D1282023

BATS

A WORLD OF SCIENCE AND MYSTERY

BATS

A World of Science and Mystery

M. Brock Fenton
Nancy B. Simmons

A Peter N. Névraumont Book

The University of Chicago Press
Chicago and London

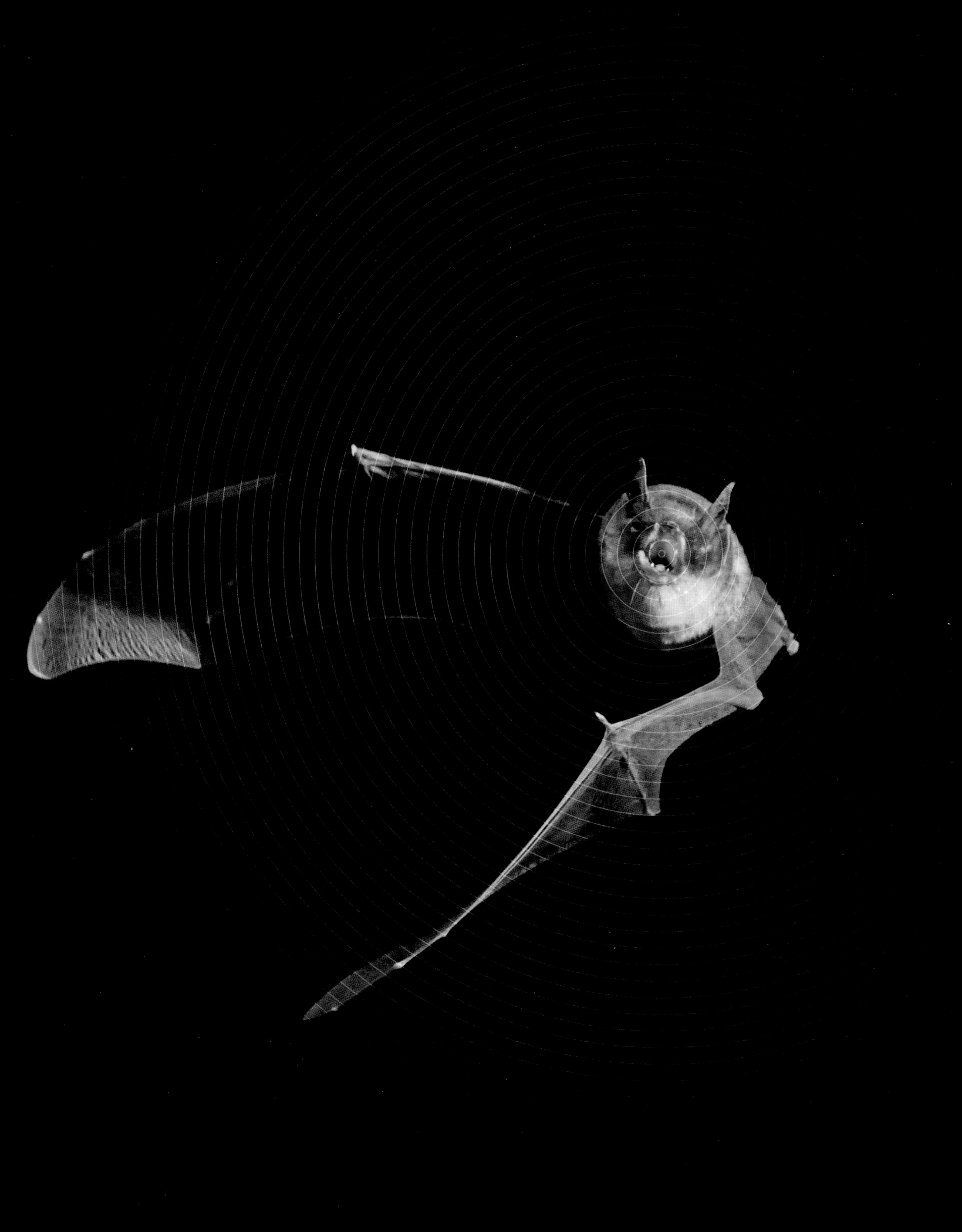

We are pleased to dedicate this book to the memory of the late Karl F. Koopman, a pioneer in the study of bats.

M. Brock Fenton
Is emeritus professor in the Department of Biology at the University of Western Ontario. He is the author or editor of several books, including *Bat Ecology*, also published by the University of Chicago Press.

Nancy B. Simmons
Is curator-in-charge of the Department of Mammalogy at the American Museum of Natural History, where she is also professor in the Richard Gilder Graduate School.

The University of Chicago Press, Chicago 60637
The University of Chicago Press, Ltd., London

Printed in China

23 22 21 20 19 18 17 16 15 14 1 2 3 4 5

ISBN-13: 978-0- 226-06512-0 (cloth)
ISBN-13: 978-0-226-06526-7 (e-book)
DOI: 10.7208/chicago/9780226065267.001.0001

Library of Congress Cataloging-in-Publication Data

Fenton, M. Brock (Melville Brockett), 1943–author.
Bats: A World of Science and Mystery /
M. Brock Fenton and Nancy B. Simmons.
pages : illustrations ; cm
Includes bibliographical references and index.
ISBN 978-0-226-06512-0 (cloth : alk. paper) —
ISBN 978-0-226-06526-7 (e-book)
1. Bats. I. Simmons, Nancy B., author. II. Title.
QL737.C5F445 2014
599.4—dc23

Contents

1

It’s a Bat!

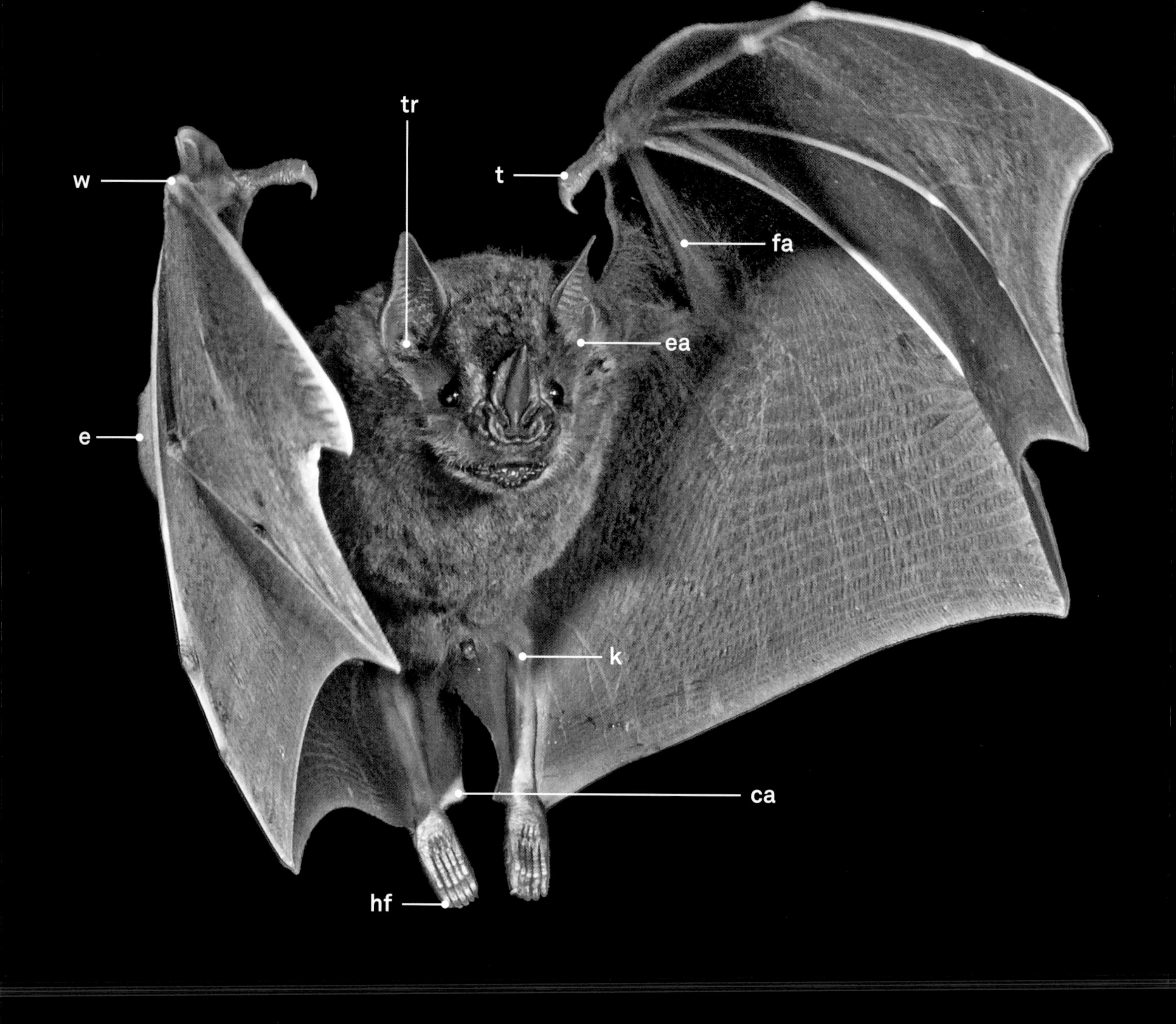

Figure 1.1.
A flying Jamaican Fruit Bat (*Artibeus jamaicensis*) showing the wrist (**w**), thumb (**t**), forearm (**fa**), elbow (**e**), ear (**ea**), knee (**k**), hind foot (**hf**), calcar (**ca**) and tragus (**tr**).

Introduction

The most distinctive features of bats are their wings and nocturnal habits. Fossils show that bats have been around for over fifty-two million years. (See Chapter 2.) If one had a time machine and could stand on the bank of an ancient stream or lake at nightfall, the flying creatures that swooped through the skies would be immediately recognizable as bats. Then, as now, bats would have appeared as quick and mysterious animals. (Figure 1.1)

People have always wondered about bats. From the time of Aesop, there have been stories suggesting that bats are otherworldly, part mammal and part bird. In some folk stories, bats are portrayed as duplicitous because they can alternate between being birds and being mammals. A recurring story recounts a ball game between birds and mammals. In one version, bats are shunned by both sides because they appear to be a mixture of the two. In another, flight allows them to score the winning point and makes them heroes, at which point they are recognized as mammals.

BOX. 1.1

Bats are Mammals

As mammals, each of us can probably think of some key features that we share with other mammals. These could include having hair or fur, giving birth to live young and feeding them with milk and having two generations of teeth (baby or milk teeth and permanent teeth). (See Chapter 7.) Bats meet all of these criteria. The basic anatomy of bats is mammalian, from skeleton to organs. Bats' hearts tend to be larger than those of other mammals of comparable size, no doubt reflecting the demands of powered flight. Bats also have some muscles lacking in other mammals, again related to their flying lifestyle. (See page 19.)

Bats are considered warm-blooded (homoeothermic) because they maintain high body temperatures when active, as do most mammals. Many bats, however, have internal thermostats that allow their body temperature to vary with ambient temperature, a specialization thought to save energy. (See Chapter 6.) This versatile approach is known as heterothermy and is a strategy that also appears in other groups of mammals, such as rodents. Bats of temperate regions especially benefit from this approach to thermoregulation.

Bats have evolved diverse dietary habits including insectivory, carnivory, frugivory, nectarivory, piscivory (fish eating) and even sanguinarivory (blood feeding). (See Chapter 5.) In this respect bats are remarkable: no other group of mammals exhibits such ecological diversity. There are no known toothless bats although a permanent evolutionary loss of teeth has occurred in some other mammals, such as anteaters. There is no evidence that bats or their immediate ancestors laid eggs. Among modern mammals, monotremes (duck-billed platypus, spiny anteater) are the only egg-layers. Bats have a placenta that facilitates exchange of nutrients and wastes between the blood of the mother and that of the fetus. Bats and all other placental mammals give birth to well-developed young. This makes placental mammals (including bats) distinct from the pouched mammals (marsupials such as opossums, kangaroos and their relatives) that bear tiny embryos that must be nurtured attached to the mother's nipple before developing sufficiently to move around alone. There is no evidence of bat-like mammals having evolved from marsupial stock.

Names of Bats

Although bats are mammals, some of the names that humans use for bats reflect the imagined dichotomy between their mammal-like and bird-like aspects. (Box 1.1) The common French name for bat is "chauve souris" or "bald mouse." In German, it is "fledermaus" or "flying mouse." But, as Denise Tupinier pointed out in her excellent book, the French have had other names for bats, such as "souris chaude" (hot mouse), "souris volante" (flying mouse), or "pissarata" (name says it all). In Scotland, bats are sometimes known as "gaucky birds", in Norway as "flaggermaus", in Holland as "viermuis." Other names for bats refer to their nocturnal activity. For example, the Greek "nycteris" refers to night, as does the Polish "nietopyr." Names such as the English "bat" do not refer to other animals or to nocturnality, instead being a unique label for a unique animal. The word "bat" is thought to be derived from the Middle English "bakke" (early 14th century.), which is probably related to Old Swedish "natbakka", Old Danish "nathbakkæ" ("night bat") and Old Norse "leðrblaka" ("leather flapper"). It is clear that humans have almost always had special names for these mysterious flying creatures of the night.

"Common" versus "scientific" names complicate the issue of naming bats. A common name is the one by which the mythical average person knows the animal. Many will recognize the common names of animals such as birds (American Robin, Bald Eagle and Nightingale). The common names of birds are standardized and relatively consistent. This is not the case for living bats, let alone fossil ones.

Scientific names of species are Latinized binomials (two-part names) that describe the organism and its general place in the overall classification of life forms. Although scientific names of larger groups such as families are written as single words in regular fonts, *e.g.*, "Pteropodidae" for Old World Fruit Bats, scientific names of species are always presented as paired names (binomials) in italics. So, *Myotis lucifugus* is the scientific name of what many people know as the Little Brown Myotis and others as the Little Brown Bat. Every species has one unique scientific name, but it may have several common names as in the case of *Myotis lucifugus*–or none at all, for example, *Onychonycteris finneyi*, a fossil bat. Using scientific names increases the precision of communicating about bats, but these names intimidate the non-technical reader. In this book we use common names wherever possible, but the first time we refer to a bat we also provide its scientific name for clarity. Biologists are much more familiar with scientific name, but variation in pronunciations of the Greek and Latin are still a challenge for them. Nobody has asked bats what they think about "common" versus scientific names. The names of all the bats in this book, both scientific and common, can be found in Bats in the Book on page 288.

Nancy Makes Up Common Names

When I was finishing up writing the Chiroptera chapter for the reference book Mammal Species of the World in 2004, I found myself facing an odd problem. The editors of the book wanted to include common names for every species—yet I found that more than fifty species of bats didn't have common names. They had been described properly in the literature with unique binomial scientific names, *e.g.*, *Myotis lucifugus*, but nobody had ever used common names for them as far as I could tell. What to do? With the permission of the editors, I simply made up names for them! Most often I coined the common name for a bat using a variant of its scientific name, *e.g.*, *Micronycteris brosseti* became "Brosset's Big-eared Bat", after the scientist for whom it was named. In other cases, the geographic range of the species helped to provide a common name, *e.g.*, *Leptonycteris curasoae* became the "Curaçoan Long-nosed Bat." It was tempting to make up silly names in some cases—I really wanted to designate a species as the "Common Baseball Bat" just for fun—but I managed to resist the temptation. I have always rather regretted that!

Table 1. The diversity and distribution of modern bats. There are twenty families and >1300 species of living bats recognized today, and about ten new species are described every year. "Laryngeal" under Echolocation means that the sounds used for echolocation are produced in the larynx (voice box).

Common Name	Scientific Name	# of Species	Echolocation	Diet	Distribution
Old World Fruit Bats	Pteropodidae	198	absent or tongue clicks	fruit, flowers, leaves	Africa, Asia, Australia, Pacific Islands
Mouse-tailed Bats	Rhinopomatidae	6	laryngeal	insects	Africa, Southern Asia
Bumblebee Bats	Crasoncyteridae	1	laryngeal	insects	Southeast Asia
Horseshoe Bats	Rhinolophidae	97	laryngeal	insects	Eurasia, Africa, Southeast Asia, Australia
Old World Leaf-nosed Bats	Hipposideridae	9	laryngeal	insects	Africa, Southeast Asia, Australia
False Vampire Bats	Megadermatidae	5	laryngeal	insects, small animals	Africa, Southeast Asia, Australia
Slit-faced Bats	Nycteridae	16	laryngeal	insects, small animals	Africa, Southeast Asia
Sheath-tailed Bats	Emballonuridae	54	laryngeal	insects	Pantropical: Africa, Southeast Asia, Australia, Tropical Americas
New World Leaf-nosed Bats	Phyllostomidae	204	laryngeal	fruit, flowers leaves, insects small animals, blood	Tropical Americas, Caribbean Islands
Moustached Bats	Mormoopidae	10	laryngeal	insects	Tropical Americas, Caribbean Islands
Bulldog Bats	Noctilionidae	2	laryngeal	insects, fish	Tropical Americas, Caribbean Islands
Smoky Bats	Furipteridae	2	laryngeal	insects	Tropical Americas
New World Disk-winged Bats	Thyropteridae	5	laryngeal	insects	Tropical Americas
Old World Disk-winged Bats	Myzopodidae	2	laryngeal	insects	Madagascar
New Zealand Short-tailed Bats	Mystacinidae	2	laryngeal	insects, fruit, flowers	New Zealand
Funnel-eared Bats	Natalidae	12	laryngeal	insects	Tropical Americas
Free-tailed Bats	Molossidae	113	laryngeal	insects	Eurasia, Africa, Asia, Australia, Americas
Bent-winged Bats	Miniopteridae	29	laryngeal	insects	Eurasia, Africa, Asia, Australia
Wing-gland Bats	Cistugidae	2	laryngeal	insects	Southern Africa
Vesper Bats	Vespertilionidae	455	laryngeal	insects, fish	Worldwide except Arctic & Antarctica

A
B

Figure 1.2.
The hand wing of bats. In **A** the wings of a flying Egyptian Rousette Bat clearly show the basic hand structure which also is obvious in **B**, the skeletal structure of the hand wing. A CT scan of a Lesser Short-nosed Fruit Bat (*Cynopterus brachyotis*) reveals the size of the wings—even when folded—with respect to the rest of the skeleton. (See also Figures 2.1, and 2.2.) CT scan (**C**) courtesy of Nancy Simmons.

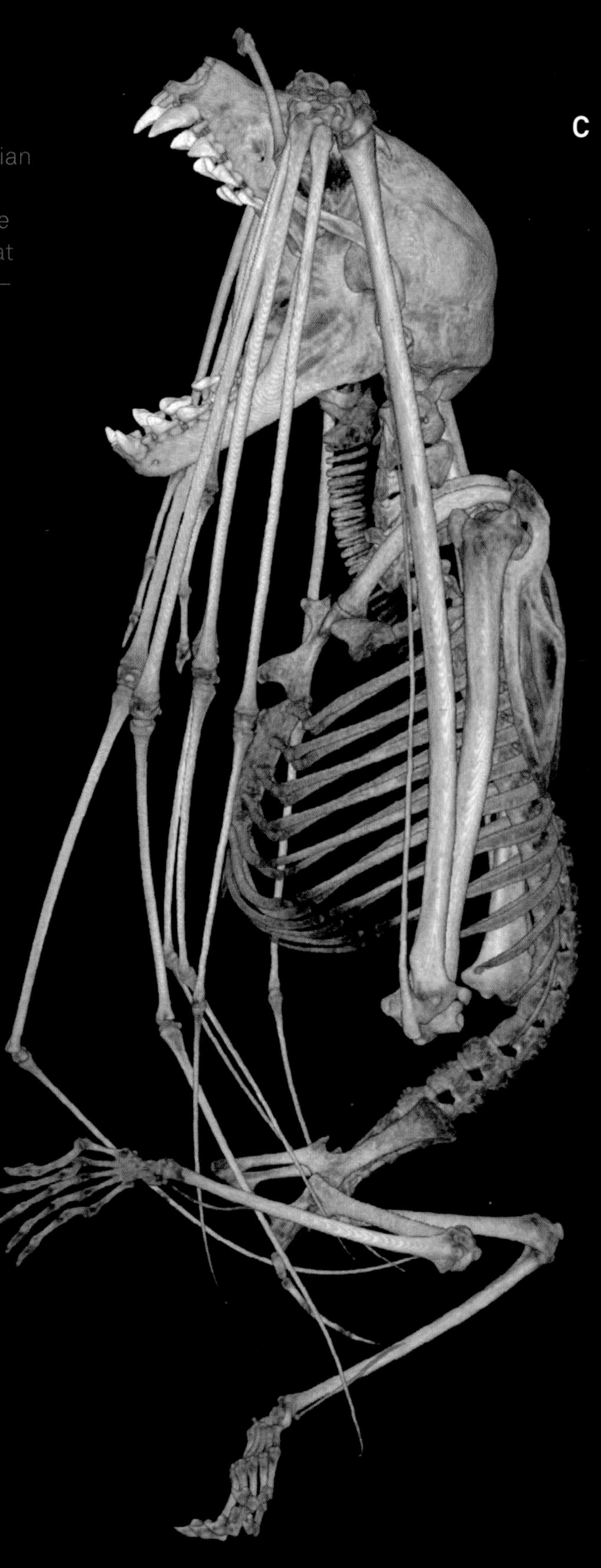

Bats constitute the order Chiroptera, from the Greek *cheiro* meaning hand, and *ptera* meaning wing. (Figure 1.2) They are the second largest group of mammals after rodents, representing about 20 percent of all classified mammal species worldwide. The >1300 species of living bats are arranged in twenty families based on their morphology, DNA and evolutionary history. Each family of bats has a scientific and at least one common name as shown in Table 1. Bats are currently classified in two suborders, the Yinpterochiroptera (yinpterochiropterans) and the Yangochiroptera (yangochiropterans). In the past scientists divided bats differently, recognizing groups called Megachiroptera (megabats = Old World Fruit Bats) and Microchiroptera (microbats = echolocating bats). These terms are no longer in use because evolutionary studies of DNA have shown that some "microbats" are actually more closely related to Old World Fruit bats than they are to other echolocating bats. Also, these terms were always misleading in the first place because some "mega-bats" are quite small, *e.g.*, Long-tongued Fruit Bats (*Macroglossus minimus*) have a wingspan of 15 centimeters (cm.) and weigh only 12 to 18 grams (10 grams = 0.35 ounces), while some "microbats" are quite large, *e.g.*, Spectral Bats (*Vampyrum spectrum*) can have wingspans of about 1 meter (m.) and weigh nearly 200 grams (g.). Also, at least two "megabats" echolocate using tongue clicks—the Egyptian Rousette (*Rousettus aegyptiacus*) and Geoffroy's Rousette (*Rousettus amplexicaudatus*)—making the distinctions even more confusing. Regardless, most scientists now use tongue-twisting names Yinpterochirotpera and Yangochiroptera—derived from "yin" and "yang" in Chinese philosophy—to describe the two main groups of bats. The geographic range of families of bats in each group is shown in Table 1.

Map to a Bat

Many of a bat's distinctive features are obvious when the animal is flying (Figures 1.1 and 1.2), but when a bat is roosting, other features become more obvious. (Figure 1.3) We have labeled some of the relatively consistent features in these figures to make them easier to interpret. All bats have two wings with flight membranes (called patagia, singular = patagium), as well as two hind legs with feet. Some bats have a membrane between the legs (uropatagium or interfemoral membrane) but others do not. Bats often have calcars, cartilaginous or bony projections from the ankle towards the tail, which allow control of the shape and stretch of the interfemoral membrane. (Figure 1.3) Tail length varies considerably in bats, spanning the whole range seen in other mammals—from very long to nonexistent.

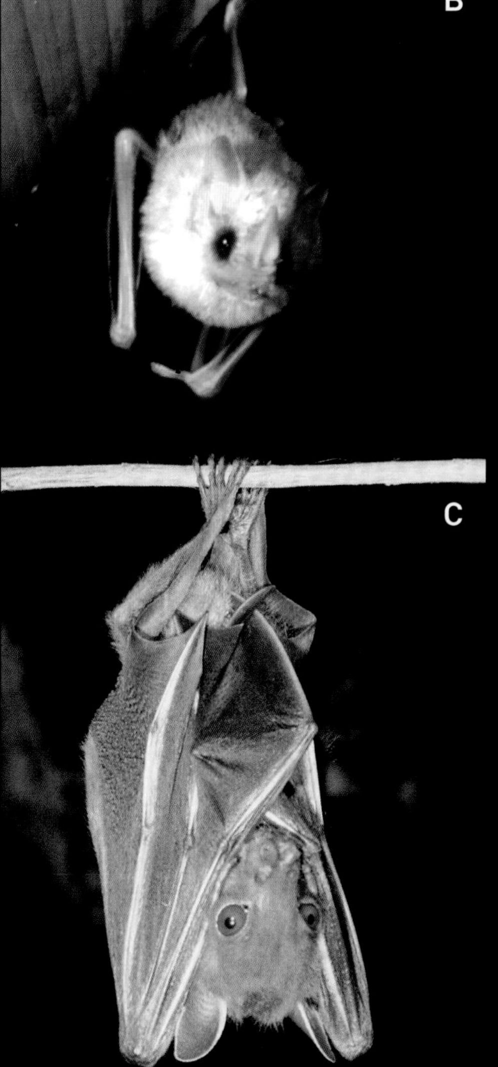

Figure 1.3.
Three roosting bats with wings that are fully folded (**A**, **C**), or partly folded (**B**). The Lesser Mouse-tailed Bat (*Rhinopoma hardwickei*) shown in (**A**) has an obvious and distinct tail; its feet and thumbs are also clearly visible. (**B**) The Honduran White Bat (*Ectophylla alba*) hangs by one foot. Its feet are obvious. (**C**) The Lesser Short-nosed Fruit Bat (*Cynopterus brachyotis*) partly envelops its body with its wings.

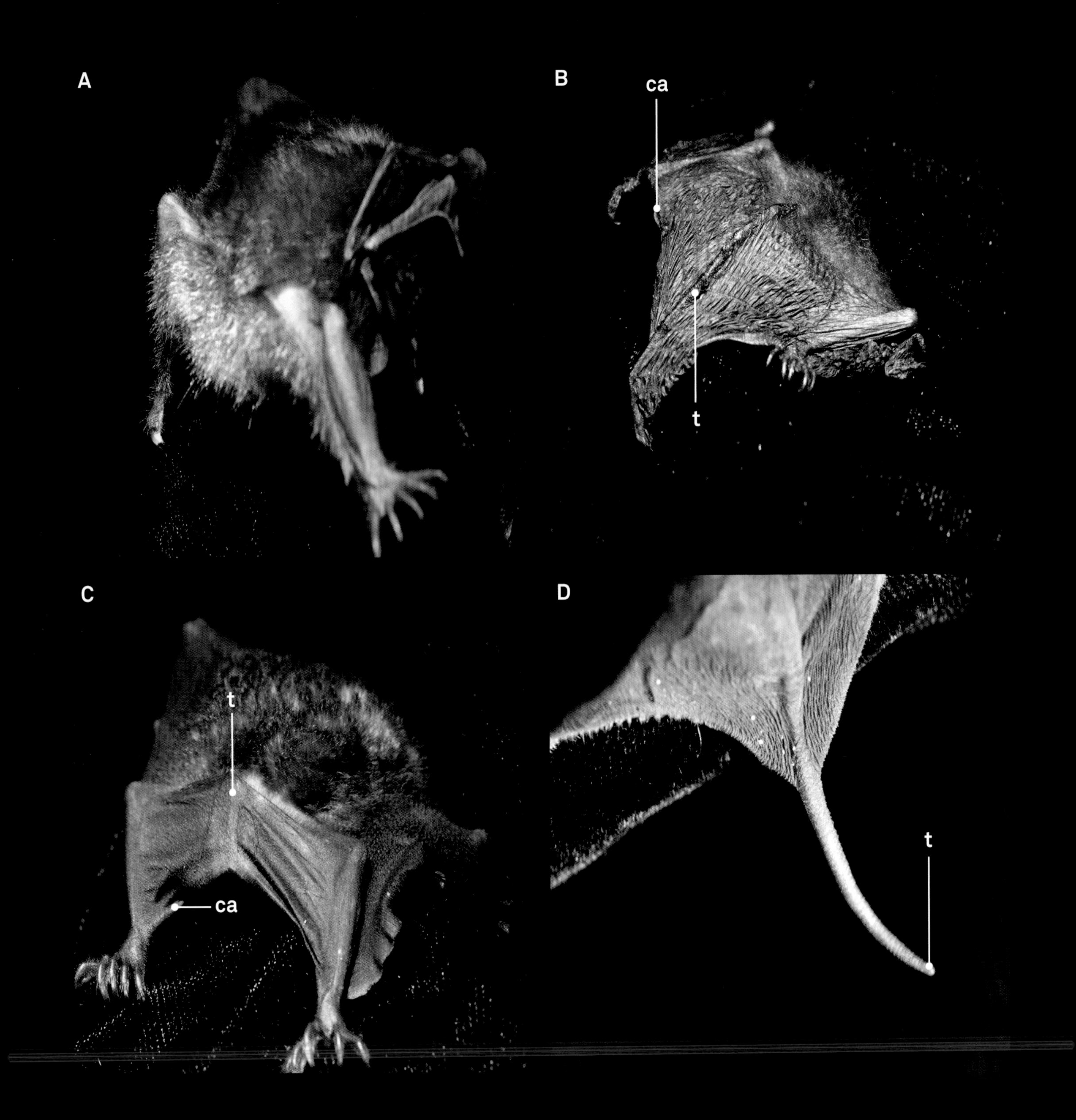
A
B
ca
t
C
t
ca
D
t

Similar Yet Different: Bats Compared to Birds and Pterosaurs

The wing skeletons of bats and those of other flying vertebrates—birds and pterosaurs—are very different from one another although they originated evolutionarily from the same basic set of arm bones common to all terrestrial vertebrates. (Figure 1.5) Ancestrally, the arm skeleton of bird, bat and pterosaur precursors contained the same set of bones that humans have: a single upper arm bone (humerus), a pair of forearm bones (radius and ulna), a group of small bones comprising the wrist (carpals), five hand bones (metacarpals) and five digits each consisting of two to three finger bones (phalanges). The arm and hand skeleton of each flying group evolved to include elongation of different parts of the arm and hand and fusion of different bones to support a wing membrane or airfoil. Note the differences in the positions of elbows and wrists between bats and pterosaurs. The humeri (upper arm bones) of pterosaurs and birds are proportionally much shorter than those of bats. In bats the wing membrane is supported by elongated arm bones (especially those of the forearm) and elongate hand and finger bones in four digits; only the thumb in bats remains relatively small. Pterosaurs, by comparison, had flight membranes supported by a single elongated hand and finger bone. (See following page.)

Figure 1.4. (opposite)
Tails of bats. (**A**) Little Yellow-shouldered Bat (*Sturnira lilium*), a tailless species, (**B**) long tailed Elegant Myotis (*Myotis elegans*), (**C**) Sowell's Short-tailed Fruit Bat (*Carollia sowellii*) and (**D**) Brazilian Free-tailed Bat (*Tadarida brasiliensis*). Arrows point to tails (**t**) and calcars (**ca**). Note the tail in Figure 1.3A.

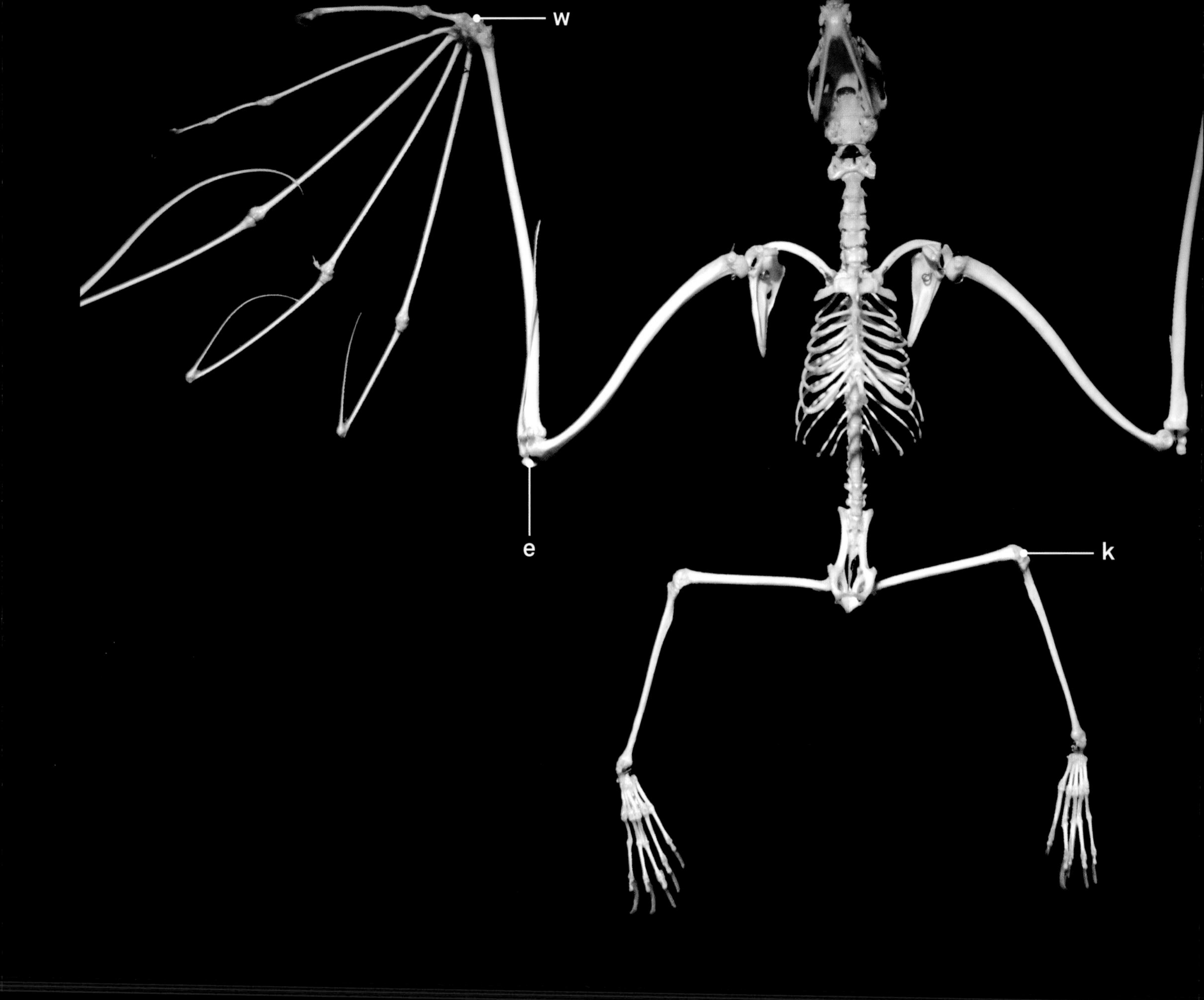

Figure 1.5.
A comparison of (**A**) the skeleton of a Flying Fox and (**B**) a reconstruction of the skeleton of a pterosaur, *Quetzalcoatlus northropi*, as well as (**C**) the wing bones of a bird. The prominent wings of bats and pterosaurs are clear. Arrows show elbows (**e**), wrists (**w**) and knees (**k**). The bird's bones show the points of anchorage of flight feathers.

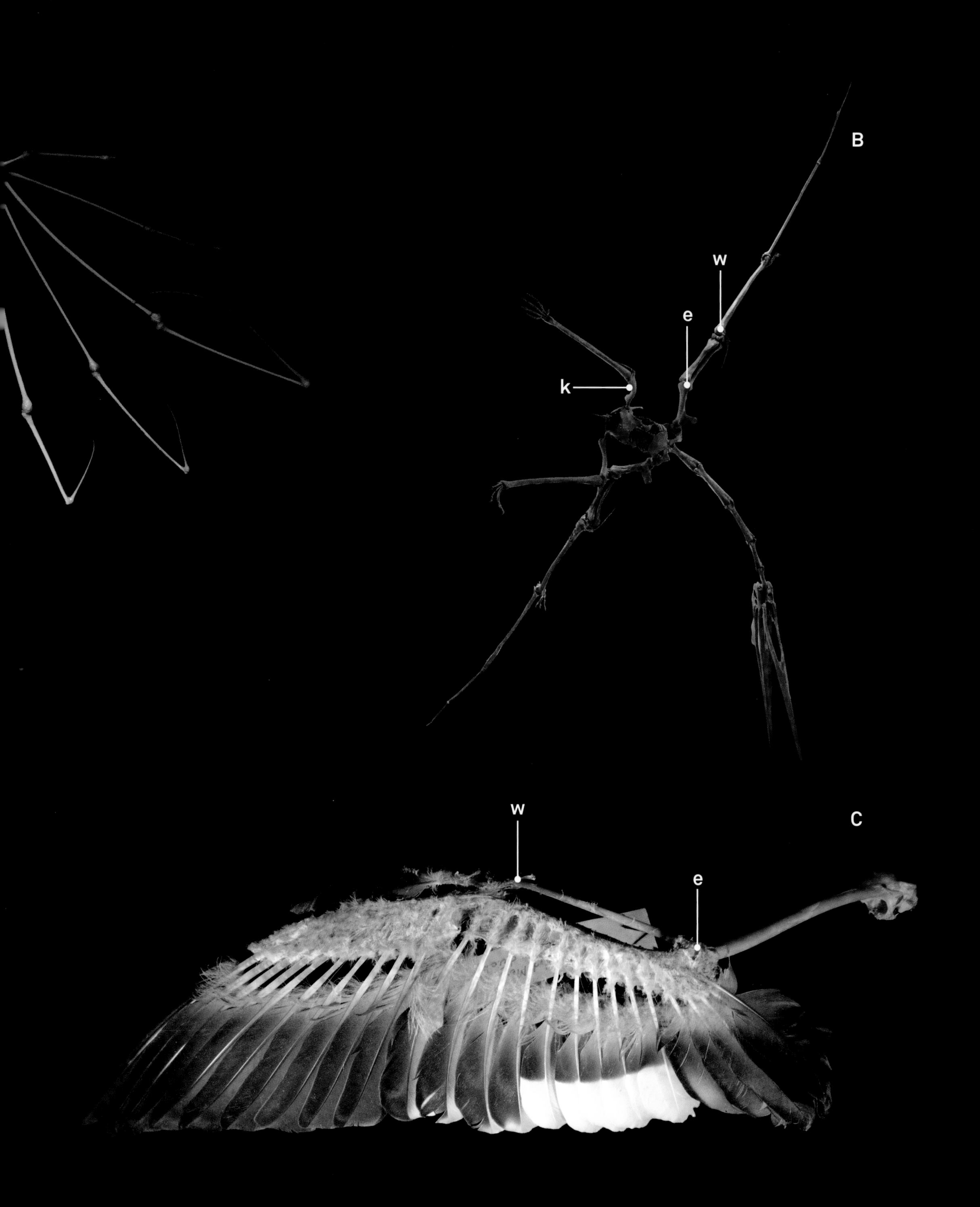
B
w
e
k
C
w
e

Pterosaurs may have had wing membranes made of skin like those of bats, but exceptionally well-preserved fossils of pterosaurs from China and elsewhere show that pterosaur wing membranes were filled with parallel fibers arranged perpendicular to the bone of the wing digit. Bat wings have many elastic strips and muscles in their wing membranes, but only some of them are arranged perpendicular to the arm or finger bones. (Figure 1.6)

The precise structure and function of the fibers in pterosaur wings remain a mystery, but doubtless these structures were important for wing function during flight. In birds, feathers comprise the flight surface and the bones of the wrist and hand have become fused to provide a robust attachment site for flight feathers. The fact that three different vertebrate groups achieved powered flight using modified forearms to produce an airfoil (the shape of the wing seen cross-section)—but did so in entirely different ways—is an excellent example of parallel evolution. (See Chapter 3.)

There are other striking differences between bats and birds in addition to their wing structure. In birds that fly, the breastbone (sternum) has a conspicuous keel, while bats typically have only a small keel that is often limited to very anterior end of the breastbone. (Figure 1.7) Birds have a wishbone (furculum) composed of fused collar bones (clavicles), whereas the clavicles in bats remain separate. Special flanges called uncinate processes produce overlap between the ribs of birds and reptiles, but these are not present in bats (or other mammals). The diaphragm and movements of the ribs are important during breathing in mammals. The absence of uncinate processes is correlated with a more flexible chest skeleton in mammals. The lungs of birds are open at either end, allowing air to flow through and resulting in more efficient breathing and cooling. Bat lungs, like our own, are dead-end sacs that cannot be as effectively ventilated as bird lungs. But the blood-gas barrier in the lungs of bats is thinner than that of other mammals, the alveoli are smaller and the lungs proportionally larger than those of mammals that do not fly. Finally, modern birds lack teeth and lay eggs, but all known species of bats have teeth and bear live young.

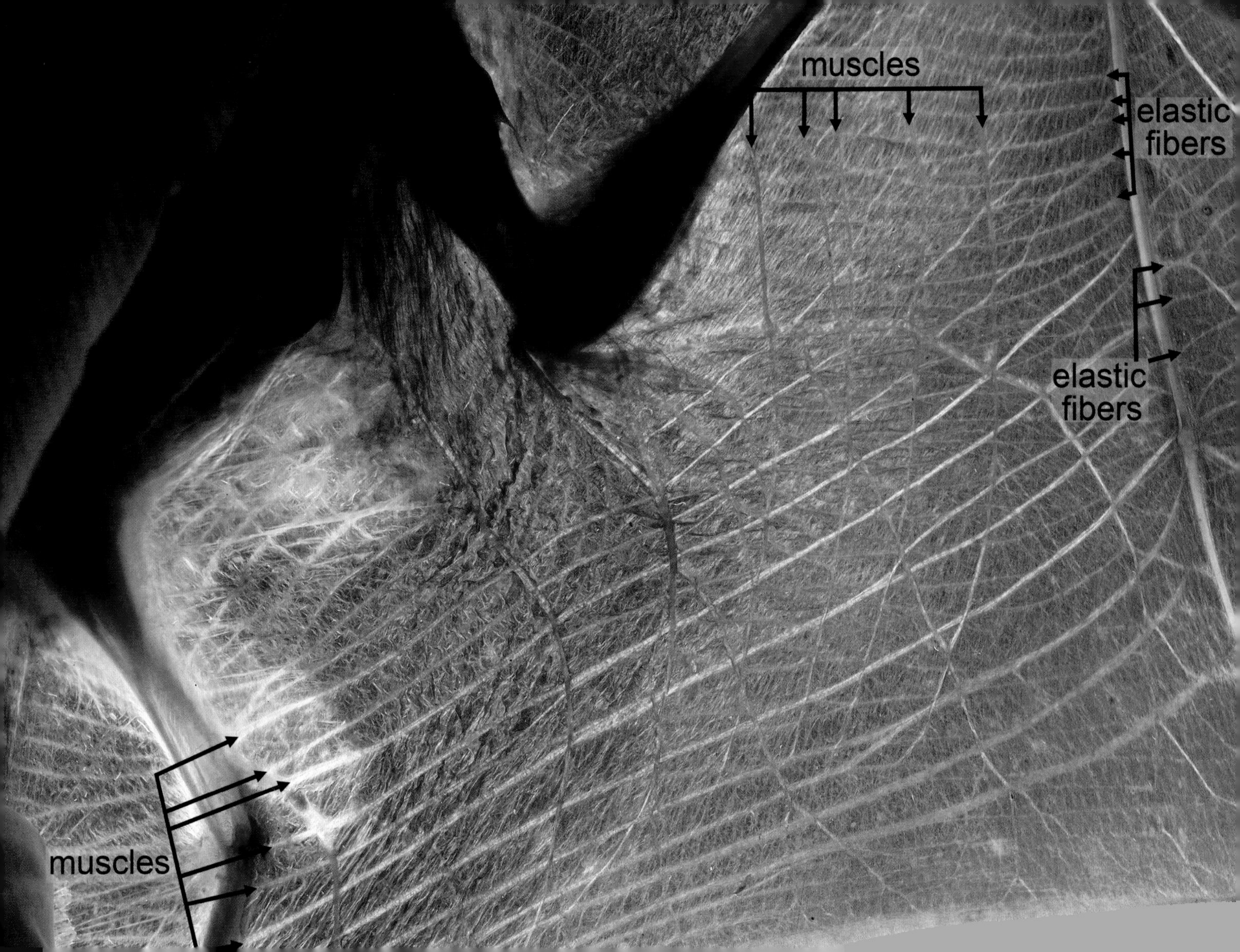

muscles
elastic fibers
elastic fibers
muscles

Figure 1.7.
A comparison of the breastbones and rib cages of a bird (**A**) Whip-poor-will (*Antrostomus vociferous*) and two bats, (**B**) Large Slit-faced Bat and (**C**) Sucker-footed Bat. Note the prominent keel (**k**) on the bird's breastbone (in **A**) compared to the arrangement of much smaller keels on the breastbones of the bats (**k** in **B** and in **C**). In the Sucker-footed Bat there is a single keel at the very top of the breastbone (**C**), while there are two keels (**k**) in the Large Slit-faced Bat (**B**) including an anterior one with two projections (two arrows) and a posterior one with a single projection. In **A**, the **u** is an uncinate process, features missing from the bats. (See page 20.)

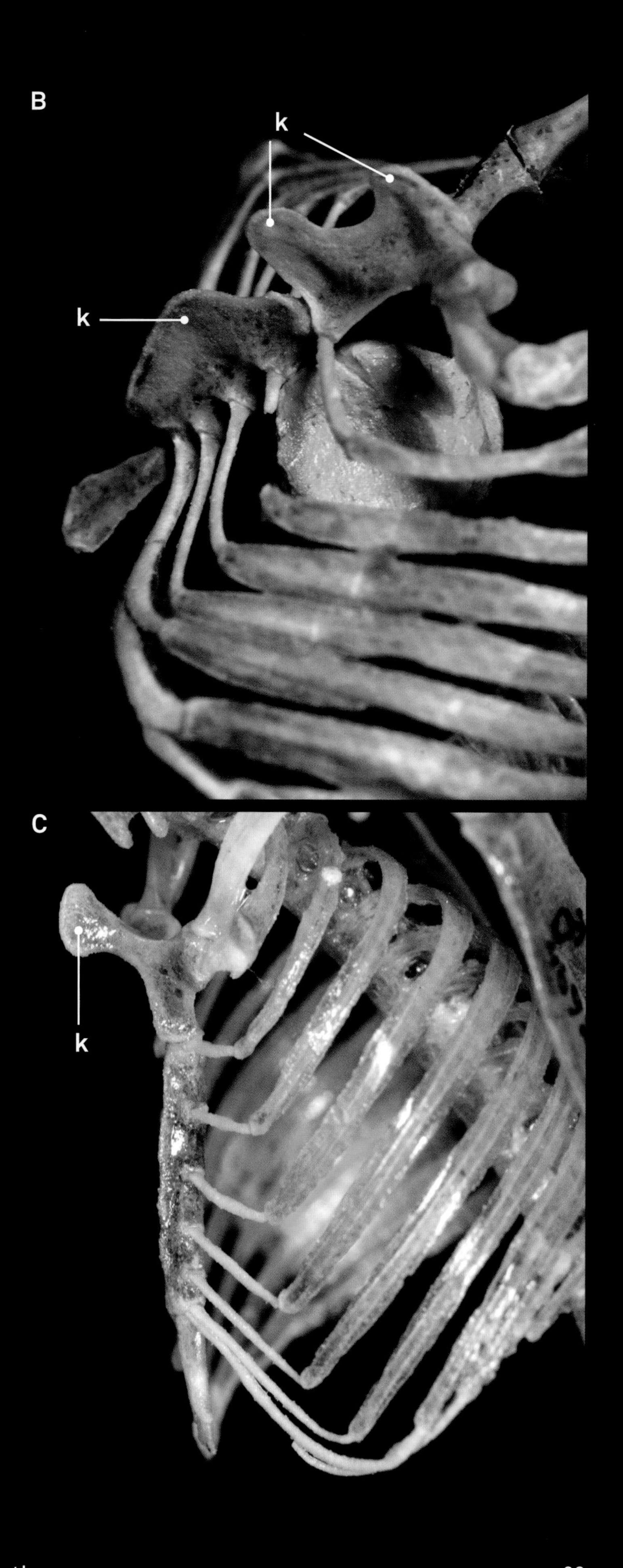
B
k
k
k
C
k

Why should living birds lack teeth and lay eggs? Egg-laying in birds is not a specialization for flight, it is a form of reproduction inherited from ancestral reptiles. Pterosaurs also laid eggs. Egg-laying is a primitive trait also seen in a few mammals (the platypus lays eggs). Most mammals give birth to live young, and this form of reproduction evolved early in the mammalian radiation. Paleontologists have identified fossils from the Early Cretaceous (125–130 million years ago) as members of the large evolutionary group of mammals that today bear live young. Within this group, the placenta, another specialization, evolved by roughly the end of the Cretaceous (~66 million years ago). The placenta allows animals to retain a fetus inside the womb until it has grown quite large. Bats inherited a placenta and live birth from such non-flying mammal ancestors. For flying animals, carrying either a large egg internally (prior to laying it) or a fetus can impose energy costs due to the additional body weight that must be supported during flight. Although it only affects females, clearly females must survive for a species to continue! Weight constraints imposed by reproduction are something that all lineages of flying animals have had to overcome to be successful.

Teeth are clearly useful for obtaining, handling and chewing food and very important in most vertebrate groups. (See Chapter 5.) Teeth, however, are made of dense materials (enamel and dentine) and are amongst the heaviest structures in the bodies of small vertebrates. Because weight matters during flight, at least some lineages of flying animals—including all modern birds and some groups of pterosaurs—lost their teeth (in an evolutionary sense) after achieving flight. Instead, birds and most pterosaurs have (or had) keratinous beaks, which are lighter. Bats, however, retained a typical mammalian dentition that has been modified in different groups to facilitate processing of different foods. (See Chapter 5.) Teeth are smaller and sometimes fewer in bats that do not need to chew their food, *e.g.*, vampire bats and nectar-feeding bats. This may be due at least in part to the energy savings accrued by reducing the body weight of the animal.

The hind legs of bats also differ from those of birds and pterosaurs. Birds and pterosaurs are bipedal animals with long bones of their hind legs robust enough to bear the weight of a walking or running animal. In contrast, the long bones of the hind legs of most bats are slender and delicate, suitable for supporting a hanging animal but not for bipedal locomotion. Furthermore, the hip and limb structure of bats has been modified so that the hind legs are rotated relative to the pelvis such that the sole of the foot to face forward rather than backwards and downward as in most other animals. In birds, pterosaurs and other mammals the sole of the foot faces down. This difference is clearly reflected in the direction of flexion of knee joints and the position of a bat's feet (Figure 1.8).

Figure 1.8.
This photograph of a flying Little Brown Myotis shows the position of the knee (arrow) as well as the hind foot. The hind limbs of bats are rotated so that the sole of the foot faces forward—in most other mammals, including humans, the sole of the foot faces backwards.

In birds that fly, two pairs of muscles largely power flight—a set of "elevator" muscles that raise the wings and a set of "depressor" muscles that bring them down. Both pairs of muscles are located on the surface of the chest (these comprise the white meat on a chicken). In contrast, in bats there are nine pairs of muscles involved in powering flight. The elevator muscles are located on the back, and the depressor muscles are on the chest. This difference probably explains why, compared to birds, bats are much thinner in profile through the chest—the flight musculature of bats requires broad areas for attachment to the rib cage, both back and front. The thin profile of the chest may have an added benefit in that it allows bats to squeeze into crevices and through small openings, giving them access to roosts inaccessible to many predators. (See Chapter 6.)

There are flightless species of birds and insects, but no known species of bats or pterosaurs is/were flightless. Flightlessness in birds and insects has often evolved on remote islands where there are few predators; the benefits of flight (in terms of providing access to resources and allowing long-distance movements) are reduced and there are additional dangers to flight, *e.g.*, individuals that fly high may be trapped by wind currents and blown away. But even bats living on remote oceanic islands in the South Pacific have retained their ability to fly. Nobody knows why this is the case. Although the structure of the forelimbs makes it difficult for most bats to walk effectively, strong walking and running behavior has evolved in some bats, *e.g.*, Common Vampire Bats (*Desmodus rotundus*). (Figure 1.9)

Brock Answers a Question

I'm often asked: "What is your favorite bat?" The problem with answering this is that, like many other bat biologists, I tend to be fickle. Today's favorite is tomorrow's also-ran. Working in the Yucatan Peninsula of Mexico in 1991, I was keen to meet some of the very neat bats that occur there. High on my list were Large-eared Woolly Bats (*Chrotopterus auritus*), Wrinkle-faced Bats (*Centurio senex*) and Tomes's Sword-nosed Bat (*Lonchorhina aurita*). On the third night, I was beside myself when a Wrinkle-faced Bat flew into the mist net. While I was busy with that bat, a Large-eared Woolly Bat flew into the net almost beside me. Within five minutes I also caught a Tomes's Sword-nosed Bat and my colleagues were kidding me about changing the battery in my pacemaker. By 9:00 pm that night, my three favorite bats were all in hand. By midnight, three other species I had not seen before were caught in the nets. So that night I had six favorite species.

Figure 1.9.
Two views of a Common Vampire Bat (**A** and **B**) on a treadmill illustrate the difference between the bat's running gait (**A**) and stance (**B**). When running, the bat reaches forward with its wrists and thumbs, swings its hind legs forward, plants them and repeats the process. Note that in **B** the bat is panting.

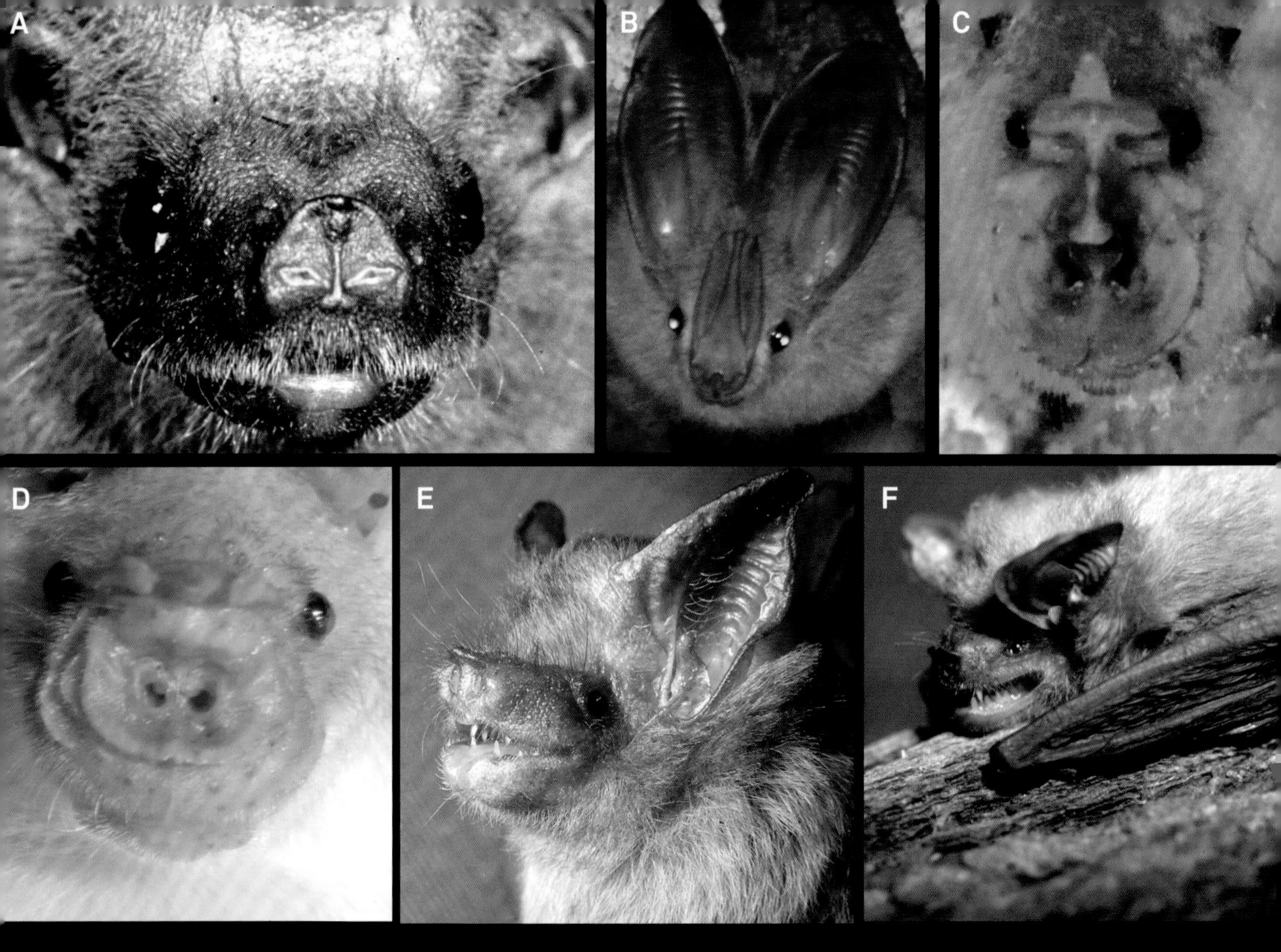

Figure 1.10.
A sample of faces and noses of bats. Included are (**A**) a Lesser Mouse-tailed Bat, (**B**) a Yellow-winged Bat (*Lavia frons*), (**C**) a Geoffroy's Horseshoe Bat (*Rhinolophus clivosus*), (**D**) a Trident Leaf-nosed Bat (*Asellia tridens*), (**E**) a Bumblebee Bat (*Craseonycteris thonglongyai*) and (**F**) a Lesser Long-eared Bat (*Nyctophilus geoffroyi*), representing Mouse-tailed Bats (Rhinopomatidae), False Vampire Bats (Megadermatidae), Horseshoe Bats (Rhinolophidae), Old World Leaf-nosed Bats (Hipposideridae), Bumblebee Bats (Craseonycteridae) and Vesper Bats (Vespertilionidae), respectively. Photographs by Brock Fenton, Robert Barclay (**B**) and Sebastien Puechmaille (**E**).

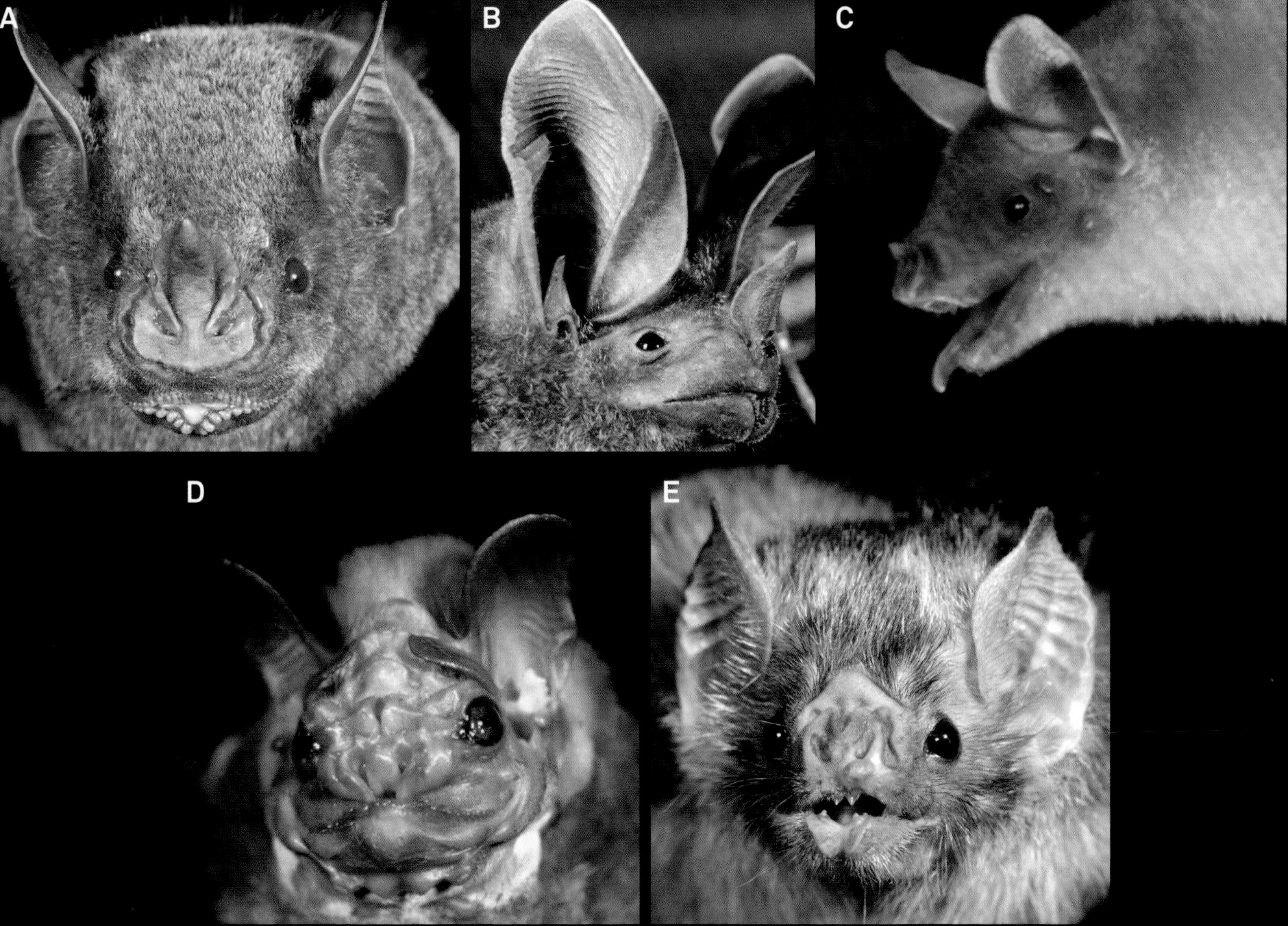

A
B
C
D
E

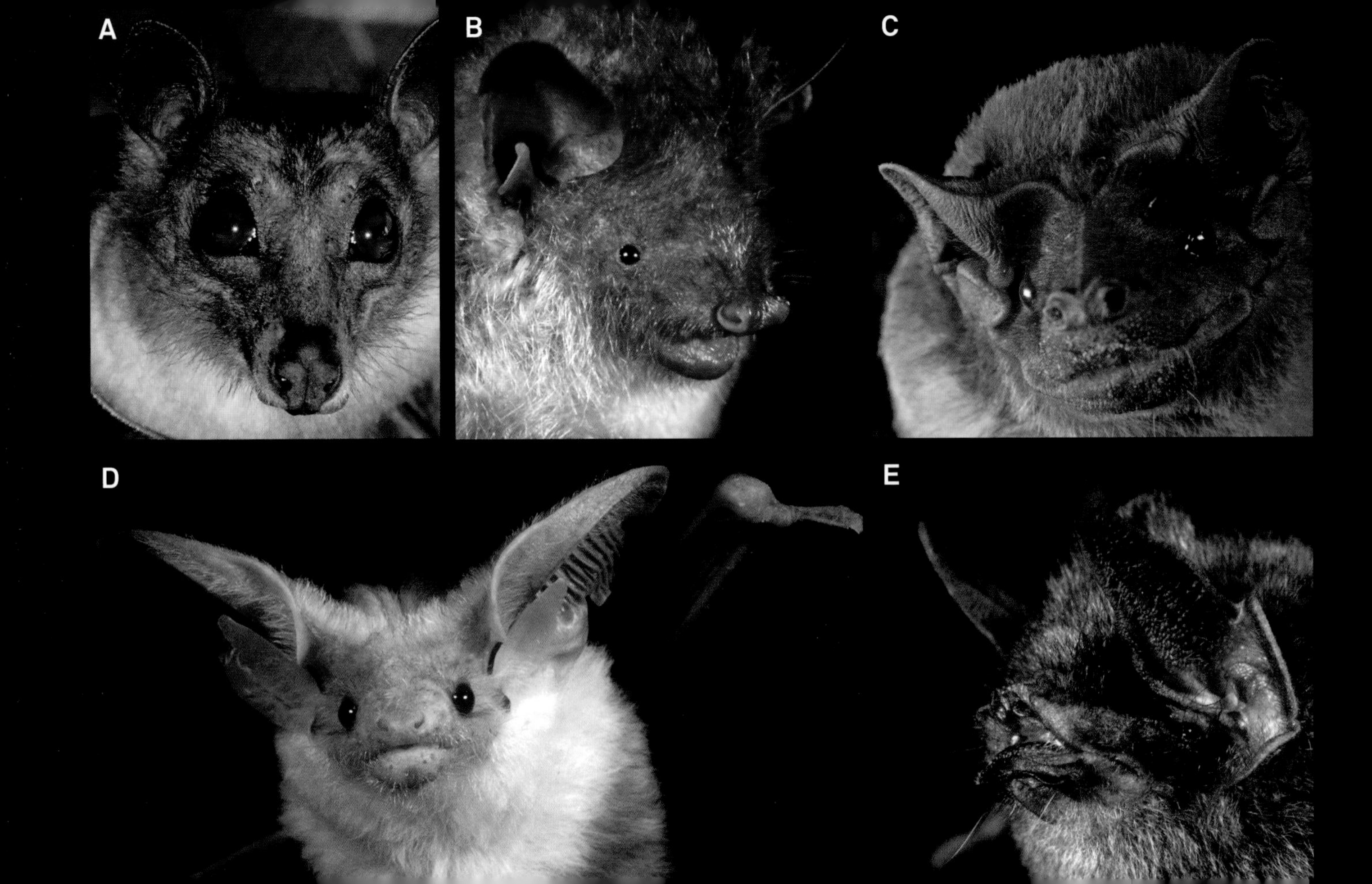
A
B
C
D
E

Bats of several families have evolved noseleaves—fleshy projections from the snout around and above the nostrils. (Figures 1.10 and 1.11) Along with noseleaves, ears and related structures such as tragi (singular tragus), an extra spike of cartilage sticking up from the base of each ear, appear to be involved in echolocation. (Figures 1.10 and 1.11 and see Chapter 4.) While noseleaves are involved in sound transmission, ears and tragi play a central role in sound reception. Flaps of skin around the mouth probably direct sounds away from the bat. (Figure 1.6E) Many but not all bat species have a tragus. (Figure 1.12) The tragus is absent in Old World Fruit Bats, Horseshoe Bats and Old World Leaf-nosed Bats. It is large and well developed in False Vampire Bats, Sheath-tailed Bats (Emballonuridae), Mouse-tailed Bats and Vesper Bats. The tragi are small in Slit-faced Bats (Nycteridae) and Free-tailed Bats (Molossidae). Other parts of the ear may serve the function of the tragus in echolocation. (See page 98.)

The Diversity of Bats

Most areas of the world support many species of bats, each of which has a slightly different anatomy and lifestyle. Most bat species occur in tropical and subtropical regions, meaning that countries closer to the equator tend to have more species of bats than countries in northern (or southern) temperate regions. Boasting ~180 species (~ = approximately), Colombia has the richest bat fauna (number of distinct species) of any country in the world. This reflects its position astride both the equator and the Andes Mountains. Colombia's equatorial tropical forests, Atlantic and Pacific drainages and desert-like north provide a rich range of climate and environmental conditions for bats.

Bat faunas in the tropics include species with a greater range of diets (fruit, flowers, animals) than bat species in temperate areas (which eat mainly insects). In temperate regions there are also fewer species of bats, for example sixteen species in the United Kingdom, nineteen in Canada and forty-five in the United States. The frigid Atlantic Labrador Current flowing down from the Arctic Ocean passes along the coast of Newfoundland, which has only two species of bats, compared to sixteen in the United Kingdom where the path of the warm Gulf Stream across the Atlantic produces a milder climate. More species of bats live along the west coast of North America than along the east coast.

Perhaps surprisingly, some tropical islands have relatively small bat faunas. There is one living bat species in Hawaii and four on the Galapagos Islands. Even large islands in the West Indies (Cuba, Hispaniola, Jamaica, Puerto Rico) each have modest bat faunas of about twenty-five species, about the same as the twenty-six species that occur on the 6,852 islands that make up the Japanese archipelago.

There is a general pattern to the distribution of families of bats. (Table 1.1) Note that while Old World Disk-winged Bats (Myzopodidae) and New Zealand Short-tailed Bats (Mystacinidae) have a very restricted distribution, Vesper Bats are widespread. Other families, *e.g.*, Horseshoe Bats, Old World Leaf-nosed Bats and False Vampire Bats, occur in the Old World (Eurasia, Africa, Southeast Asia), while others, *e.g.*, Bulldog Bats (Noctilionidae), Moustached Bats (Mormoopidae), New World Leaf-nose Bats, Funnel-eared Bats (*Natalus* spp.), Smoky Bats (Furipteridae) and New World Disk-winged Bats (Thyropteridae), occur only in the Western Hemisphere (the Americas).

Although there are a few records of bats from the Arctic, *e.g.*, Southampton Island in Hudson's Bay, there are none from the Antarctic. Mind you, it is only a matter of time before fossil bats are found in both areas, reflecting changes in the distributions of bats that in turn reflect changes in prevailing climates. Over geologic time, climate change has often been brought about by shifts in the positions of the continents relative to each other. We expect that fossil bats may someday be found in Antarctica, which was originally part of Gondwanaland—a continent that included South America, Africa, India and Australia. Gondwanaland began to break apart well before bats evolved, but Antarctica remained connected to South America until the Oligocene (~30 million years ago). During the first twenty million years of bat evolution, parts of Antarctica probably had a much more hospitable climate than today. In Scandinavia, bats occur well north of the Arctic Circle (60° N), surviving and perhaps thriving under the midnight sun. Areas north of the treeline may provide few roost opportunities for bats, perhaps explaining the lack of bats in treeless tundras.

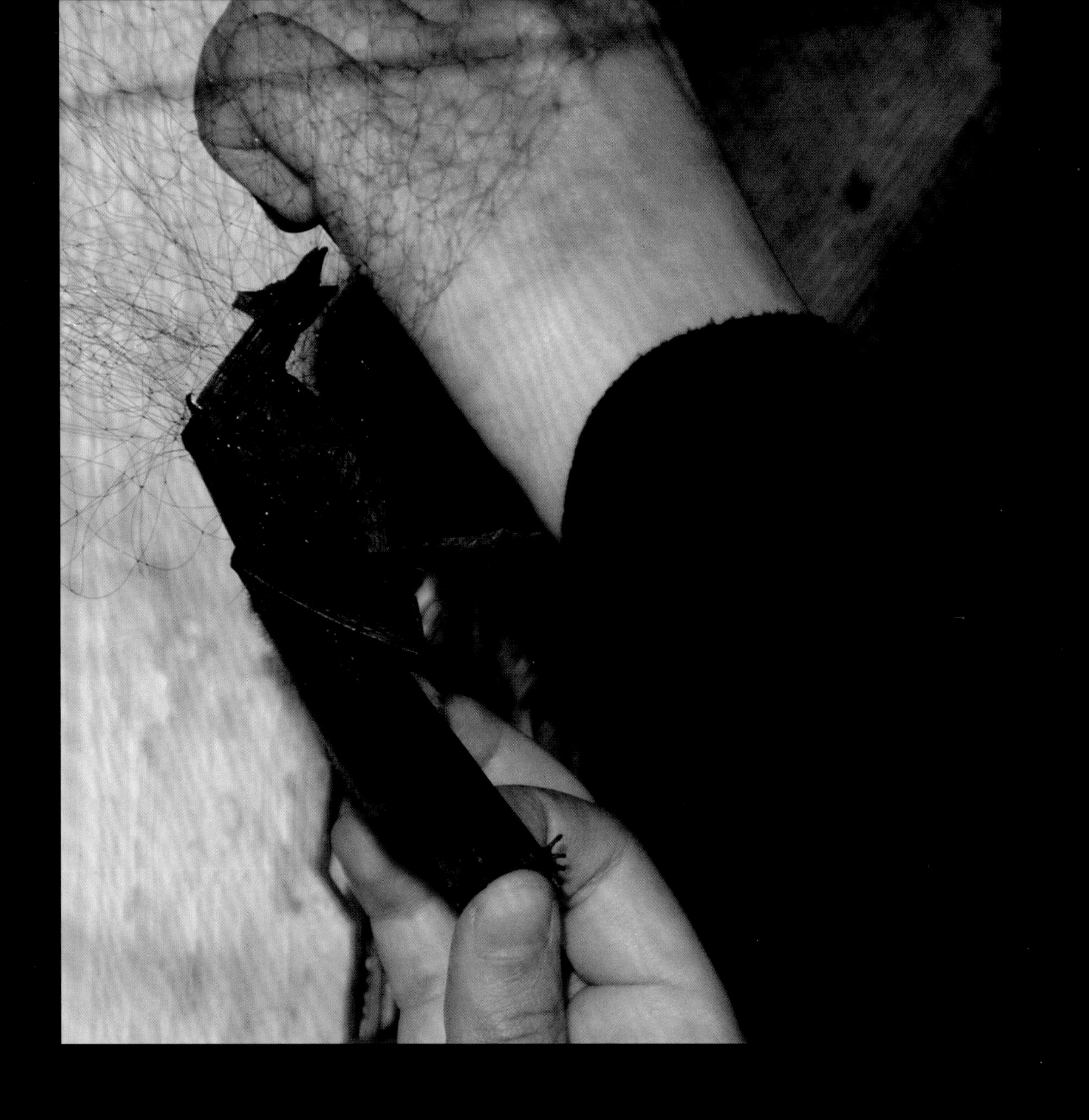

Figure 1.13.
Removing a Little Brown Myotis from a mist net. The fine mesh is obvious against the hand and white wall background.

Figure 1.14.
Nancy Simmons and Deanna Byrnes at a harp trap. Deanna holds a bat, while Nancy records data about its species, gender and age. The vertical monofilament lines are obvious. Set in a bat flyway, bats colliding with the lines slide down them into the bag below. Stand-alone harp traps were developed by Bat Conservation International founder Merlin Tuttle and revolutionized the business of catching bats. To many bat biologists they are Tuttle Traps.

How Brock Was Hooked on Bats

At about age six years, I met my first bat at a family cottage north of Toronto in Canada. It was August and the owners of the neighboring cottage were painting its exterior. The Little Brown Myotis that roosted behind the shutters were displaced and one of them ended up in our cottage. After flying about for a while, it landed on the stone mantelpiece.

I decided to catch the bat, but did not realize it was hanging upside down as bats routinely do. So grabbing it low was a mistake—I practically put my fingers into its mouth. I was bitten and immediately released the bat, which took off and continued to fly around the cottage. The amusing part of this was that the only room in the cottage with a low ceiling was the bathroom. So while I watched the bat, my mother and sister took refuge there. These dramatic events, at least to my six-year old eyes, sparked my ongoing curiosity about bats.

Eleven years later, when I was in my second year at Queen's University in Kingston, I had the opportunity to accompany Roland Beschel, whose wide-ranging scientific interests made him an ideal mentor, on a search for hibernating bats. This meant exploring local caves and looking for bats that had been banded in the summer somewhere else. I found bats, but to my disappointment any that were banded had been tagged in the cave where I found them.

At first I was torn between the lure of bats and the lure of caves, but before long the bats won out. How far do bats travel between their summer and winter quarters? How do insects protect themselves against marauding bats? Why are there eight species of bats in Ontario, rather than just one, or twenty?

In the years since then, bats have never disappointed me. They repeatedly challenge me and force me to recognize that what I had thought to be true about them is wrong, or more complicated. In short, the unanswered questions about bats drew me into science and they continue to hold me there. Bats really are the gift that keeps on giving.

Nancy Meets Her First Bats

As a postdoctoral researcher at the American Museum of Natural History, I began working on bat anatomy in 1989. My goal was to develop a better understanding of the evolutionary origins of bats and their amazing specializations for flight and echolocation. The research was fascinating, but it involved only museum specimens—skeletons and dried skins of bats collected by bat researchers in the past, sometimes over a hundred years ago. I had never held a live bat in my hands when I began working on bat evolution. But that all changed in 1991 when my husband and I started a faunal inventory project at Paracou, a forestry research station in French Guiana.

The goal was deceptively simple: document all the mammals that lived in one patch of rainforest. My husband, an expert on rodents and marsupials, would handle those parts of the fauna as well as the other larger mammals; the bats were my job. I researched everything I could find about capturing bats, bought mist nets (imagine giant volleyball nets made of nylon thread) and poured over books and papers about how to identify Neotropical bats. I bought field equipment and plane tickets, and off we went to spend two months in the rainforest.

My first night of netting bats was a comedy of errors. I didn't know how to cut poles for the nets, and I had no idea how tangled a mist net could become if you dropped it. (Figure 1.13) But I learned quickly, and immediately became hooked on the mystery and excitement of catching bats. I put up a series of mist nets along a trail and waited for the bats to arrive. Seeing that first small, struggling form in the net was a thrill I'll never forget. Carefully grasping the bat in one hand while using the other to untangle the threads, I was unprepared for how beautiful and beguiling a live bat could be. Warm and wiggling, with soft and flexible wings, and its tiny heart beating a mile a minute! And the diversity in that small patch of rainforest was completely unexpected (at least to me). So many species—all different—and I loved the challenge of trying to figure out the identity of each bat we caught. If I hadn't been totally hooked on bats before then, I was after that first field trip, and have spent my life ever since studying these endlessly intriguing animals. Recently I have started using harp traps which capture bats without entangling them, and which therefore can be left open all night. (Figure 1.14) I always feel a thrill of anticipation when I walk out to check a harp trap, wondering what new bats I may find in it. It's rather like Christmas morning with the undiscovered bats being like unopened presents!

2

Ancient Bats

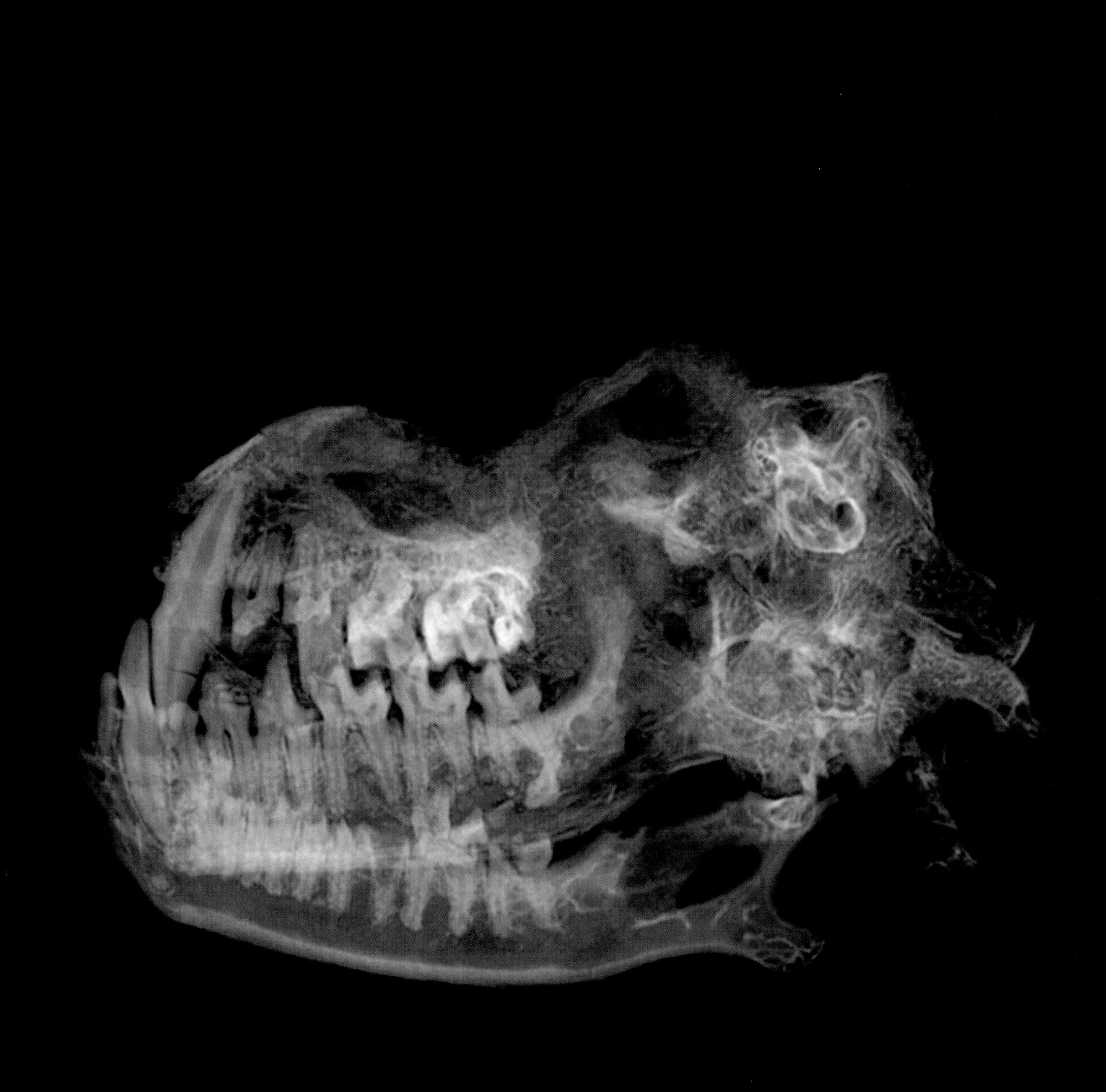

Origin and Evolution of Bats

One of the most amazing things about fossils of the oldest bats is that they are so similar to modern bats. Wings supported by elongated hand and finger bones (hand wings) are unique among vertebrates and are the most unmistakable feature of living bats. (See Chapter 1.) The wings and other skeletal structures important in flight are clearly visible in many Eocene fossils and differ little from those of living bats, as can be seen by comparing the skeleton of the modern bat in Figure 1.5 to those of the fossils in Figure 2.1 and elsewhere in this chapter. Notable among features of the bat skeleton that can be seen in fossils are elongated digits, a long forearm, a robust humerus (upper arm bone), broad shoulders, a strong clavicle (collar bone) and a wide rib cage. The sturdy sternum (breastbone) often has a small keel for attachment of muscles that power flight.

Figure 2.1.
One of only two known specimens of *Onychonycteris finneyi*, the most primitive known bat. Note the tiny claws on the tips of the fingers—these claws are evolutionary leftovers from the terrestrial ancestors of bats. Fingertip claws are absent from the long wing digits in all extant bats, probably having been lost to reduce the weight of the bones of the fingers and thus reduce the energy that bats need to flap their wings. The skull of *O. finneyi* was 23 mm long. Photograph courtesy of Joerg Habersetzer.

Nancy Describes a Fossil

I will never forget the day that I first met *Onychonyceris*, one of the oldest and best-preserved fossil bats. *Onychonycteris finneyi* is a 52.5 million year old bat that I described with colleagues in the journal Nature in 2008. (Figure 2.1) This fossil was found by commercial collectors in Wyoming, and my first hint that it existed was a photograph shown to me by a friend. I had studied many fossil bats before, and I knew instantly that this one was something completely new and different. Clearly it was a bat—with the long fingers perfect for supporting a skin flight membrane or patagium. I could see immediately that its shoulder, rib cage and sternum preserved all the specializations for powered flight seen in living bats. (Compare Figures 1.5 and 2.1.) *Onychonycteris*, however, also had several primitive features, including relatively large hind limbs and tiny claws on all five fingers of the wing. Living bats usually only have a claw on the thumb, and a few ancient bats and living flying foxes (Old World Fruit Bats) also have a claw on the index finger. But no bat that I had ever seen before—fossil or living—had claws on all five fingers. Tiny remnants inherited from their non-flying ancestors, these claws were lost during evolution of bats as their hands became more specialized for flight. But there they were, clear as day—five tiny claws on each wing of this new fossil bat! Even though it was just a photograph, I could tell that day *Onychonycteris* was more primitive than anything anyone had seen before. Wow! My heart raced and I was so excited that I danced a little jig in my office. It would be a few years before I would have an opportunity to study the fossil in person—it was subsequently donated to the American Museum of Natural History—but I have never forgotten the thrill of seeing it for the first time in that photograph and understanding its importance. Not quite a "missing link" between bats and non-flying mammals, it nevertheless gives a snapshot of an earlier stage in bat evolution than previously known.

Remarkably, bats appear in the fossil record on several continents at about the same time—the Early Eocene, roughly 47.5 to 55 million years ago. The oldest known bat—as of 2014—is *Archaeonycteris praecursor*, known only from a single tooth from Europe. It may be as old as 55.5 million years old. This tooth is very similar to teeth of somewhat younger fossil bats known from complete skeletons, yet is it subtly different in size and form. Accordingly, it has been recognized as a distinct species and given its own scientific name. Teeth are the most diagnostic parts of the mammalian skeleton and new species are frequently described based on teeth or fragments of jaws.

Fossil bats are known from the Early Eocene of Europe, North America, South America, India and Australia. The most anatomically primitive bats—(Figure 2.1) *Onychonyceris finneyi* and (Figure 2.2) *Icaronycteris index*—come from North America, but given the incompleteness of the fossil record, paleontologists do not know exactly where bats originated. This is because many extinct bats that are known only from teeth cannot be placed with certainty in a particular part of the bat family tree. It is possible that fragmentary bat fossils from a continent other than North America will someday be shown to be closer to the root of the bat tree than are *Onychonycteris* and *Icaronycteris*.

Recent studies of the evolutionary relationships of the major groups of mammals have concluded that the closest living relatives of bats are mammals that look and act nothing like bats: hoofed mammals and whales. The oldest members of that lineage, identified primarily from fossil teeth and jaws, were small-bodied terrestrial animals that lived just after the end of the Age of Dinosaurs, approximately 66 million years ago. Since the bat lineage must be as old as its closest fossil relatives, bats must have originated at the same time—around 66 million years ago. This leads Nancy and her colleagues to conclude that about ten million years of evolutionary history of bats is not yet known from fossils. The Paleocene ancestors of bats must have existed, but what were they like and where did they live? Nobody yet knows, and that is one of the enduring mysteries of the early evolution of bats. As of 2014, all fossils that can be definitively identified as "bat" come from rocks that are Eocene or younger. It is likely, however, that some fragmentary Paleocene fossils (bits and pieces of jaws and teeth) now sitting in museum drawers may eventually be identified as bats once more complete specimens are discovered.

Depending on when they originated, the first bats would have lived in a world full of birds and perhaps other feathered dinosaurs. Pterosaurs, the earliest known vertebrates to have evolved powered flight, were extinct by the end of the Cretaceous period. They probably did not overlap in time with the earliest members of the bat lineage—but scientists cannot tell for sure.

Figure 2.2.
This specimen of *Icaronycteris index*, perhaps the most famous fossil bat in the world, was described in 1966 by the late late Princeton University paleontologist Glen Jepsen. It reigned as the oldest and most primitive bat skeleton until *Onychonycteris* was described over forty years later by Nancy. Hopefully scientists will not have to wait another forty years for the next big discovery! The skull of *I. index* was 20 mm long. Photograph courtesy of Joerg Habersetzer.

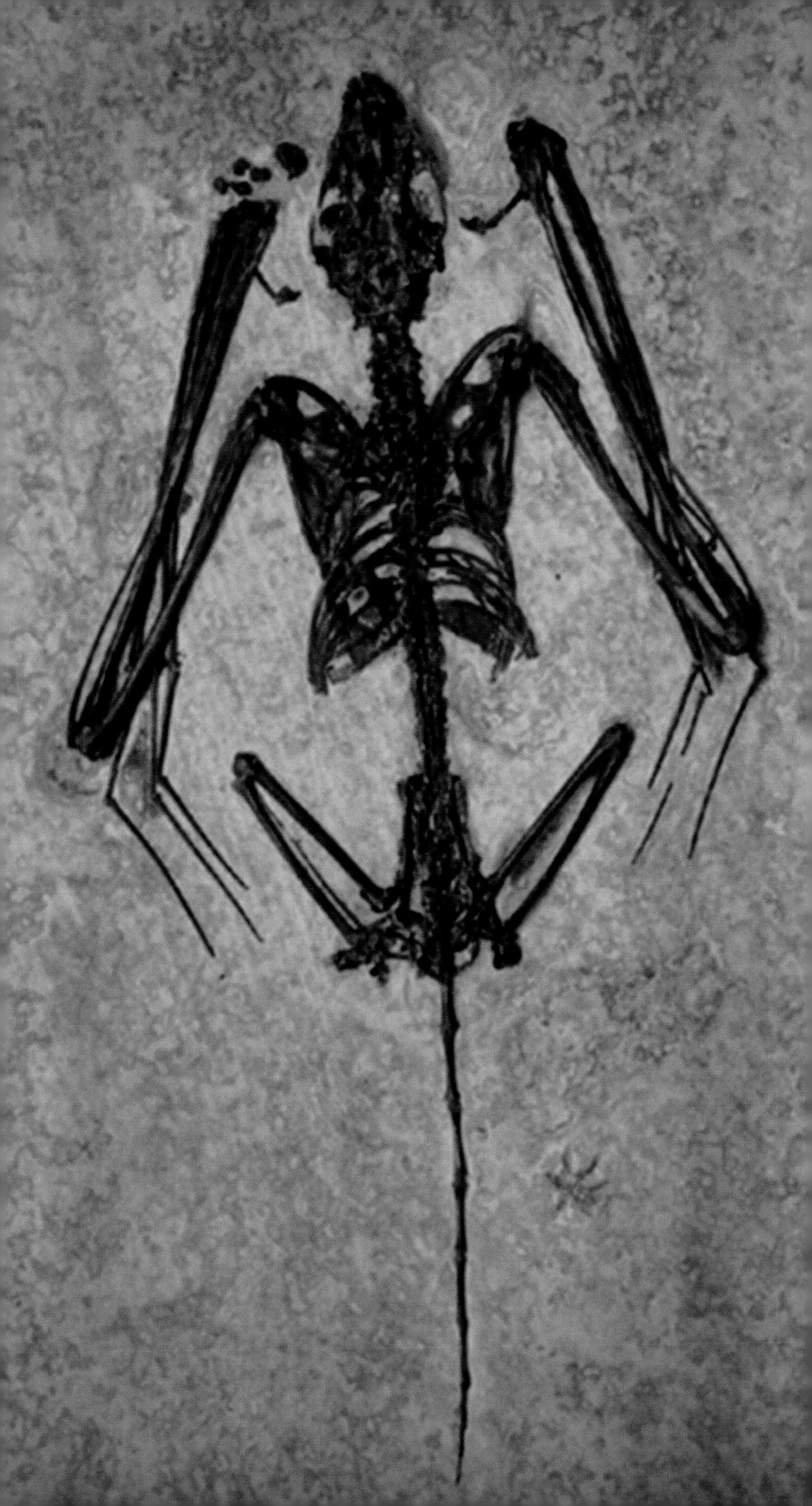

Birds appear in the fossil record in the Upper Jurassic, about 160 million years ago—nearly 100 million years before bats. Having survived the extinction event at the end of Cretaceous, birds were diverse and well established worldwide by the time bats came on the scene. Evolutionary biologists wonder what made it possible for bats to evolve in the face of potential competition from birds for food, airspace and roost sites. Nocturnal activity may have been an important factor. Virtually all birds are diurnal, but many mammals—especially small-bodied forms—are nocturnal. All living bats do most of their flying at night, although some bat species regularly emerge before dark to hunt flying insects at dusk. Most researchers presume that the earliest bats were nocturnal too, although there is no fossil evidence to support this conjecture. Being active at night may have prevented overt competition for food between the earliest bats and birds and may have protected early bats from predation by raptors and other diurnal animals. Being nocturnal also helps living bats avoid overheating, which may have been important in the early stages of the evolution of flight as well. Everything, however, has a cost—and when they took to air, early bats had to solve the problem of how to avoid obstacles and find food while flying under low light conditions.

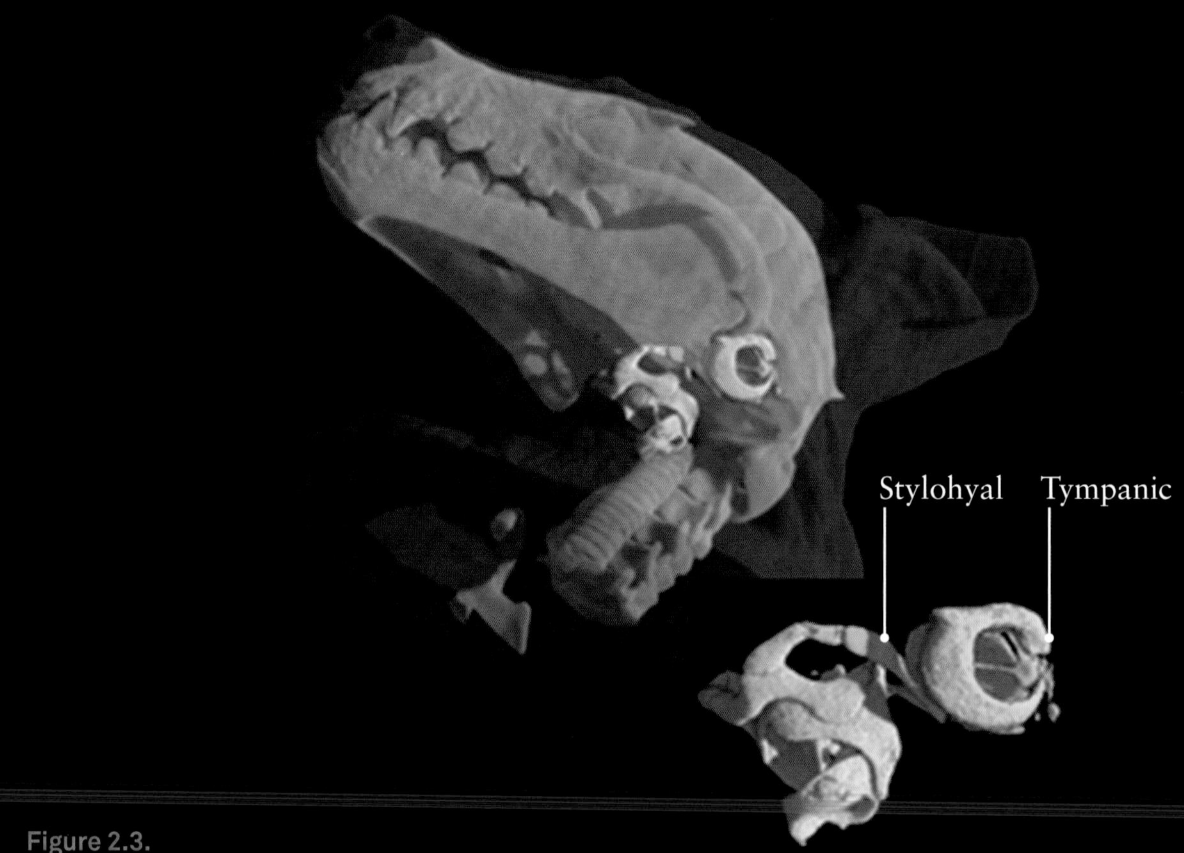

Figure 2.3.
A micro CT scan of the head and throat of a Blanford's Fruit Bat (*Sphaerias blanfordi*) showing the stylohyal and tympanic bones in situ and in detail. This species of bat does not use echolocation and the stylohyal does not directly contact the tympanic. Image courtesy of David McErlain and David Holdsworth.

Echolocation and Bat Diversity

Echolocation is another key factor in the evolutionary success of bats. (See Chapter 4.) Anatomical evidence suggests that most Eocene bats were echolocators. Using echolocation would have made early bats more effective at detecting and tracking flying insects. Taken together, flight, echolocation and nocturnal habits were a recipe for evolutionary success. But how can we tell if an extinct bat could echolocate? Reconstructing the behavior of extinct animals is notoriously difficult. In this case, features of the head and neck skeleton related to hearing and vocalization in living bats are correlated with echolocation. These same features can be examined in well-preserved fossils. These traits include the size of the cochlea (the bony housing of the inner ear) and the connection of the ear to the throat region via the stylohyal bone (Figure 2.3), a tiny bone that is part of chain of bones that support the throat muscles. (Figure 2.4) Bats that produce their echolocation calls in their voice box (larynx) have a relatively large

In echolocating bats, the part of the inner ear responsible for detection of high-frequency sounds is enclosed in an enlarged cochlea, suggesting that increased reliance on hearing high frequencies—critical for echolocation—may have caused this evolutionary change. The function of the stylohyal modifications is less clear, but Brock and his colleagues have hypothesized that having a direct bony connection between the throat (where echolocation calls are produced) and the ear (where they are detected) helps bats keep track of the timing between producing sounds and hearing returning echoes.

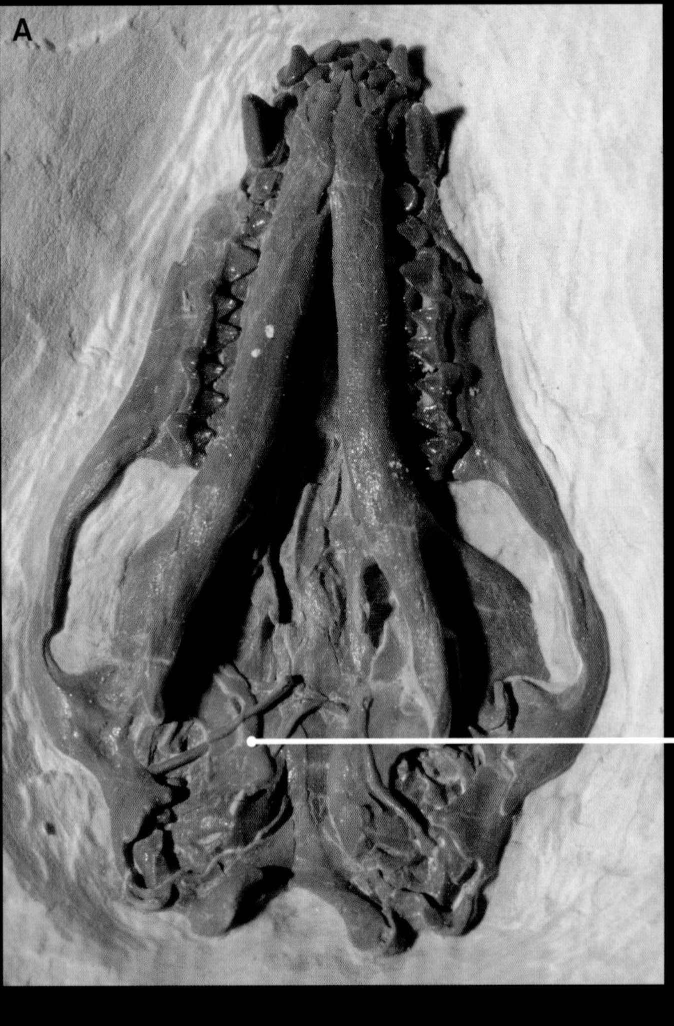

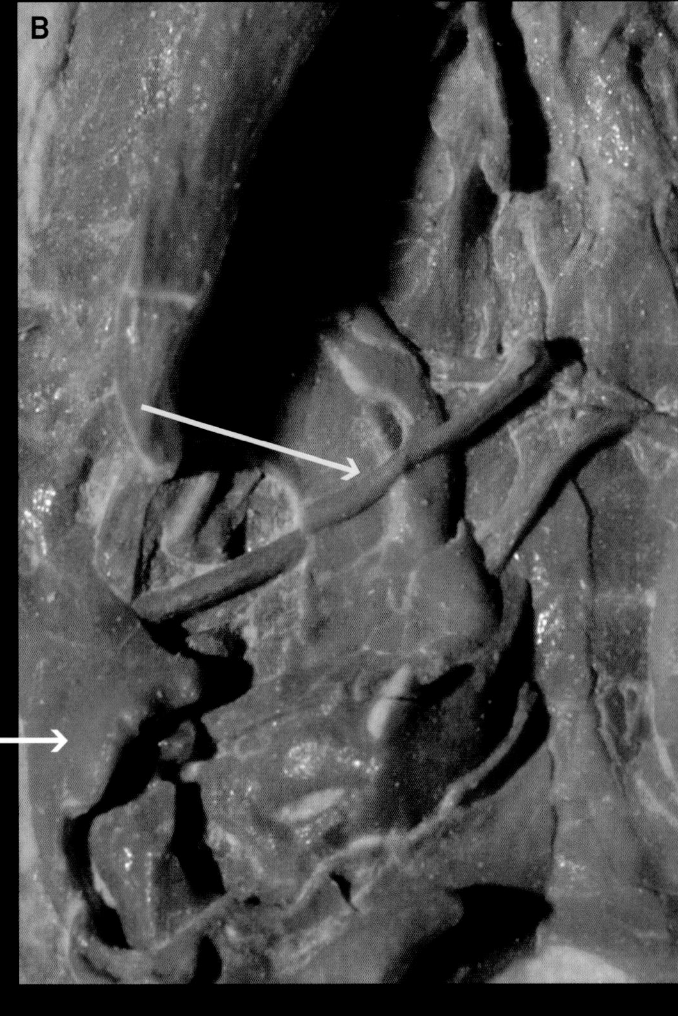

Figure 2.4.
The stylohyal bone (yellow arrow) is part of a chain of small bones connecting the ear to the throat. It is a long rod in *Onychonycteris* (**A** and **B**). In a Pearson's Horseshoe Bat (*Rhinolophus pearsoni*) the long rod extends from a flattened join with the tympanic bone over towards the hyoid (**C** and **D**). The stylohyal is part of a chain of bones that connects the voice box to the tympanic bone in living bats that produce echolocation calls in their voice box (larynx). The connections are not clear in *Onychonycteris*. White and Purple arrows show enlargements (**B** and **C**) of areas on the skulls (**A** and **D**).

Fossil bats like *Palaeochiropteryx* and *Icaronycteris* show clear evidence that they were echolocators. With the help of x-rays and CT scans, Nancy and her colleagues have been able to "see" into the skulls of these animals and measure the relative size of their cochlea. Stylohyal elements are likewise preserved in some specimens. Skeletal features of mammals can be very revealing about the habits of these animals. Most fossil bats are not preserved well enough for scientists to evaluate their ears, but scientists can use evolutionary trees to reconstruct the probable abilities of poorly-known fossil taxa based on their relationships to species that are better preserved. This is the technique that Nancy and her colleagues have used to conclude that most or all Eocene bats were

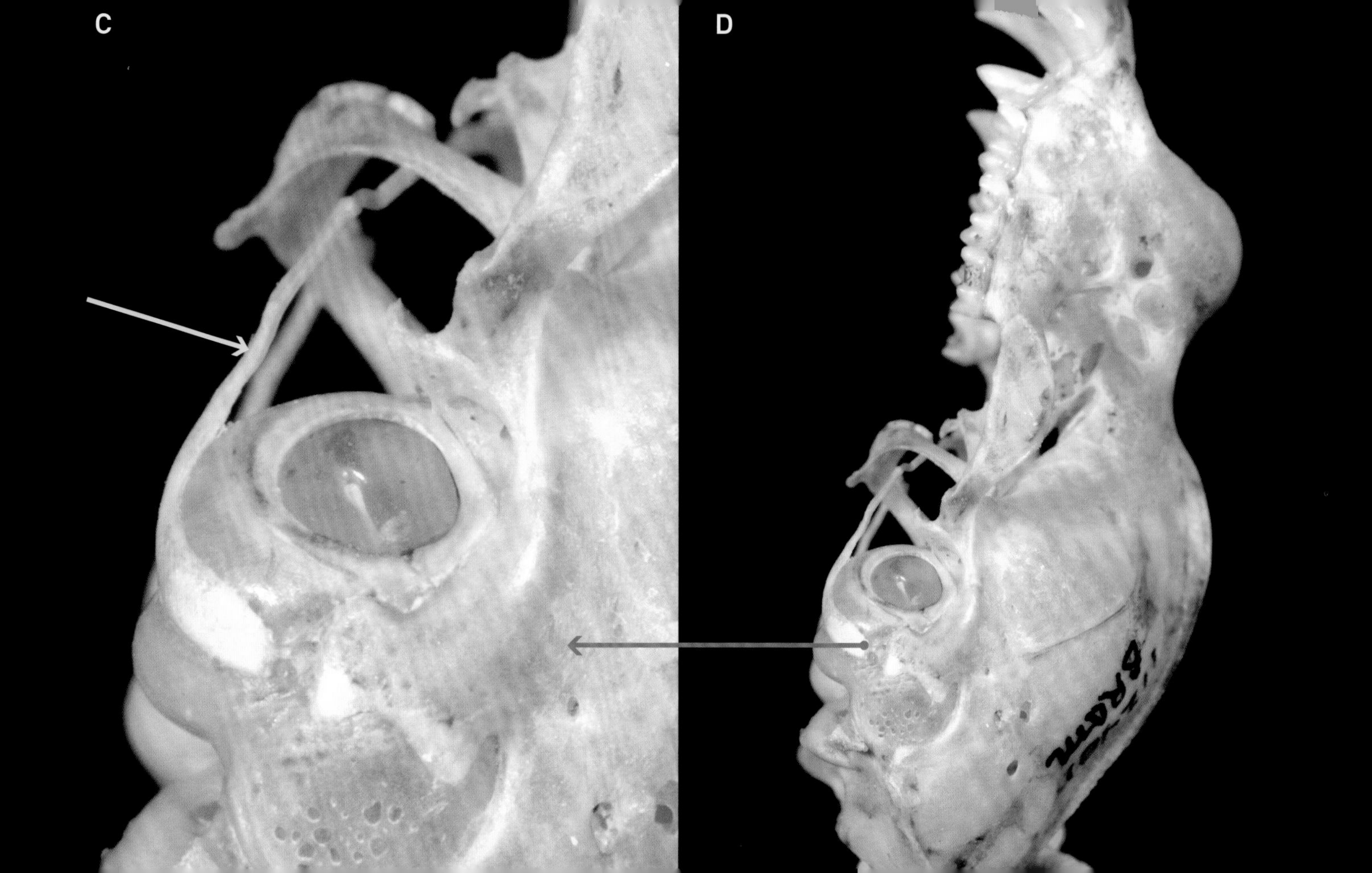
C
D

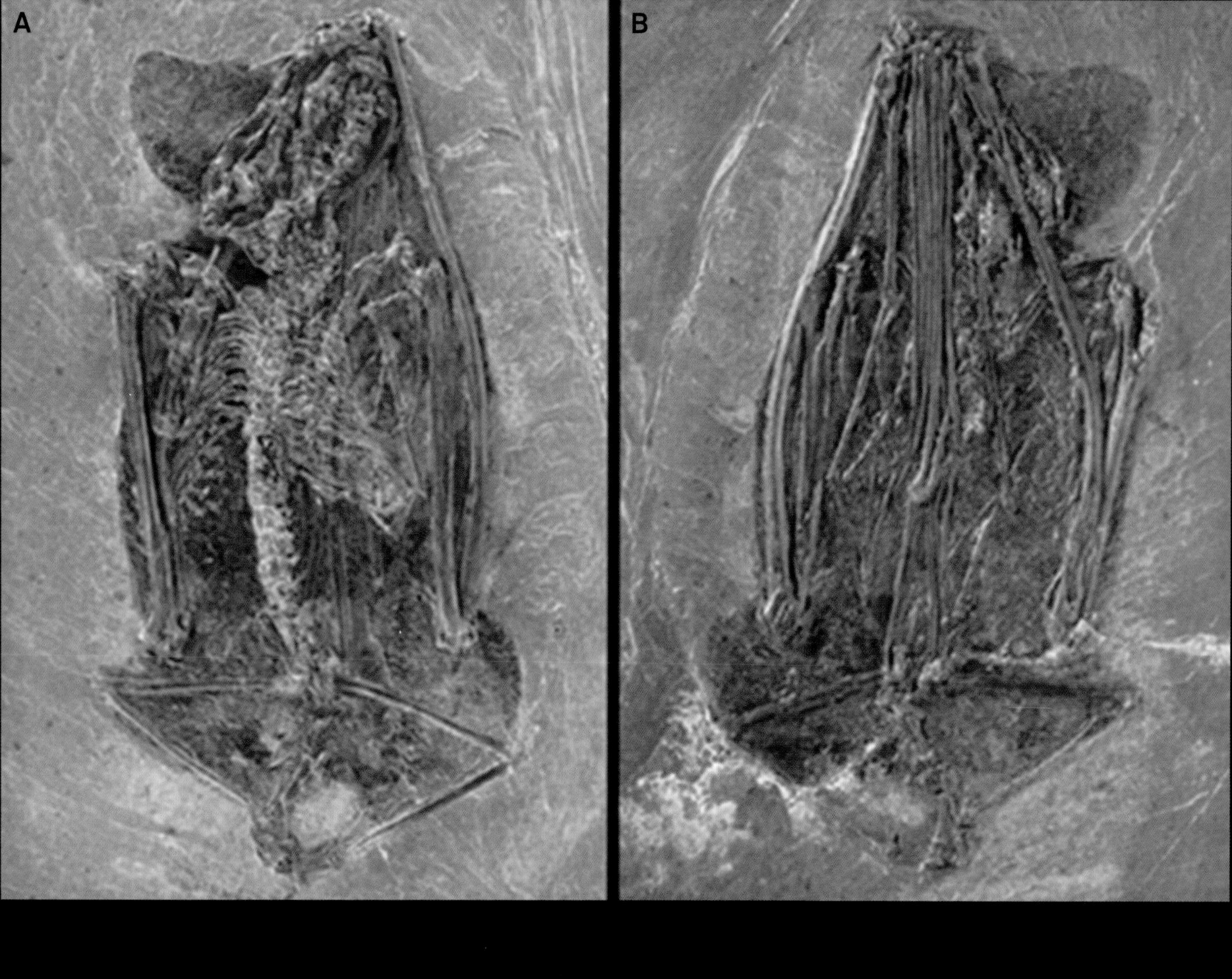

Figure 2.5.
Two views of the fossil of *Tachypteron franzeni* (**A** and **B**) from the Middle Eocene of Messel, Germany. This 47.5 million year old fossil is the oldest member of the family Emballonuridae, Sheath-tailed Bats. Shown for comparison is a living member of the family, (**B**) a Shaggy Bat (*Centronycteris maximilliani*). Fossil photographs by S. Tränker courtesy of the Senkenberg Messel Collection, Frankfurt; Shaggy Bat by Brock Fenton.

C

Today living bats are arranged in twenty family groups, nineteen of which use laryngeal echolocation—echolocation with sounds produced in the larynx (voice box). (See Table 1.1.) Many of these families have members that date back to the Eocene period. For example, the oldest known Sheath-tailed Bat, a living family that includes Shaggy Bats (*Centronycteris maximillianai*), is *Tachypteron franzeni* from the middle Eocene of Messel, Germany. (Figure 2.5) The paleontologists from the Senckenberg Research Institute, who described this spectacular fossil in 2002, recognized it as a Sheath-tailed Bat based on the form of its teeth, skull and the bones of its wings, especially the long and curved bones of its forearms.

In addition to ancient members of living lineages, such as *Tachypteron*, scientists recognize ten extinct families of bats. These include Onychonycteridae, Icaronycteridae, Archaeonycteridae, Palaeochiropterygidae and Hassianycteridae. Each of these names is a mouthful, but each name reflects the name of the first and most famous member of the family—for example, Onychonycteridae is the family that contains *Onychonycteris* and is named after that taxon. Most families include other species as well, often from different parts of the world than their namesakes. Palaeochiropterygidae, for example, contains nine species including fossil bats from Europe, India and China. Similarly, Icaronycteridae includes species from North America, Europe and India. While Onychonycteridae was named based on two specimens originally found in Wyoming in 2008, in 2012 it was shown that this family also includes species from England and France. Archaeonycteridae and Hassianycteridae both include species from Europe and India. The trend in Eocene bats seems to be for wide distribution of members of multiple evolutionary lineages. Only a few of the more poorly-known families are limited to only one continent, and that may be an artifact of preservation—simply very little is known about those groups, *e.g.*, Tanzanycteridae, known from only a single specimen, *Tanzanycteris mannardi*, from Tanzania.

Treasure Troves of Fossils

Many of the best-known fossil bats come from spectacular fossil treasure-troves known as lagerstätten (German, from Lager "lair or den" and *Stätte* "place"), sedimentary rock deposits containing large numbers of exceptionally well-preserved fossils. There are about sixty lagerstätten known worldwide but only about a dozen of these occur in rocks of appropriate age to preserve bats. Three of these deposits preserve exceptional bat fossils and faunas: the Eocene Green River Formation (ca. 53.5 to 48.5 million years old) the United States, the Messel Oil Shale (ca. 47.5 million years old) of Germany and the Oligocene Riversleigh Deposits (ca. 25 to 4 million years old) in Queensland, Australia. One of the special things about lagerstätten is that they offer a snapshot in time of entire faunas, not just individual species. Much of what is known about the earliest bats comes from these unique deposits.

The Green River Formation consists of fine-grained sandstone, mudstone and siltstone layers that were deposited in a series of freshwater lakes in what are now the states of Wyoming, Colorado and Utah. The deposits containing bat fossils, known as the Fossil Lake beds, are located in Wyoming and are approximately 52.5 million years old. The rocks themselves are pale laminated mudstones that beautifully preserved the dark brown bones of vertebrates. Most of the fossils from the Green River Formation are fish—over a million specimens representing twenty-five species are known. Other animals, however, were sometimes fossilized, including bats, insects and other arthropods, salamanders, frogs, turtles, crocodiles, lizards, snakes, birds and early members of the carnivore and ungulate lineages. Paleontologists think that the climate in Wyoming at the time these animals lived was moist- temperate to subtropical—rather like Florida today.

Two species of bats occur in the Green River beds, *Onychonycteris finneyi* and *Icaronycteris index*. (Figures 2.1 and 2.2) Only a few specimens of each are known—about a dozen *Icaronycteris* skeletons and two *Onychonycteris*. Some of these fossils, however, are quite well preserved, and Nancy and her colleagues have surmised a lot about these animals from comparisons with living bats as well as other fossil mammals. The two types of Green River bats differed in sizes and may have used different foraging techniques. *Icaronycteris* was smaller (probably ~25 grams based on the diameter of the shaft of the humerus), had wing proportions like many modern bats and seems clearly to have been capable of echolocation. In contrast, *Onyconycteris* was larger (~40 grams) and had shorter, broader wings. Rather than continuously flapping, it may have used a flutter-glide technique. The scientists who described this bat, including Nancy, believe that it was not capable of echolocation because it lacked an enlarged cochlea and there is no evidence that it had the stylohyal modifications seen in echolocating bats. Other scientists, including Brock, have argued that it might have had a stylohyal connected to the ear region, and thus might have been capable of primitive echolocation. Regardless, it was clearly a very different bat from *Icaronycters*, which apparently lived at the same time and place.

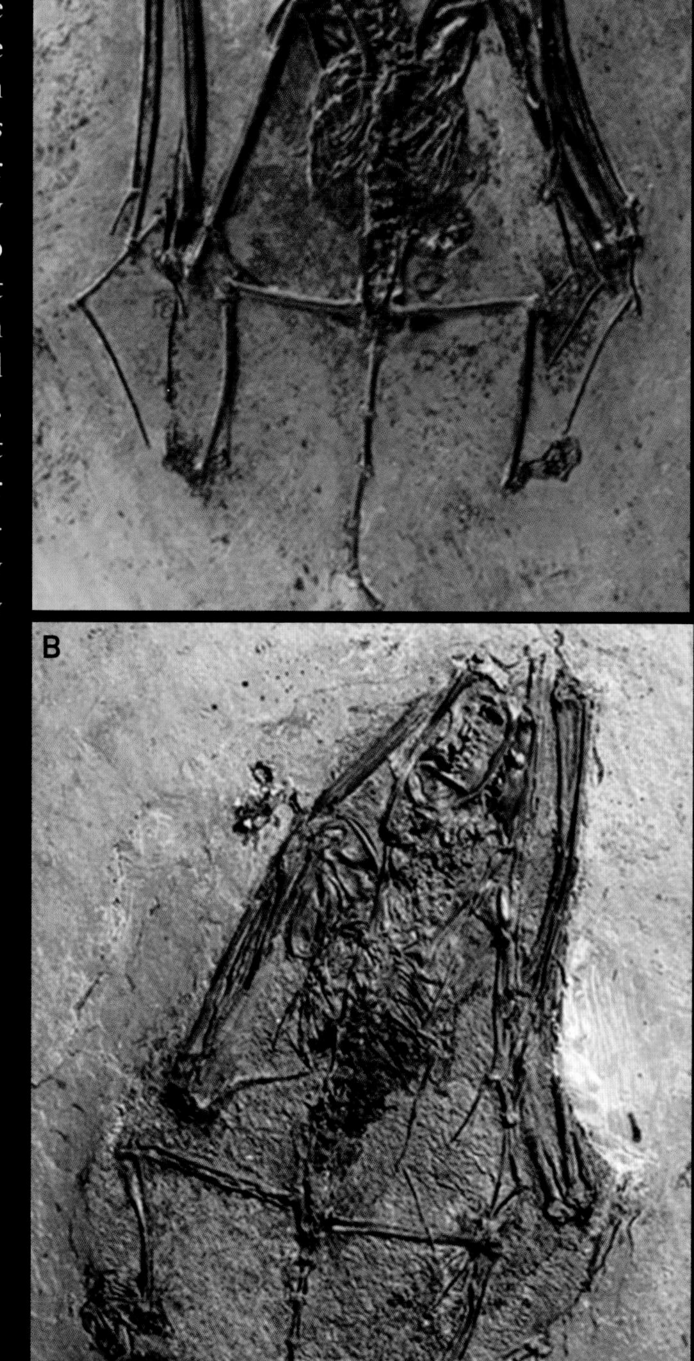

Figure 2.6.
Two of the fossil bats from Messel: *Palaeochiropteryx tupaiodon* and *Hassianycteris messelensis*, the latter (**B**) showing stomach contents (black area under the rib cage). Photographs A. Vogel (top) and E. Haupt (bottom), courtesy of the Senkenberg Messel Collection, Frankfurt.

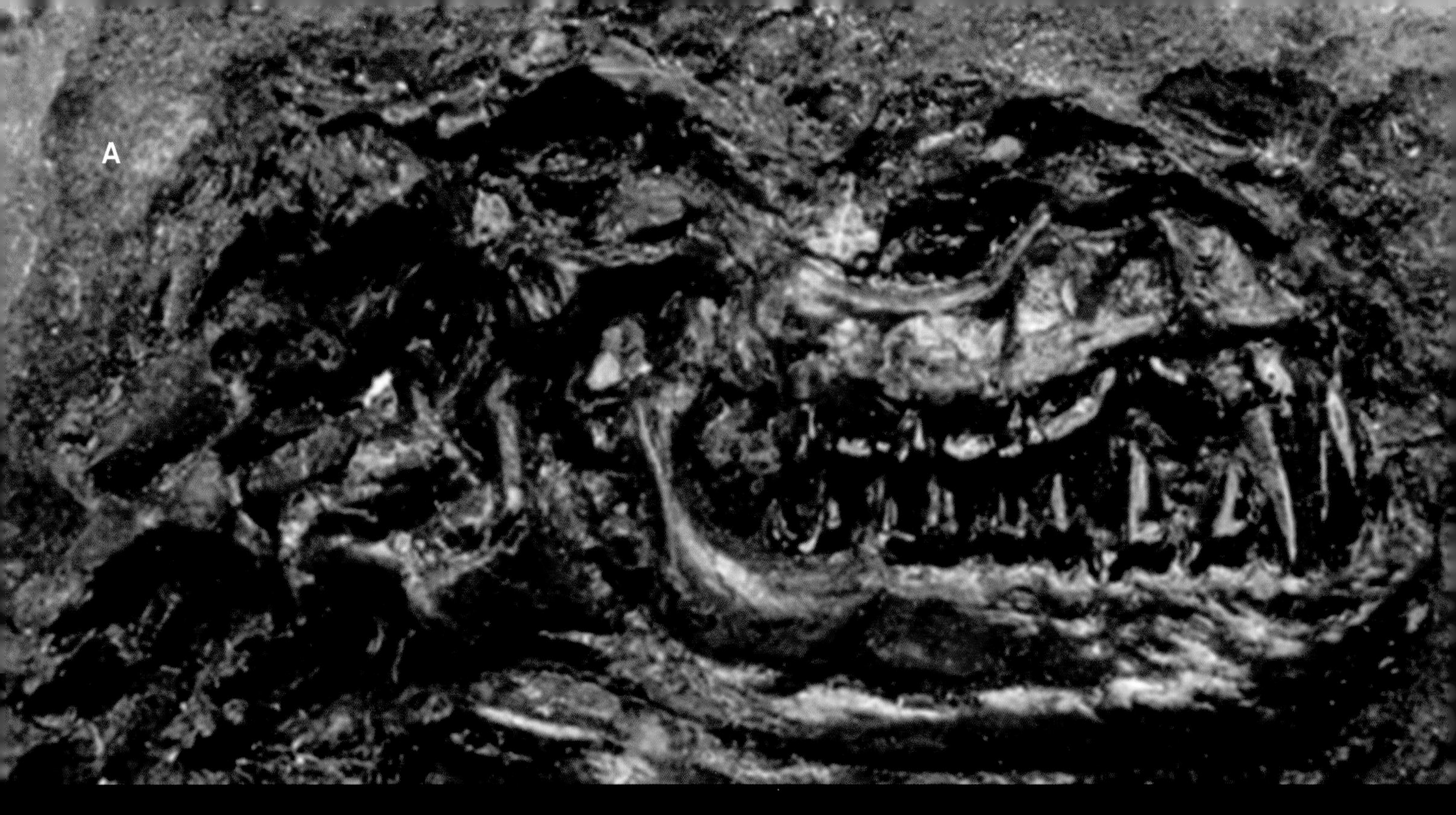

Messel Pit, a World Heritage Site located about 35 km. southwest of Frankfurt, Germany, contains even more species of extinct Eocene bats. Unlike the Green River, where bat fossils are exceedingly rare, Messel is rich with bats—over 700 individual bat fossils have been collected there. The Messel rocks are very fine-grained oil shales deposited in an ancient lake about 47.5 million years old. The pit that forms the Messel Oil Shale site used to be a quarry, but it is now protected because of the rich fossil fauna that it preserves. Thousands of fish, insects (some with distinct coloration still preserved), amphibians and reptiles have been found at Messel. It is, however, the mammal fauna that is most striking. It includes pygmy horses, large mice, a primate, a marsupial, hedgehogs and pangolins in addition to the bats. The climate at Messel was probably subtropical to tropical at the time these animals were alive.

A total of eight species of bats has been described from the Messel Pit, including the oldest member of the Sheath-tailed Bats, *Tachypteron franzeti* (Figure 2.5), two species of *Archaeonycteris*, two species of *Palaeochiropteryx* and three species of *Hassianycteris* (Figure 2.7). This bat fauna included both the smallest of the Eocene bats (*Palaeochiropteryx tupaiodon*, ~10 grams) and the largest (*Hassianycteris messelensis*; ~90 grams). Based on their skull anatomies, all of these bats were probably echolocators. They seem to have been very different in other ways, however, including how they foraged and what they were eating.

Figure 2.7.
Skull and X-ray of skull of *Hassianycteris messelensis* showing the robust jaw and teeth that it used to crunch the hard shells of beetles. The X-ray (**B**) clearly shows the teeth and their deep roots. The coiled cochlea also is clearly evident, the rounded white structure with a dark opening preserved in the upper left corner of the skull. Images by Joerg Habersetzer.

and the ground or water. The size of their cochleae (the bony housing of the inner ear) suggests that they were good echolocators. *Palaeochiropteryx* probably hunted small insects on the wing (a habit above the ground. *Hassianycteris* probably did the same in the Eocene skies around Messel, hawking large moths and insects over the forest and the lake.

The intermediate-sized bats at Messel—*Archaeonycteris* species—seem to have specialized on beetles. Their stomachs contain no moth scales or bits of caddis flies but are instead full of pieces of beetle armor, some of which still preserve iridescent colors. The wing shape of Archaeonycteris is similar to that of *Palaeochiropteryx*, suggesting that these bats could fly near the ground and vegetation. *Archaeonycteris*, however, is characterized by a cochlea of only moderate size, suggesting that while these bats could echolocate, they may not have used echolocation for detecting or tracking flying insects. Perhaps they gleaned prey from surfaces after detecting the sounds of beetles hitting the vegetation.

Riversleigh Bats

The Riversleigh deposits of Australia provide paleontologists with other snapshots of bat history, in this case on a southern continent and later in time—the Lower Oligocene/Upper Miocene through the Pliocene (25 to 4 million years ago). Another World Heritage Site, Riversleigh is the most famous fossil locality in Australia. Fossil-bearing rocks there record over twenty million years of evolutionary history during a period when rainforests were being slowly transformed into more arid grasslands. Fossils of numerous mammals, birds and reptiles have been recovered from limestones deposited in pools and caves at Riversleigh, including the oldest known monotremes, koalas, marsupial lions, bandicoots and rat-kangaroos. [Figure 2.8]

Figure 2.8.
Giant sperm of a 17 million year old seed shrimp (ostracod crustaceans) in a phosphatic preservation. The conditions in a Miocene bat cave in the Riversleigh deposits in Australia were ideal for this kind of preservation due to the high levels of phorphorus derived from the bat guano. The spiralization of the sperm is clear in this longitudinal view. The scale is 10 μm long. This discovery is the unexpected benefit of finding fossil bats and the roosts they used. Citation: Matzke-Karasz, R., Neil, J. V., Smith, R. J., Symonová, R., Mořkovský, L., Archer, M., Hand, S.J., Cloetens P. & Tafforeau, P. (2014). Subcellular preservation in giant ostracod sperm from an early Miocene cave deposit in Australia. *Proceedings of the Royal Society B: Biological Sciences*, 281(1786), 20140394.

A total of thirty-five species of fossil bats is known from Riversleigh sites, including twenty-three species from the Oligocene/Miocene deposits. All of these are echolocating bats—curiously, there are no Old World Fruit Bats preserved at Riversleigh. This is notable because the fauna of Australia today includes a dozen species of Old World Fruit Bats. Where were the ancestors of these bats during the mid-Tertiary? Nobody knows—but clearly they were not at Riversleigh, or at least did not fall victim to whatever circumstances lead to preservation of other bats there.

The bat fauna in the oldest deposits at Riversleigh includes many members of living bat families: early Old World Leaf-nosed Bats, False Vampire Bats, Sheath-tailed Bats, Vesper Bats, Plain-nosed Bats and Free-tailed Bats. Judging from specimens collected so far, bat species diversity was greater during the earlier periods (Oligocene/Miocene) than in the later deposits (Pliocene in age) at Riversleigh. This may be due to changes in climate and vegetation. Modern bat faunas are much more diverse in tropical forests than in grassland environments, so perhaps the differences in diversity seen over time at Riversleigh reflect ecological changes. Conversely, preservation may be biasing the composition of these faunas. For example, the Pliocene bat fossils come mostly from a single fossil cave at Riversleigh known as Rackham's Roost. Any local bat species not roosting in that cave when the deposits were laid down remain undiscovered. This might be one explanation for the absence of Old World Fruit Bats at Riversleigh—most living species roost in trees and avoid caves.

Bat Ancestors

The family tree (phylogeny) of bats has changed dramatically in the last decade. Phylogenetic trees today are based primarily on genetic evidence with fossils inserted based on anatomical analyses, but new discoveries can change prevailing views. The currently accepted tree indicates that all bats evolved from a common ancestor that could fly, and that powered flight evolved only once in mammals. Gliding mammals—such as flying squirrels and colugos—are not closely related to bats. There is currently no positive evidence that bats passed through a gliding stage in their evolution, although many specialists in bat flight mechanics believe that they did. The details of how bats achieved powered flight remain unclear because no fossil intermediates between *Onychonycteris* and the non-flying ancestors of bats are known. In short, there are no missing link fossils—or they haven't been recognized yet.

The ancestors of bats were probably small, nocturnal insect-eaters that lived mainly in trees. Scrambling around in the branches, they would have been agile and had keen senses, including good hearing and low-light vision. Like living insectivores such as shrews and moles, they probably had a relatively long muzzle with molar teeth that had sharp cusps and "W" shaped crests for efficient puncturing and slicing small, armored arthropod prey. Jaw fragments and fossil teeth showing these features are common in fossil collections around the world—indeed, they are frequently found in Late Cretaceous (~100 to 66 million years ago) and Paleocene (~66 to 56 million years ago) fossil sites. Perhaps paleontologists already have fossils of enigmatic "pre-bats" in their collections but haven't yet recognized them for what they really are! For now, the most primitive known member of the bat lineage is *Onychonycteris*—already clearly a bat.

Other Features of Fossil Bats

Hanging upside-down by their hind feet is a behavior of most bats. Specialization of the forelimbs as wings may be an underlying practical reason for this roosting habit. Hanging upside down with wings folded may serve to protect the flight membranes, plus it can help provide a layer of insulation around the body when the bat is resting. This posture also facilitates take-off when a bat leaves its roost. To initiate flight, most bats need only let go and spread their wings. While there is no direct evidence, scientists think that hind-limb hanging probably evolved early in the bat lineage because fossil bats have feet like living bats. One of the bones of the big toe is elongated so all the toes are approximately the same length, which means that all five claws on the feet can be used together for grasping the substrate, be it a branch, wood, rock or some other surface.

Small body size is a consistent feature of most bats. As adults, living bats range in size (body mass) from 2 grams to over 1500 grams but most species weigh less than 50 grams. The fossil record provides no evidence of much larger bats, suggesting that they have always been relatively small mammals. Reconstructions of ancestral body size suggest that the primitive body size for bats was 13 to 18 grams. This means that both larger and smaller body sizes represent evolutionary specializations, probably associated with changes in prey type, habitat, echolocation behavior, roosting habits and/or flight behavior. In general, small body size probably reflects the fact that most bats eat insects, most of which are relatively small. The only living insectivorous mammals that achieve large body sizes are those that eat ants and termites, such as anteaters and the aardvark. Their prey occurs in very large numbers, neatly organized into colonies that can be raided by mammals with large claws for tearing their way into the nest. Insectivorous bats, lacking such claws, must rely on insect prey spread less densely around the landscape. (Though see Figure 6.12.) As will be discussed in later chapters, the physics of sound—a key aspect of echolocation—may also constrain body size. Regardless, it is interesting to think that the largest and smallest bats that have ever lived may be alive today.

Where Bats Fit in the Mammal Family Tree

Figure 2.9 illustrates the position of bats among living mammals. This "family tree" (or phylogeny) proposes that bats' closest relatives among living lineages are shrews, hedgehogs, whales, cows, horses, carnivores and scaly anteaters—a diverse group that collectively is called Laurasiatheria. Somewhat surprisingly, among all of these groups it is the horse + cow + whale lineage that is the most closely related to bats. The evidence supporting this arrangement is a combination of genetic markers and morphological features. Some aspects of this tree are very different from traditional views of bat relationships. Biologists used to think that bats were most closely related to either the group of small insectivorous mammals that includes shrews and their relatives or to primates and their kin—colugos among them. The latter relationship made sense because colugos are gliding mammals that have membranes between their fingers, an arrangement that one could imagine having been ancestral for bats. There is now, however, compelling evidence that bats are more closely related to carnivores and ungulates (cows, horses, whales) than to colugos or shrews.

Figure 2.9.
Cladogram showing the evolutionary relationships of bats to other mammals. Bats belong to a group of mammals known as Laurasiatheria (gray box), which encompasses such extant creatures as dogs, whales, cows, carnivores, scaly anteaters and some insectivorous mammals such as shrews and hedgehogs.

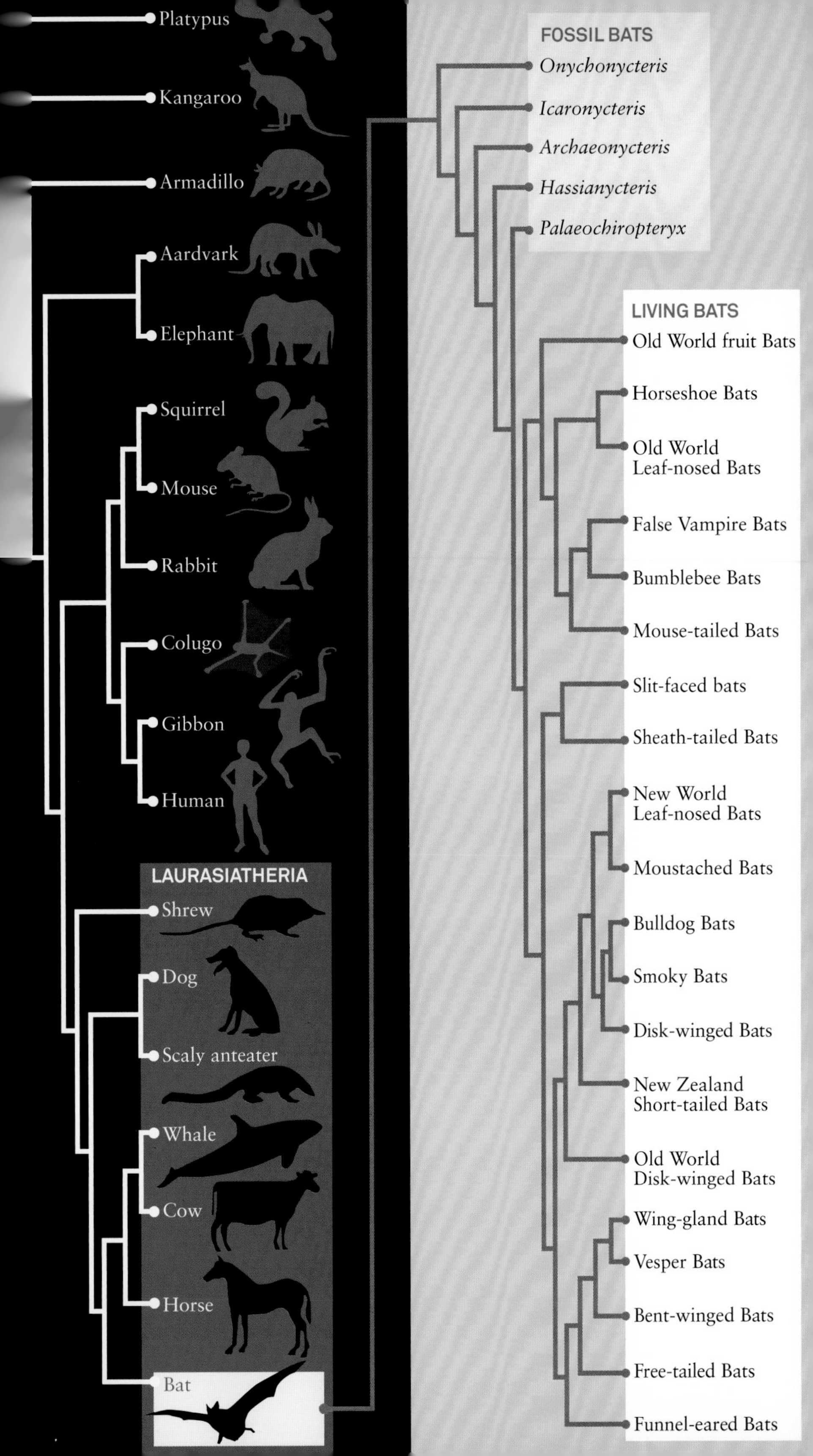

Platypus
Kangaroo
Armadillo
Aardvark
Elephant
Squirrel
Mouse
Rabbit
Colugo
Gibbon
Human
LAURASIATHERIA
Shrew
Dog
Scaly anteater
Whale
Cow
Horse
Bat
FOSSIL BATS
Onychonycteris
Icaronycteris
Archaeonycteris
Hassianycteris
Palaeochiropteryx
LIVING BATS
Old World fruit Bats
Horseshoe Bats
Old World Leaf-nosed Bats
False Vampire Bats
Bumblebee Bats
Mouse-tailed Bats
Slit-faced bats
Sheath-tailed Bats
New World Leaf-nosed Bats
Moustached Bats
Bulldog Bats
Smoky Bats
Disk-winged Bats
New Zealand Short-tailed Bats
Old World Disk-winged Bats
Wing-gland Bats
Vesper Bats
Bent-winged Bats
Free-tailed Bats
Funnel-eared Bats

Figure 2.10.
A fragment of an upper jaw with two molar teeth from the Upper Eocene of Egypt. This fossil represents a bat species never before seen by scientists. Nancy is currently working with paleontologist colleagues to formally describe and name this new taxon, which will also be the basis for a new bat family. The large, bulbous hypocone cusp (indicated by the yellow arrow) is a unique feature of this animal. Based on comparisons with living species, we think that this bat was probably omnivorous because broad teeth and rounded cusps are seen in living species that include plant material (like fruit) in their diet. Photograph by Paul Velazco, American Museum of Natural History.

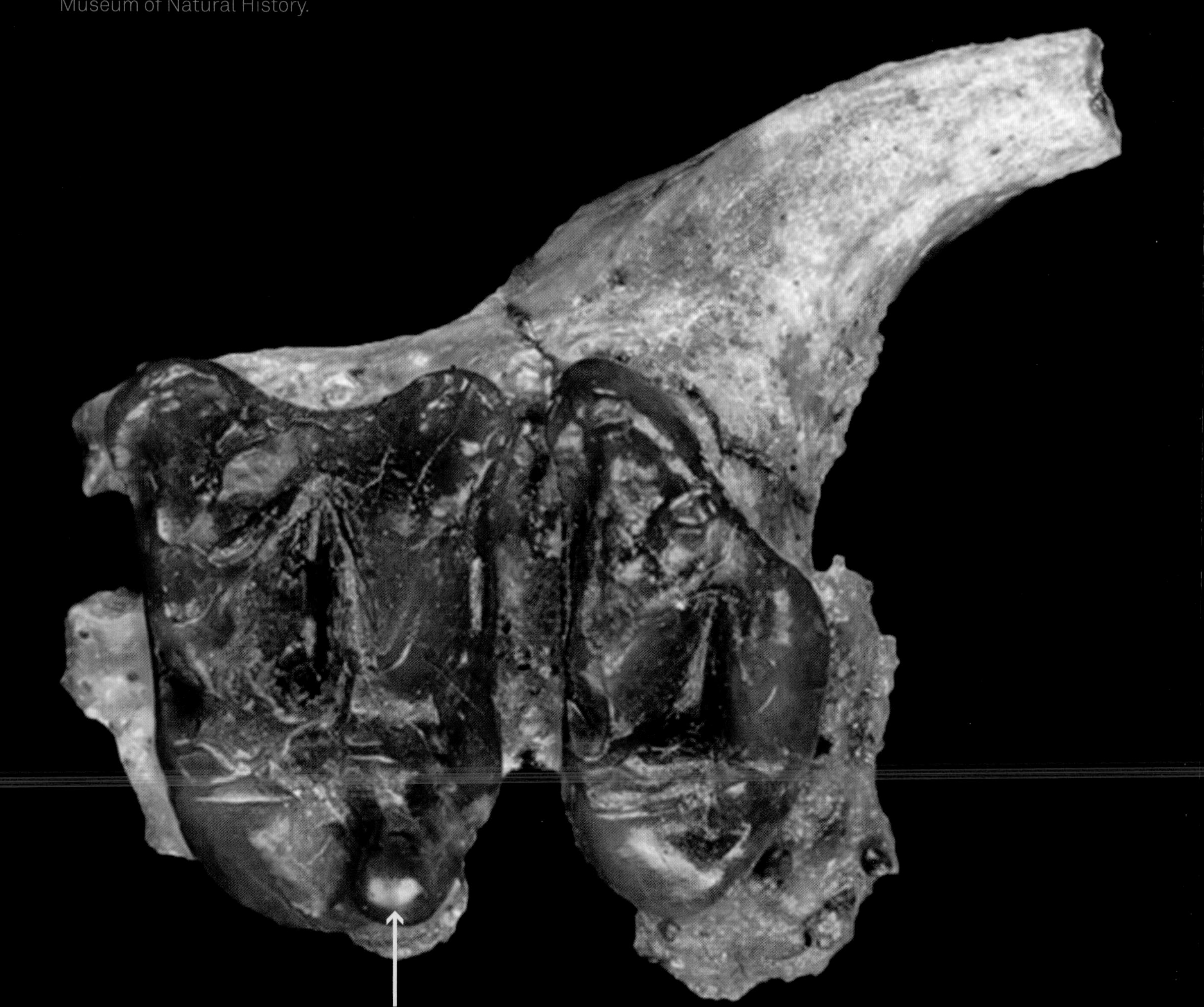

Relationships among the various lineages of laurasiatherian mammals have proven difficult to resolve, but modern genetic techniques are providing vast amounts of new data. In 2013, a group of biologists from Queen Mary University in London published a new tree for bats based on comparisons of complete genomes of twenty-two mammal species. Their results unambiguously placed bats in the position shown in Figure 2.9. What would the common ancestor of such diverse animals as bats, cats, cows and whales have looked like? It is hard to say, but most likely this common ancestor was a small insectivorous mammal of some sort. Clearly, the living members of the laurasiatherian sister-group of bats, *e.g.*, cats and cows and whales, are all highly-derived animals specialized for very different lifestyles that do not include flight. The closest relatives of bats may actually be laurasiatheres whose lineages have gone extinct. To find out, researchers need to unearth fossils even closer to the origin of bats than *Onychonycteris*.

More and More Fossils

There are over 280 extinct species of bats known (as of 2014) and many new bat fossils are discovered every year. Few fossil bats, however, are found as complete skeletons. Most bat fossils are bits and pieces of teeth and jaws, like fragment shown in Figure 2.10. Even though this specimen is very incomplete, the shape of the teeth and arrangements of cusps and crests indicates that it represents a new species—probably the first known member of a whole new family of extinct bats. By comparing this specimen with other living and fossil bats, Nancy and her colleagues will determine how similar to or different it is from previously known species. Much can be learned even from fragmentary fossils if the right parts are preserved. Teeth are particularly informative because they fossilize well, are indicative of diet and are usually species-specific (which means that every species has unique dental traits that differentiate it from all or most other species). As noted earlier, paleontologists can often tell if they have a new bat species from just a few teeth.

The humerus (upper arm bone) is another bone that is very useful for distinguishing one bat species from another. The humerus is a key element in the flight apparatus, and its form probably reflects differences in flight capabilities and habits. Details of the shapes of surfaces at the two ends of the humerus where they connect with other bones differ in different lineages of bats, often being characteristic of individual species. Some fossil bats are known only from isolated humerus bones. One such species is *Mormoops magna* from the Caribbean. Living species of *Mormoops* occur throughout much of the Neotropics, ranging southern Arizona to Peru and on many Caribbean islands. The large extinct *Mormoops*—bigger than any living member of the group—was described in 1974 based on two isolated humeri recovered from Pleistocene Cuban cave deposits. Interestingly, in 2012 Nancy and her colleagues found two more specimens of *Mormoops* magna in fossil cave deposits from the Dominican Republic—and they were also isolated humeri! (Figure 2.11) She hopes that future collections of fossil bats from caves in the Caribbean will eventually lead to more complete specimens of this bat. In the meantime, *Mormoops magna* remains a "headless bat" and she can only conjecture about what the rest of it looked like.

Nancy Receives a Photograph

One of the exciting things about working on bat fossils is that there are always chances for new fossil sites to be discovered, new animals to be encountered and new insights to be gained into bat history and evolution. Every year brings advances in our knowledge about bats. In 2012 a friend sent me a picture of the floor of a flooded cave in the Dominican Republic. (See Figure 2.11.) While surveying the cave for fossils, scuba divers associated with the Museo del Hombre Dominicano had found an area littered with bat bones. Just looking at the picture, I could tell that there were literally dozens of fossil bats just waiting to be picked up. The scuba divers collected the specimens and sent them to me in New York for analysis. I compared them with skeletons of modern bats and found that most species in the cave are still alive today. Two now-extinct species, however, were also found in the collection, including the giant Ghost-faced Bat (*Mormoops magna*). (See page 57.) These finds document recent changes in the bat fauna of the island of Hispaniola, perhaps linked to climate change or stochastic events such as hurricanes.

Figure 2.11.

A photograph taken underwater in Oleg's Bat Cave in the Dominican Republic. The dark objects are mostly bat bones—dozens of fossil skulls and limb bones, all resting on the sandy floor of the cave. These specimens, presumed to be from the Late Pleistocene (ca. 125,000 to 11,700 years ago), were collected by scuba divers and subsequently identified by Nancy and her colleagues. They found that the assemblage included eleven species, two of which are now extinct on the island. Photograph courtesy of the Dominican Speleological Society and Phillip Lehman.

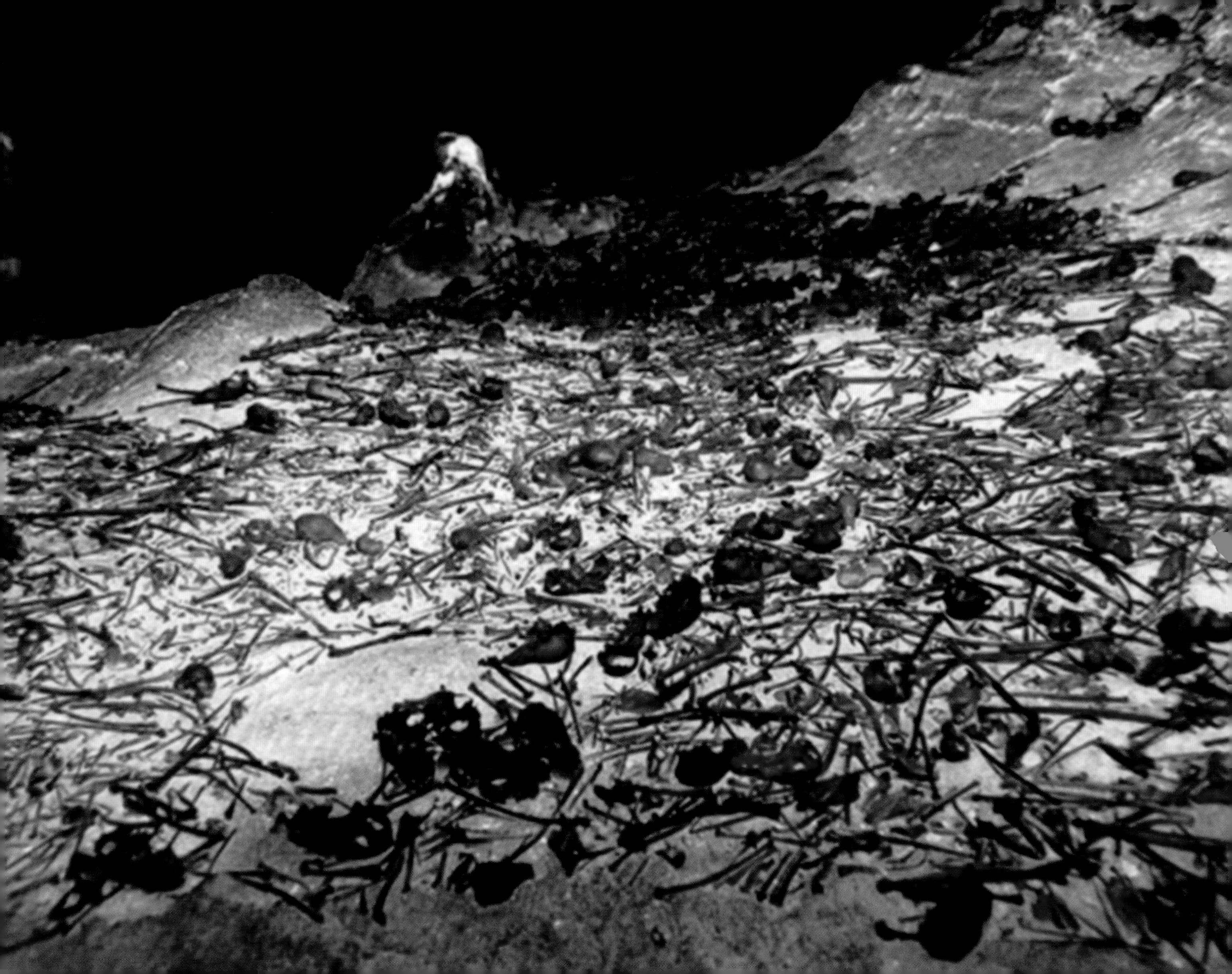

3

Taking Off

Figure 3.1.
A flying female Lesser Yellow-shouldered Fruit Bat with wings poised for the downstroke.

Bat Abilities

The flight of a Little Brown Myotis released in an enclosed room demonstrates the abilities of a flying bat. The bat usually flies around quickly, apparently looking for a way out. Typically, it starts at the ceiling and works its way around the perimeter, inspecting the walls, windows and doors. In the absence of an opening, the bat works its way ever lower in the room until it is flying just above the floor, negotiating its way amongst the legs of furniture and observers. When there is enough space, the bat will fly out under the door. If not, the animal lands and tries to crawl out under the door. If it cannot find any possible avenue of escape, it typically roosts in the darkest available space, often behind a curtain.

Indian False Vampire Bats (*Megaderma lyra*)—carnivorous bats more than four times larger than Little Brown Myotis—show the same general behavior. When there is an opening in the wall, an Indian False Vampire Bat will go through it. When the width of the opening is larger than its wingspan, the bat just flies through. When the opening is smaller than the bat's wingspan but large enough to admit its body, the bat flies directly at the opening and, just as it arrives, tucks up its wings and dives through. If a mischievous experimenter puts fine threads across the opening, the bat bulls its way through. When fine wires rather than threads obstruct the opening, the bat looks elsewhere for a way out. These observations represent impressive demonstrations of flight performance and control, as well as of the sensory abilities of bats.

An Exclusive Club

Although gliding has evolved numerous times in mammals, only bats among mammals are capable of sustained flight powered by flapping wings. (Figure 3.1) Flight places bats in an exclusive club of animals that includes (or included) birds, pterosaurs and insects. (See Chapter 1.) Flight confers mobility as well as ready access to resources not available to animals that walk, run, burrow or climb. These resources may include food and/or places to roost. (See Chapters 5 and 6.) Being able to fly also can help bats to avoid and escape would-be predators, and also allows some bats to migrate long distances.

Like most other flying animals, bats tend to have small body sizes, which permit them to subsist on smaller amounts of food than larger animals. In terms of total quantity, the Big Brown Bat (*Eptesicus fuscus*) living in your attic will eat much less than your dog or your cat. The bat, however, eats proportionally much more per gram of body mass than does a dog or cat, or even a mouse the same size as the bat. Flight is energetically expensive if you measure energy burned per unit time in flight, not by unit distance covered. As a result, the metabolic rates and concomitant energy needs of bats, birds and insects are generally much higher than those of similarly-sized animals that cannot fly. Likewise, an actively flying bat has a higher metabolic rate than the same individual when not flying. The actual costs of flight depend upon flight speed, wind conditions and the size of the animal.

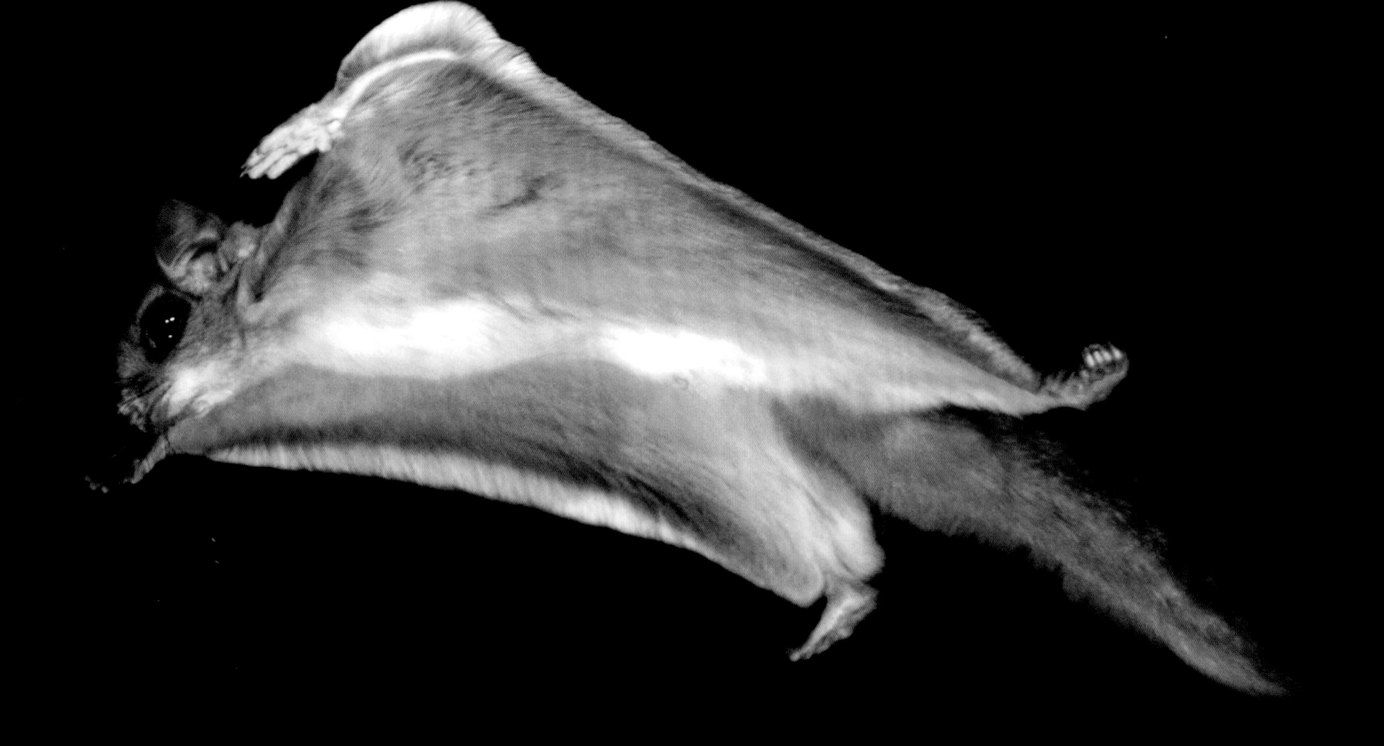

The flapping flight of bats (and birds, insects and pterosaurs) involves generating lift and thrust or propulsion through repetitive wing movements, which distinguishes it from gliding and parachuting. (Figure 3.2) The wings of all flying vertebrates are modified forelimbs, but the anatomical details of the wing, *e.g.*, structure of the bones, muscles and wing surface, differ among these groups. (See Chapter 1.) Among animals, flight evolved independently four times beginning with flight evolving in insects by about 320 million years ago, in pterosaurs by about 220 million years ago, and in birds by 160 million years ago. As noted in Chapter 2, flight was well developed in bats by 52 million years ago but must have originated somewhat before that time. Arguably, the evolution of flying insects set the stage for the appearance of other flying animals because they provided an abundant food source most effectively exploited by a flying predator.

Lift

Flight requires production of lift and thrust (propulsion), a combination that allows an object (or animal) to rise in the air and move forward through the air. In a fixed-wing aircraft, lift is generated by movement of air across an airfoil—the wings—and thrust (propulsion) is generated by the engine(s). Lift is generated because air traveling across the upper surface of an airfoil has farther to travel than air moving across the lower surface. The faster moving air across the upper surface of the airfoil results in lower air pressure above the wing than below the wing, which effectively sucks the wing upward in the air. Meanwhile the air going below the wing is moving slower, which generates more pressure and effectively pushes the wing up. Hence an airplane with air moving over its wings is pulled up from above and pushed up from below at the same time, forces that together constitute lift.

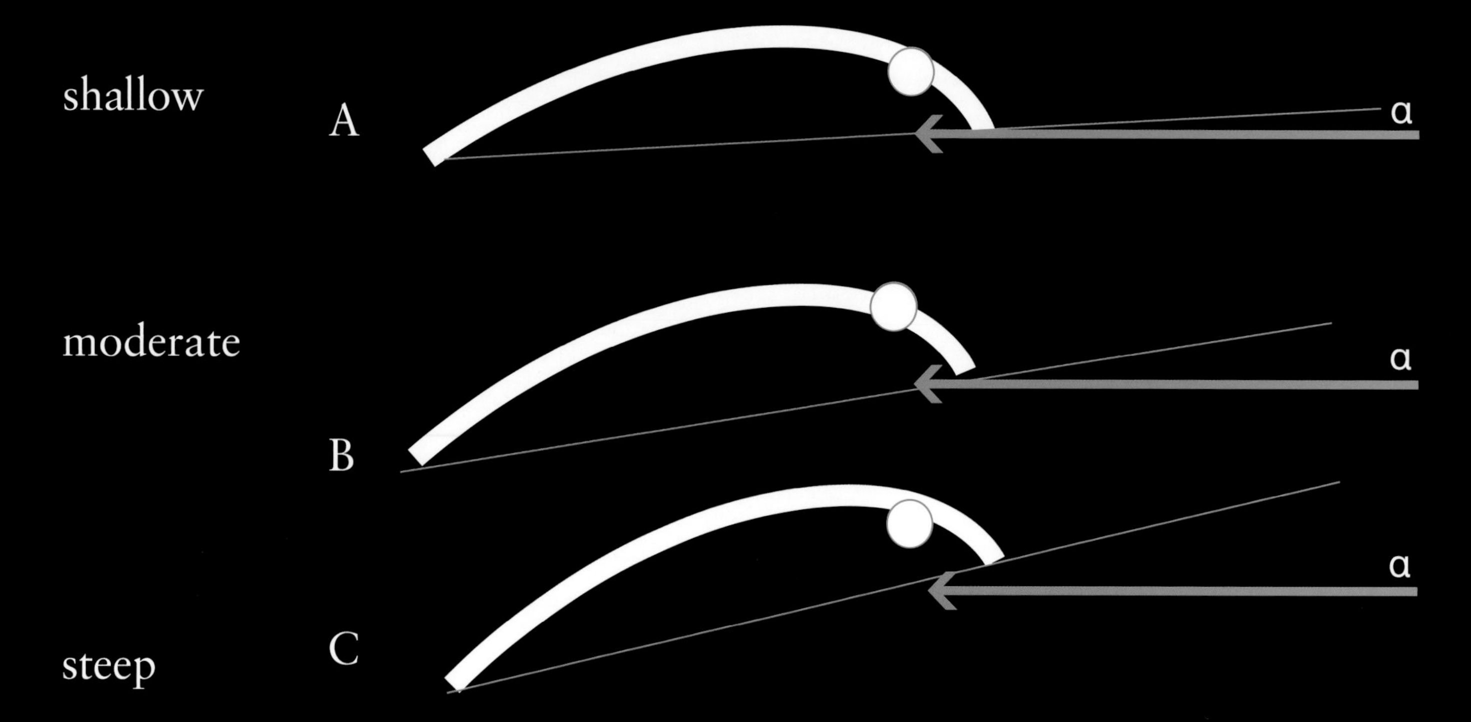

Figure 3.3.
Three diagrammatic cross-section views of a bat wing; the white circle is the forearm. Angles of attack are shown as **α**. Compared are shallow (**A**), moderate (**B**) and steep (**C**) angles of attack. Here α is relative to the movement of air (arrows).

The more curved the airfoil, and the greater the speed of the airflow over the upper surface of wing and hence the greater the lift, providing the degree of curvature does not impede the flow of air. The angle at which the airfoil moves through air (the angle of attack) strongly influences the outcome. (Figure 3.3) If the angle of attack is too shallow, less lift is produced. (Figure 3.3A) If the angle of attack is too steep, air tends to be shed from the upper surface of the wing and stalling may occur. (Figure 3.3C)

Bats (and birds and pterosaurs) are/were not fixed wing aircraft. In bats, the airfoil section of the wing is located primarily between the fifth finger and the body, and this area generates lift during part of each upward flap of the wingbeat cycle. (Figure 1.1) Bat wings are compliant, fluttering to some extent as the wing moves. This fluttering apparently does not interfere with generating lift, but may affect the amount of lift generated at any given moment.

Thrust

Thrust or propulsion is the force that propels a flying machine in the direction of motion. In aircraft, engines produce the thrust that overcomes the drag generated as an aircraft moves through the air. In fixed wing aircraft, thrust is generated by the engines operated by pulling (propeller) or pushing (jet) the airplane. In bats, the downward flapping of the wings produces thrust.

The traditional view that flying bats generate most of energy for thrust with the distal parts (those situated away from the center of the body) of their wings is less accepted today. Rather, we know that the movement of air across the top of the wing produces vortices. These circulating patterns of rotating air appear to generate most of the aerodynamic force required for thrust. A bat flying on a straight and level course continually changes the positions of its body and limbs. Examine images of flying bats, *e.g.*, Figures 3.4 and 3.5, and note changes in the bats' postures as well as the details of the positions of their thumbs, hind limbs and, where appropriate, tails. When the bat is maneuvering, it constantly adjusts the positions and attitudes of wings and the structures that support them, from hind limbs to thumbs, hind feet to calcars. (Figure 3.5) The details of position of arm and hand bones are critical, and subtle changes in flexion and extension at various joints can produce large changes in flight direction. Among flying animals, bats have high angles of attack. As this angle increases it generates more lift. Stalling occurs at the point when generation of lift stops. In horizontal flight, this can happen when a flying bat approaches a wall. The bat tends to slow down and quickly reaches the tipping point when it either lands or turns and dives to regain airspeed and resume horizontal flight.

Drag

Maintaining the pattern of air flow across the wings is vital to flying animals. The body posture of a flying Egyptian Rousette (Figure 3.5) provides a further variation on the themes illustrated by a flying Big Brown Bat (Figure 3.4). For aircraft, accumulation of ice on the upper surface of the wing interferes with the pattern of airflow, and thus reduces lift. Accumulation of ice on airplane wings also increases the drag caused by friction generated by resistance to air moving across the surface of a moving body. Drag increases the energy required to maintain speed and stay aloft. Drag helps explain the slick, body-hugging costumes and competition styles of human athletes from speed skaters to cyclists, as well as the designs of racing cars.

Figure 3.5.

Figure 3.5.
A flying Egyptian Rousette frozen at the top of the downstroke. In addition to the position of the wings and of the thumbs, notice how the hind feet are held, facing forward with toes together (a position that facilitates landing).

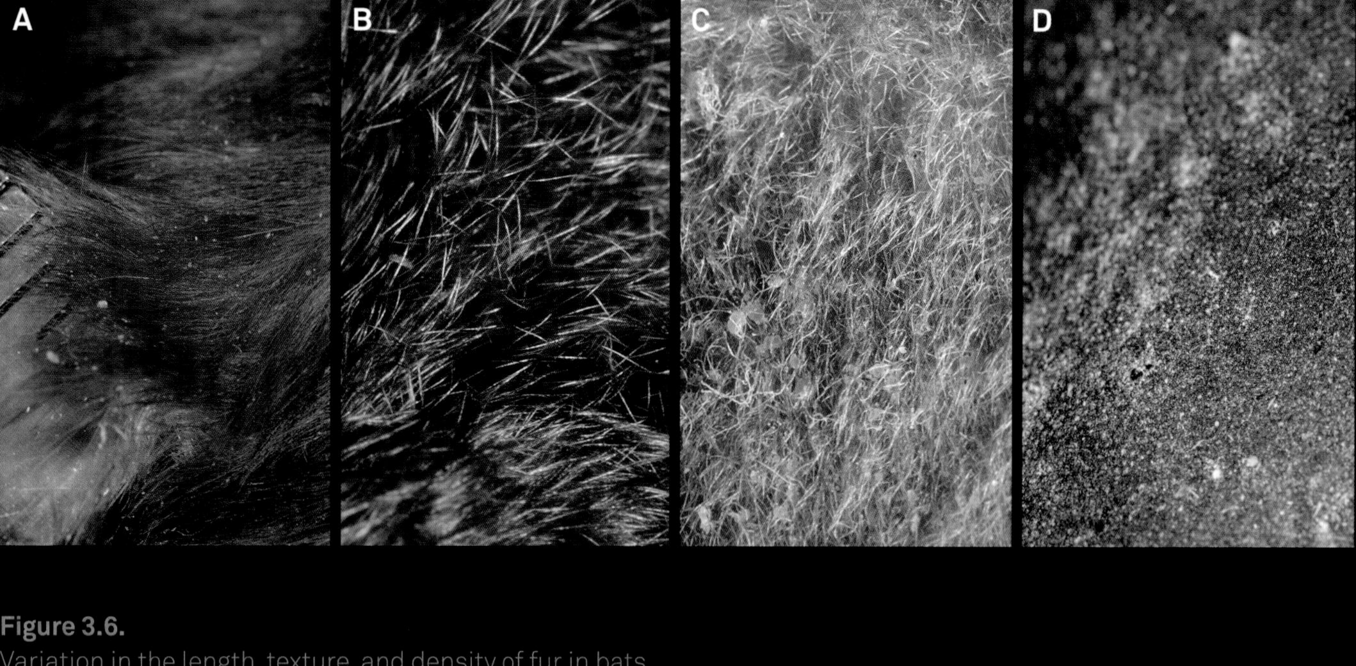

Figure 3.6.
Variation in the length, texture, and density of fur in bats, from the long and silky fur of (**A**) a Western Bonneted Bat (*Eumops perotis*) to the woolly fur of (**B**) a Cyclops Round-leafed Bat (*Hipposideros cyclops*), (**C**) the very short fur of a Greater Bulldog Bat (*Noctilio leporinus*) and (**D**) the bare skin of a Hairless Bat (*Cheiromeles torquatus*).

In 1994, Brock was part of a group that measured the drag associated with the fur of bats. (Figure 3.6) Bats show considerable variation in fur length from species to species, ranging from woolly, *e.g.*, Cyclops Round-leafed Bat (Fig. 3.6B) to long and silky, Western Bonnetted Bat (Figure 3.6 A), very short, Greater Bulldog Bat (Figure 3.6C) or hairless, Hairless Bat (Figure 3.6D). Does fur length influence the cost of flight? Placing a piece of bat skin fur-side-up in a small wind tunnel allows us to measure the force (in grams) of drag associated with air passing over the bat's skin (and fur). Compared to Hairless Bats, other bats had higher levels of drag associated with air passing over the skin. By turning the bat fur sample around so that air moved against the lie (nap) of the fur, we measured the difference in drag. This reversal experiment demonstrated that a bat flying backwards would incur higher costs of drag than one flying forwards.

To further complicate the picture, some other species appear to have naked backs because their wing membranes meet down the middle of the back. (Figure 3.7A) Naked-backed bats include some species of Old World Fruit Bats (genus *Dobsonia*) and some Moustached Bats such as Davy's Naked-backed Bat. (Figure 3.7) In these bats, the wing membranes meet in the middle of the back, covering the fur beneath. The biological significance of this arrangement remains unclear, although the enlarged area of the flight membrane may contribute additional lift during flight.

The reduced drag associated with bare skin raises a question: Why do bats have fur? Clearly they evolved from furred ancestors, which explains the presence of a fur coat, but why have they not reduced or lost their fur? In fact bats show considerable variation in their fur, both in texture and length of individual hairs. (Figure 3.9) Bulldog Bats and Free-

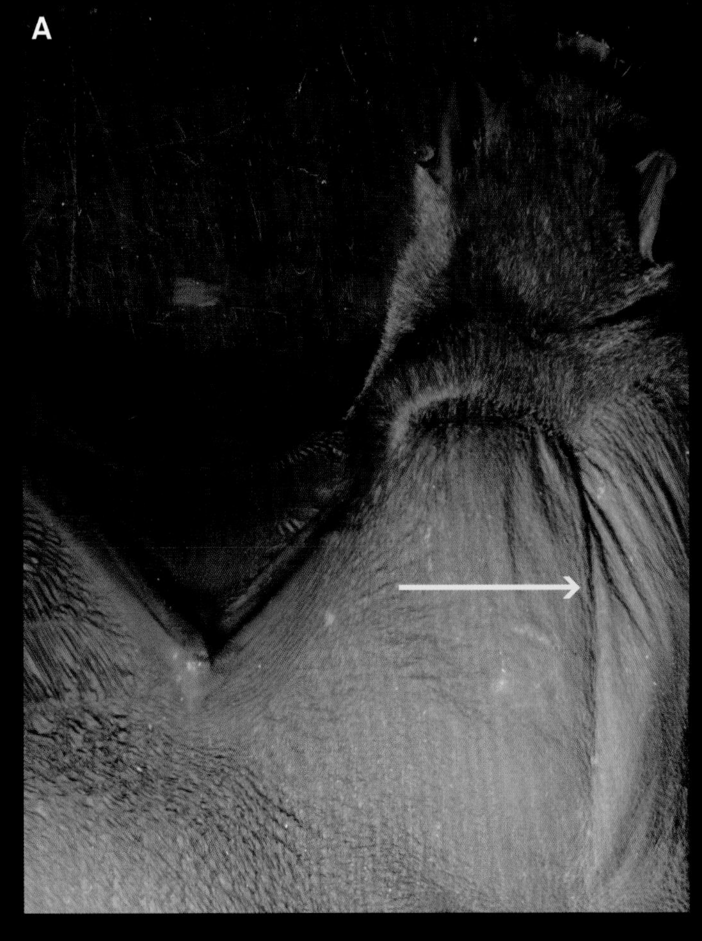

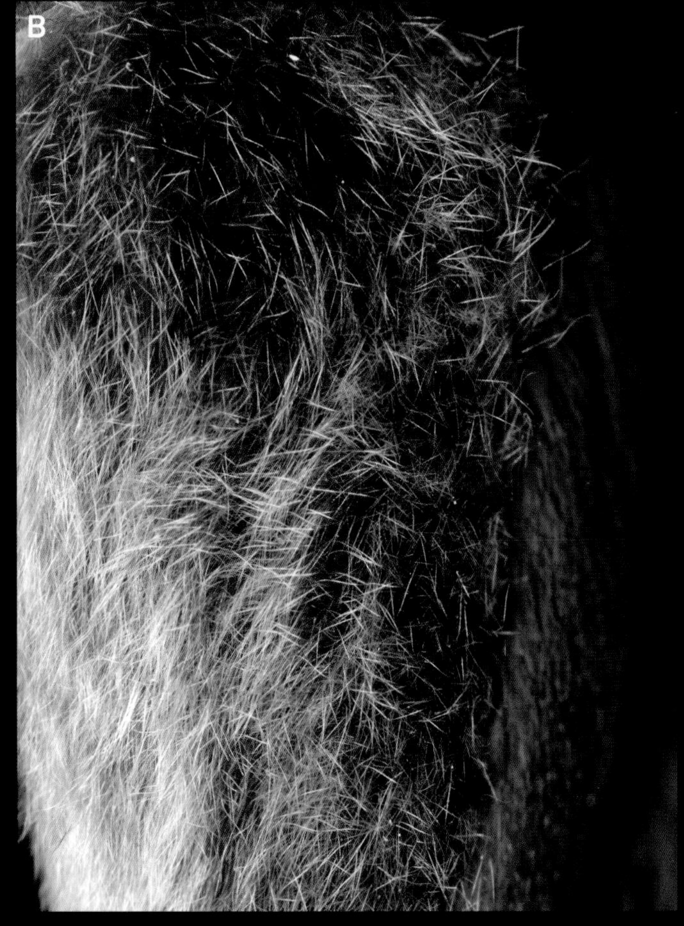

tail Bats tend to have the shortest fur. A virtually complete lack of fur occurs only in Hairless Bats, large cave-dwelling Free-tail Bats from Southeast Asia. (Figure 3.8) Bats with longer, narrower wings tend to have shorter fur than those with shorter, broader wings. It seems clear that fur plays many other roles in the lives of bats from insulation to disruptive coloration, camouflage

Figure 3.7.
In this dead Davy's Naked-backed Bat, the wing membranes meet in the middle of the back (**A**–arrow). Under the back wing membrane the fur is long and silky (**B**), more like the fur on the animal's belly than the fur on the head and shoulders.

Figure 3.8.
A Hairless Bat from Malaysia. These bats are large (nearly 200 grams) and live in the tropics, so they are capable of effective thermoregulation despite lacking fur. Photograph by Tigga Kingston.

Power and Speed

Muscles power the flight of bats, with the upstroke generating lift and the downstroke providing propulsive power or thrust. Bats are mammals and most of their bones and muscles have counterparts in other mammals. At least five muscles, however, are unique to bats. These muscles directly control the tautness of the wing membranes, contributing to the stiffness of the skin and the degree of curvature across the airfoil section (the camber). The first of these muscles connects the back of the bat's head to its hand and runs along the anterior edge of the wing. The role of this muscle (the occipitopollicaris) in bat flight remains unclear. The same is true of the coraco-cutaneous, which anchors networks of elastic fibers to the armpit area. The humeropatagialis is another unique muscle that occurs in some but not all bats. This muscle tightens and braces the distal parts of the wing membrane. The tensor plagiopatii consists of two parts, both of which anchor and brace the trailing edge of the wing membrane. In bats with a tail membrane, the depressor ossis styloformis helps to spread the interfemoral membrane between the hind legs by swinging the calcar away from the leg and towards the tail or bat's midline.

How fast does a bat fly? Decades ago, releasing bats and timing their flights along long hallways was a traditional way to measure the flight speeds. Another method involved having a bat fly in a wind tunnel and determine experimentally how they respond to air moving at different speeds. There are questions about how relevant such measurements of speed are to understanding the life of bats in the real world, just as timing an athlete running inside a small gymnasium may not provide an accurate view of his or her speed on a track in an open field. More recently researchers have used radar or stroboscopic photography to measure bats' flight speeds in nature. Other measurements come from data collected by arrays of microphones. Computer software converts times of arrival of individual pulses of sound (echolocation calls emitted by a bat) at different microphones and, from this, reconstructs the flight path of the flying bat in the wild. These approaches to studying flight behavior have provided scientists with data about flight speeds in bats, at least for selected species.

The answer to the simple question "How fast do bats fly?" depends upon the bat and the situation. Hunting Little Brown Myotis fly about 18 kilometers per hour (kph.). Hunting Eastern Red Bats (*Lasiurus borealis*) fly at 23 kph., Hoary Bats (*Lasiurus cinereus*) at 27 kph. and Brazilian Free-tail Bats at 34 kph. Flight speeds of bats hunting flying insects typically fall between about 10 and 34 kph. Little Brown Myotis approaching roosting sites (caves and mines) fly at about 29 kph., much faster than the 18 kph. they fly as they actually enter the site. Migrating Silver-haired Bats (*Lasionycteris noctivagans*) fly over 50 kph. At the other extreme, Little Brown Myotis flying low over water move more slowly, about 8 kph. In other words, there is no simple answer to the question of how fast bats can fly. Biologists need more accurate data about flight speeds of bats to better appreciate their versatility. Smaller satellite tags will hopefully be developed in the future to provide more specific details of the flight speeds of more bats. These data may open our eyes to flight speeds at both ends of the spectrum, but probably will inform us more about higher rather than lower flight speeds because bats flying in cluttered habitats have more erratic flight paths than those flying in the open.

Cost and Fuel

Heartbeat (pulse) rates can provide information about the relative energetic effort that mammals expend at different activities. Measure your pulse rate when you are sitting and relaxed. Then exercise for thirty minutes (run, walk, cycle, swim, dance) and measure your pulse again. Notice the increase between the two readings. The heart rate of an alert Little Brown Myotis hanging in a roost at room temperature is about 200 beats per minute. This jumps to over 1000 beats per minute when the same individual is flying at room temperature. When hibernating at 5°C, this bat's pulse rate will be about five beats per minute. These examples illustrate the dynamic range of the metabolic rate and relative costs of different activities (as measured by pulse rates) of Little Brown Myotis.

Biologists measure the metabolic rate and the relative costs of different activities using naturally occurring stable isotopes of hydrogen (^{2}H, known as deuterium or D_2, and ^{1}H) and oxygen (^{18}O and ^{16}O). The measurements take advantage of the fact that isotopes of hydrogen and oxygen clear the body at different rates. Oxygen is lost as water (H_2O) and as carbon dioxide (CO_2), whereas the hydrogen isotope deuterium is lost only as water. In the experiment, biologists first capture and take a blood sample from a bat. Next they inject water labeled with known amounts of these isotopes (^{2}H and ^{18}O) into the animal, and then later recapture the bat and obtain a second blood sample. They measure the amounts of the two rare isotopes in the two samples. The difference between the injected amounts and the subsequent sample provides an indication of the energy consumed during respiration (oxygen in and CO^2 out). More recently, measuring isotopes of carbon (^{12}C and ^{13}C) in exhaled bat breath has allowed researchers to determine the source of the carbon used in fuelling flight.

Results of experiments such as these have shown that flying bats consume energy about twelve times faster than non-flying bats. Although hovering is more expensive than horizontal flight, it actually costs some flower-visiting bats less than it costs either hovering Hawk Moths or hummingbirds. But how do bats fuel flight? A Pallas' Long-tongued Bat (*Glossophaga soricina*) obtains about 78 percent of the cost of hovering during feeding from the nectar ingested as it feeds. In terms of source of fuel, the bats are not as efficient as hummingbirds, which cover 95 percent of the cost of hovering from the nectar they ingest, but bats are much more efficient than other mammals at turning sugars into energy. Both the nectar-feeding bats and hummingbirds do much better, for instance, than humans. We can use about 30 percent of the energy in the chocolate bar that we eat just before exercising, while we exercise. This difference in efficiency boils down to whether the animal uses endogenous fuel (lipids from fat stores in the body, for example) or exogenous fuel (obtained from the food ingested during foraging, such as the energy in nectar). The typical path for nutrients from food first is absorption into the cells lining the gut. From there the nutrients move into the lymph and blood. Some bats and other animals, however, move nutrients directly through spaces between cells lining the intestine and into the body. This is known as paracellular absorption, and it at least partly explains how bats obtain virtually immediate access to the energy in the food they consume. Insectivorous bats such as Lesser Bulldog Bats (*Noctilio albiventris*) also use exogenous fuels to cover the cost of flight, and this may prove to be true for other insectivores. Furthermore, use of exogenous fuels may help to explain rapid food passage through bats. (See Chapter 5.)

Migrating bats are expected to fuel flight with body fat not stored in muscles. Although many species of bats may migrate, we have few details about their physiology. We do know that male and female Hoary Bats (*Lasiurus cinereus*) differ in their amounts of stored fats and the levels of activity of enzymes used to bind fatty acids. Migration in birds and in bats poses important physiological challenges and research in this area promises to yield more interesting results.

Analysis of isotopes appears to support the expectation that Horseshoe Bats that hunt from perches have lower flight costs than those spending more time on the wing. Flight time is one element in this calculation, but a more important factor is the cost of maneuvers that require active changes in wing position and movement.

Control

While bats that eat mainly fruit appear to use flight mostly as a means of transport, bats that pursue flying insect prey and those that feed at flowers to obtain nectar and pollen face additional challenges. (See Chapter 5.) It is more expensive to pursue flying evasive prey or to hover at feeding sites than it is to just fly in a straight line because of greater demands on energy and control. These flight patterns require different abilities to maneuver, which are reflected in differences in wing morphology ranging from wing shape to the design of wing tips, and in skeletal differences particularly around the joints. The shape of the bone of the upper arm (humerus) and shoulder blade (scapula) show great diversity among bats, probably reflecting differences in their flight abilities. (Figure 3.9)

Figure 3.9.
Posterior (**A**, **C**, **E**) and anterior views (**B**, **D**, **F**) of the upper arm bones (humeri) of three bats illustrating skeletal variation associated with flight. The bats shown include (**A**, **B**) a Short-faced Fruit Bat, (**C**, **D**) a Gervais's Fruit-eating Bat (*Artibeus cinereus*) and (**E**, **F**) an insectivorous Hairless Bat. In these photographs **h** = the head of the humerus (where it articulates with the shoulder blade), **gt** = the greater tuberosity, **lt** = the lesser tuberosity and **mr** = medial ridge.

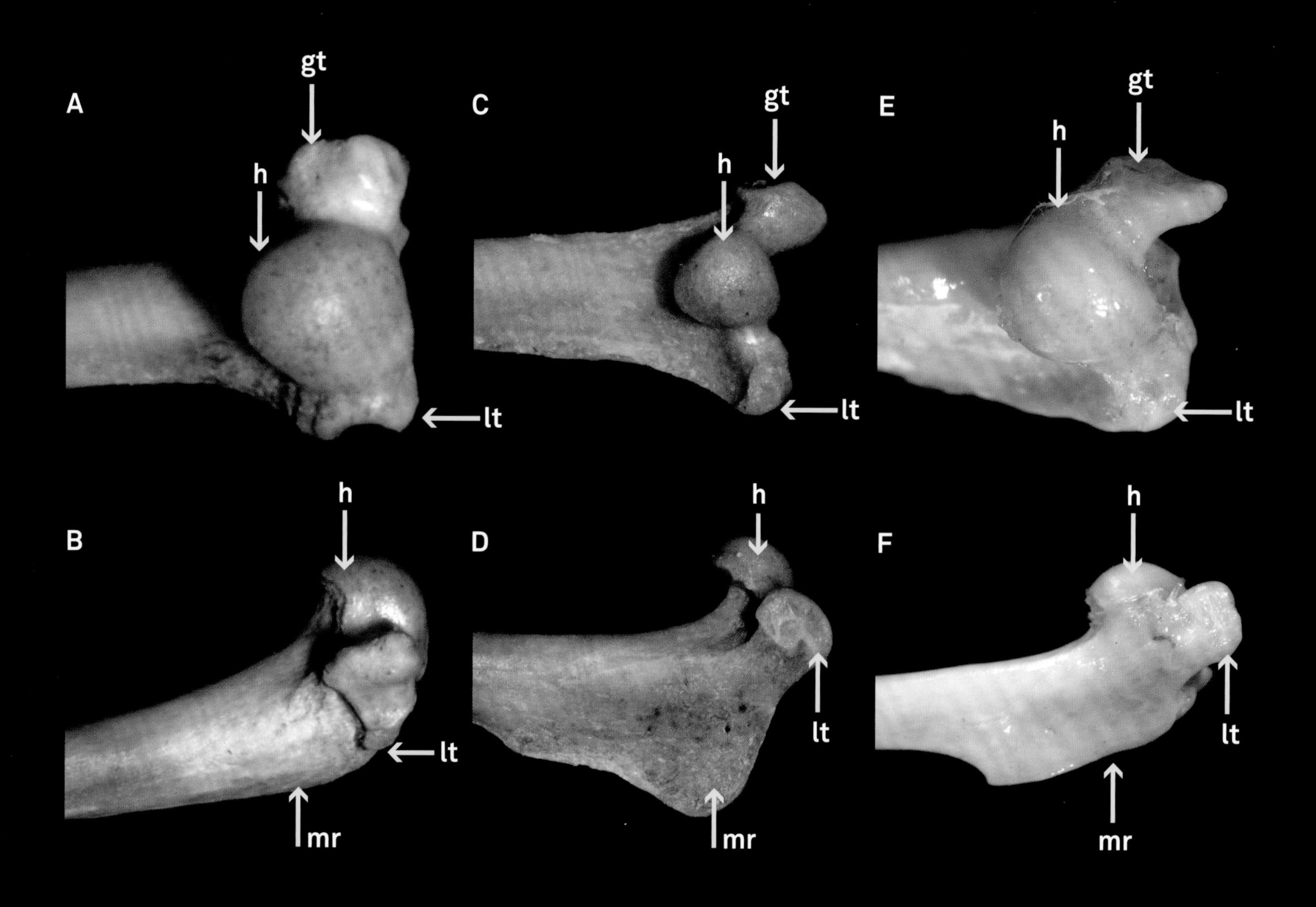

Figure 3.10.

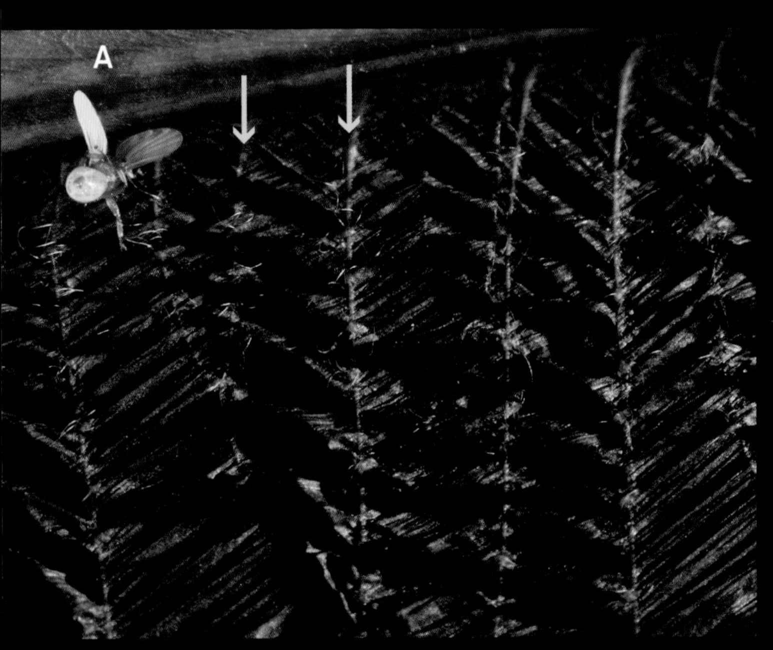

Figure 3.11.

Obstacle courses have been used to measure a bat's ability to maneuver in flight. (Figure 3.10) A successful bat does not collide with any conspicuous obstacles. Changing the distances between obstacles in a row and staggering obstacles between rows makes it impossible for an animal to fly straight through the course without hitting an obstacle. When obstacles are one wingspan apart, the bat has to actively maneuver to avoid hitting one. Obstacle courses can be used to measure predictions about maneuverability based on the morphology (wing shape, size, wing tip) of bats wings

In 2009 Brock was involved in a study that used an obstacle course to test the effects of alcohol on fruit-eating New World Leaf-nosed Bats. Frugivorous and nectarivorous bats sometimes consume alcohol, the product of fermentation that naturally occurs either in fruit or in flowers. When Egyptian Rousette Bats (an Old World Fruit Bat) have consumed alcohol, they do not fare well when trying to fly through an obstacle course. For reasons that remain unknown, some fruit- and nectar-eating New World Leaf-nosed Bats with blood alcohol levels of 0.3 percent (higher than the DWI level in most states!) flew through the obstacle course without collisions. Part of this may be due to differences in sensory systems—Rousette Bats use a combination of vision and tongue-click echolocation, whereas the New World Leaf-nosed Bats are laryngeal (voice box) echolocators. However, we still do not know for sure.

Figure 3.10.
Two tents joined by a flight corridor (**A**) were used to challenge the flight maneuverability of bats. The flight corridor housed an obstacle course (**B**), in which lengths of plastic chain served as obstacles. The distances between obstacles were adjustable, while banks of obstacles were 1 meter apart. The obstacles were conspicuous to the bats.

Figure 3.11.
A comparison of close-up views of the wings of (**A**) a Mesoamerican Moustached Bat (*Pteronotus mesoamericanus)* and (**B**) a Gervais' Fruit-eating Bat. The bone at the top of **A** is the bat's forearm (with a small bat fly—see Chapter 7). Note the rows of sensory hairs (arrows in **A**) that are absent from **B**. The bones in **B** are converging on the bat's wrist. There are some small hairs (arrow) between two of the hand bones in **B**.

Close examination of the wing surfaces of bats reveals lines of small hairs, each associated with a sensory cell. (Figure 3.11) In the 1790s, Lazzaro Spallanzani had suggested the possibility that bats had very sensitive wings and flew "by feel." (See Chapter 4.) In 2011, two papers published by the National Academy of Sciences provided experimental evidence of the sensitivity of bat wing surfaces, and particularly the role of sensory hairs. In one experiment, Cindy Moss and a team at the University of Maryland used a depilatory cream to remove the sensory hairs from the wings of captive Big Brown Bats. The hairs grew back, giving the experimenters the opportunity to examine and measure flight performance with hairs, without hairs, and then after the hairs had grown back. The results demonstrated that the sensory hairs on the wings of these bats play an important role in how the bat senses air flow over its wings, which in turn is critical for maneuverable flight. A close look reveals a variety of patterns of sensory hairs on the wings of bats. (Figure 3.11) Neural input from these hairs and associated sensory cells provide information about air movement across a bat's wings.

A second experiment conducted by a team led by University of Maryland neuroscientist Susanne Sterbing-D'Angelo revealed that the wing membranes of Big Brown Bats are very sensitive to touch. Putting this in perspective, their sensitivity is comparable to that in our finger tips. Bats can literally feel changes in air pressure and speed over the skin of their wings. In addition to vision, echolocation and the skeletal morphology of a bat's wing, these findings illustrate that nerve endings and sensory hairs on the wing surface are crucial to flight performance.

Drinking

Although many bats obtain the water they need from their food, others fly low and drink by dipping their tongues into a lake or stream. (Figure 3.12) This type of drinking behavior has been reported from bats in different families, from Old World Fruit Bats to Free-tailed Bats to Vesper Bats. In fact, an effective way to capture some species of bats is to set nets over open water.

Some bats use echolocation to recognize water surfaces. In 1988 while teaching a field course about bats, Brock and his students observed some Little Brown Myotis flying in a room with a shiny linoleum floor. They noticed that a few bats flew close to the floor and appeared to attempt to drink from it. They actually drank if there were pools of water on the floor.

Hugh Aldridge, currently at the University of California, Berkeley, studied bat drinking behavior in more detail. He reported that when flying close to the water's surface, bats adjusted their wingbeats by not using their full range of wing movement. At the bottom of the abbreviated wing stoke, the bat's wing tips were about 2 mm. from the water surface. Had they used a full downstroke, the wing tips would have been under water. This would have produced a dramatic braking effect and caused the bat to crash. When flying low over a water surface, Hugh found that Little Brown Myotis increased their wing beat frequency compared to flight in a tunnel (15.05 versus 9.23 beats per second) and flight speed (3.99 versus 2.66 kilometers per second), and reduced induced power output (2.05 versus 4.35 watts per kilogram). The resistance that would have been generated by putting much more than the tongue tip in the water probably explains the details of changes in flight mechanics and behavior during in-flight drinking.

Large Bulldog Bats that use their hind feet to gaff small fish or other aquatic prey also changed their wingbeat cycle, which may reduce overall costs of flight while improving flight control. Bat biologists still do not know how other bats such as Large Slit-faced Bats (*Nycteris grandis*) and Indian False Vampire Bats take fish or frogs from the water surface. To date there are no reports that any bats behave like kingfishers and dive into the water to attack prey.

Nancy Takes Advantage of Thirsty Bats

When I was working in French Guiana in the 1990s, one of the problems I faced was trying to catch Free-tailed Bats. In rainforest habitats these bats, which are specialized for fast flight but not for maneuverability, roost high in the trees and rarely come anywhere near the ground. Rather than braving flight through vegetation in the forest understory, they hunt in the unencumbered spaces above the forest canopy or in clearings. I was trying to survey the bat fauna to determine how many and which species of bats lived in the area, but I never caught any Free-tailed Bats. I knew that they were there, but how could I catch some? After a big rainstorm one day, I had an idea. I noticed that a bulldozer working on a nearby road had accidentally scraped out a hollow area about ten meters wide and thirty meters long, and that this depression filled with water when it rained. It was only about a foot deep in the middle, but this giant puddle might be a place that Free-tailed Bats came to drink. The open spaces along the road cut provided an ideal flight path for space-loving Free-tailed Bats. Putting on my rubber boots, I strung a mist net across the puddle and waited for dark. I was not disappointed. Thirsty bats appeared as dusk fell, zooming along the road and dipping down at the puddle to drink. Soon I had many bats in my net, and yes—there were Free-Tailed Bats! I eventually captured over 100 bats—including three species of Free-tailed Bats—over that single puddle. Ever since that time, I always keep an eye out for places that bats might come to drink. Where natural open fresh water sources are scarce, a manmade "puddle" can turn out to be great place to see and catch bats.

Figure 3.12.
A Trident Leaf-nosed Bat coming in to take a drink is reflected in the water's surface. Jens Rydell took this excellent picture at a pond in Israel.

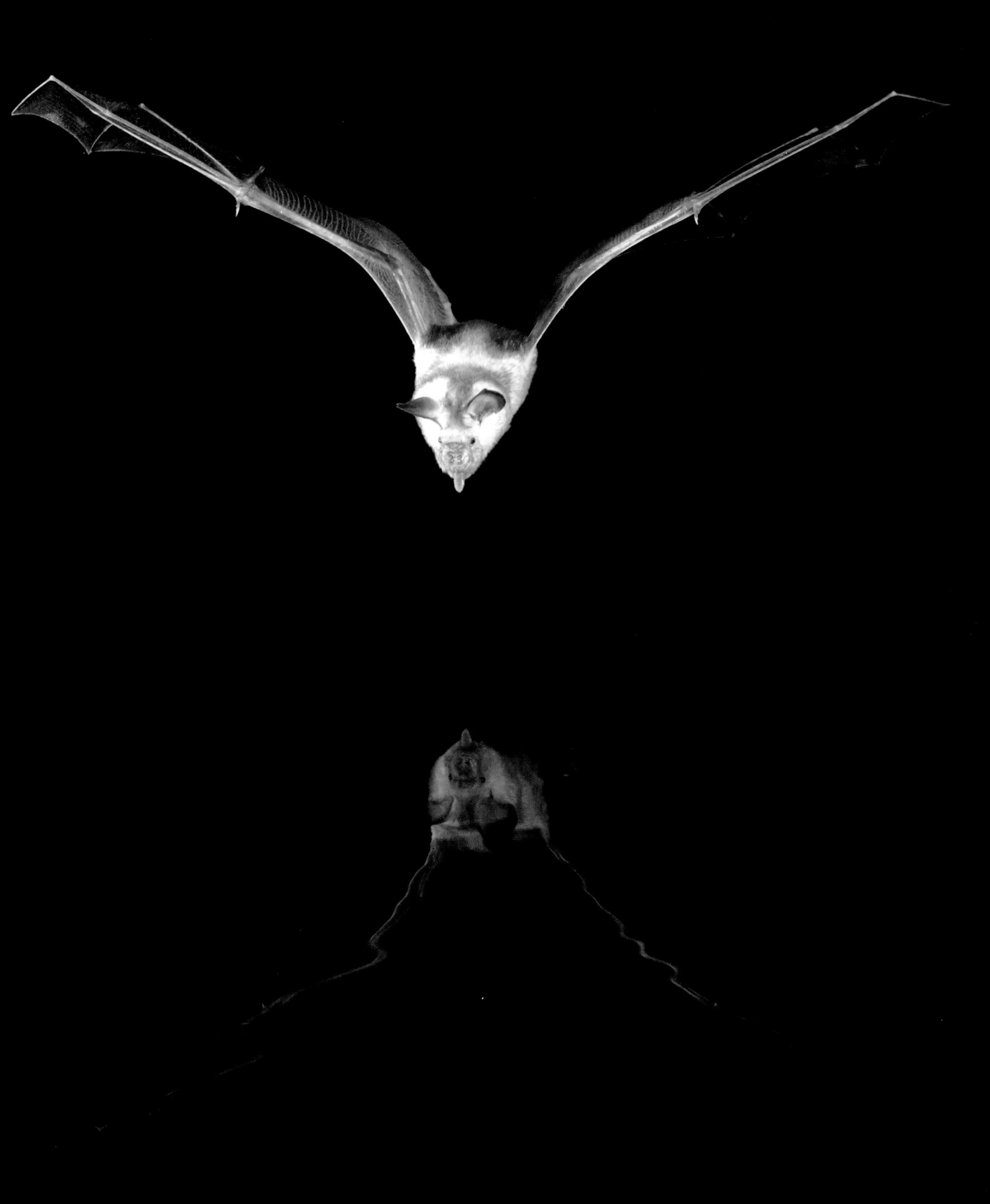

Take-off and Landing

Most species of Free-tailed Bats are very entertaining during take-offs. These bats roost in high places and use an initial long drop to build up airspeed and then take flight. This means that a bucket trap set below the entrance to a roost almost always catches these bats. It also means that if you put a Free-tailed Bat on the ground, it will quickly climb up your legs (or those of someone else) and end up on the top of your head. If you raise an arm, the bat will climb higher to reach the drop that it wants. This is an unusual way to initiate people to the ways of bats. Most species of Free-tailed Bats have long and narrow wings. While this wing form restricts the sorts of places that they can roost, it allows Free-tailed Bats to fly particularly far and fast.

Why do bats hang upside down? The most obvious reason is that they are "lazy" (although by that measure any resting animal might be considered lazy). Hanging upside down for bats means that taking off only involves letting go and spreading your wings—gravity does the rest. The main reasons, however, for hanging behavior probably have more to do with the fact that the structure of the forelimbs (wings) makes it difficult for most bats to walk effectively. Some bats have unique tendons and tendon sheaths (bands of tissue that hold the tendons in place) that allow their feet to "lock" closed when the bat is roosting, allowing the bat to hang without exerting any energy. Muscular energy is needed to let go, not to hold on. Nancy and her colleagues have documented these tendon locks in bats of many different bat families.

Foraging Common Vampire Bats face different problems with take off than most other bats. Vampires often feed on the ground or close to it. They ingest large amounts of blood (at least two thirds of their body mass) and that weight can be hard to lift off the ground. Vampire bats urinate copiously to rid themselves of the weight of the water in blood plasma, but taking off is still a challenge. (See Chapter 5.) To deal with this, Common Vampire Bats use their very long thumbs in two ways. First, the thumbs extend the wing membrane, and second they are used like throwing sticks to provide extra leverage to launch the bat from the ground. Common Vampire Bats also crouch low to the ground and then, with a powerful contraction of muscles usually used in the downstroke of flight, they launch themselves up into the air, gaining additional leverage from the long thumbs. Biologists still do not know of any other bats that use the Common Vampire Bat's approach to take off, including either Hairy-legged Vampire Bats or White-winged Vampire Bats. These other vampires face the same weight problems with take-off, but they are primarily arboreal, so they may be able to "solve" the take-off problem simply by leaping off the closest branch. It has been suggested that some pterosaurs took flight in the same way as Common Vampire Bats.

For a flying animal, landing can be perilous because the touch-down must be gentle enough to avoid injury. Bats are no exception. Some bats land on ceilings, overhead and more-or-less horizontal surfaces. A detailed analyses of the landing behavior of three species, conducted by Brown University biologist Sharon Swartz and her colleagues, revealed two patterns, including the use of reverse summersaults. Lesser Short-nosed Fruit Bats made four-point landings (two thumbs and hind feet). Pallas' Long-tongued Bats and Seba's Short-tailed Bats (*Carollia perspicillata*) made two-point landings (with the hind feet). The four-point landings (Lesser Short-nosed Fruit Bats) produced larger peak forces (over three body weights), while two-point landings (Short-tailed Fruit Bats and Pallas' Long-tongued Bats) produced an impact of about 0.8 body weight. Both of the New World Leaf-nosed Bats made left- and right-handed landings. Some bats use a summersault as they come into land, directing their feet at the landing surface.

Brock Observes Some Fancy Flying

The adroitness of bat flight can be impressive. In 1978 in the Chiricahua Mountains in eastern Arizona, Gary Bell, then a graduate student, and I watched a Southwestern Myotis (*Myotis auriculus*) hunting along the lighted eaves of a building at a research station. The bat flew along and never landed, but with its mouth deftly grabbed insects that had landed on the eaves. We provided the bats with more targets, carefully pinned insects. As Gary was in the middle of placing another moth, a bat grabbed it from the pin while pushing off against his fingers with wrists and hind feet. The bat did this without actually landing.

4

How Bats See with Sound

Figure 4.1.
An echolocating Lesser Mouse-tailed Bat (*Rhinopoma hardwickei*) emerging from its cave roost in Israel. Note the open mouth and forward-pointing ears. Photograph by Jens Rydell.

Sounds and Echoes

The picture of a flying bat (Figure 4.1) with its mouth open today suggests echolocation. Today, almost everyone knows about echolocation or biosonar. Bats and some other animals orient by using sound. Specifically, they use echoes of sounds they produce to locate objects in their path. Bats and other echolocators listen for differences between what they said (sounds they produced) and what they hear (echoes of their calls) to collect information about their surroundings. (Box 4.1) The differences between pulse and echo are data for the echolocating bats.

MUST BOX 4.1

Other Echolocators

Echolocation allows animals to operate under conditions of uncertain lighting or in complete darkness. Anyone who has watched a fly walk across patterned wallpaper will have observed how the fly appears and disappears according to the background. In this case, vision is not as reliable as echolocation would be. The ability to orient and detect objects successfully without using vision opens up numerous opportunities in conditions where vision is either unreliable or problematic. Echolocating birds and bats can nest (birds) or roost (birds and bats) in the total darkness of caves where they may be less vulnerable to predators. In addition to sight and smell, some small mammals such as shrews and tenrecs (hedgehog-like mammals from Madagascar and Africa) appear to use echolocation as a way to collect information about their surroundings, particularly when moving in unfamiliar settings. Many echolocating toothed whales use echolocation to detect, track and assess potential prey such as swimming fish or squid.

Echolocation is varied in its distribution among vertebrates—it not a characteristic of all bats (Order Chiroptera) and other animals also echolocate. Other echolocators include toothed whales (odontocete cetaceans) such as dolphins, some shrews, some tenrecs and at least two groups of birds—Oilbirds (*Steatornis capripensis*) and several species of Swiftlets (genera *Aerodramus* and *Collocalia*).

The majority of Old World Fruit Bats do not echolocate. Two species that do, the Egyptian Rousette (*Rousettus aegyptiacus*) and Geoffroy's Rousette (*Rousettus amplexicaudatus*), produce echolocation signals by clicking their tongues, while another, the Dawn Bat (*Eonycteris spelaea*), may echolocate using sounds produced by clapping the tips of its wings together. All other bats that echolocate produce their calls in their larynx (voice box). Echolocation using signals produced in the voice box is typical of insectivorous bats in both major evolutionary lineages, the Yinpterochiroptera and the Yangochiroptera. (See Figure 2.9.)

The first detailed experiments that indicated that there are sounds people cannot hear were performed in 1794 on bats and owls by Lazzaro Spallanzani, an Italian priest and physiologist. His findings led him to the conclusion that "bats could see with their ears." His suggestion was not well received and Spallanzani was mocked by some of his contemporaries. The French naturalist Baron Cuvier is reputed to have said "Mr. Spallanzani, if bats see with their ears, do they hear with their eyes?" Then, as now, suggestions of phenomena beyond our experience often generate disbelief and ridicule.

Spallanzani was trying to solve the mystery of how bats and owls find their way in the dark. Using animals he had caught in the wild, Spallanzani showed that owls would not willingly fly in a totally dark room. If forced to fly in the dark, the owls collided with objects in the room, including furniture and walls. In contrast, the bats he tested continue to fly unimpeded in the dark. Frustratingly, however, Spallanzani could not see

Figure 4.2.
The series of echolocation calls from (**A**) a flying Little Brown Myotis and (**B**) a flying Hoary Bat (*Lasiurus cinereus*), illustrating the differences in cadence or the pattern of timing. The time sequence in A is just over 1 second (1000 ms.) long; in B 934 ms. Time between Little Brown Myotis echolocation calls is about 70 ms., between Hoary Bat calls, 273 and 248 ms., respectively.

Figure 4.3.
Three ways to look at an echolocation calls: (**A**) the time-amplitude display, (**B**) the spectrogram and (**C**) the power spectrum. Note that in A and B the horizontal axis is time. In **C** the horizontal axis is frequency (the vertical axis in **B**). The vertical axes in **A** and **C** are relative energy. By analyzing calls of bats in multiple ways, scientists can better understand the structure of each call and assess the information that may be obtained from returning echoes.

Figure 4.2.

Amplitude

50%

A

0%

-50%

2100 2200 2300 2400 2500 2600 2700 2800 2900 3000 ms

Amplitude

50%

B

0%

-50%

3500 3600 3700 3800 3900 4000 4100 4200 4300 4400ms

Figure 4.3.

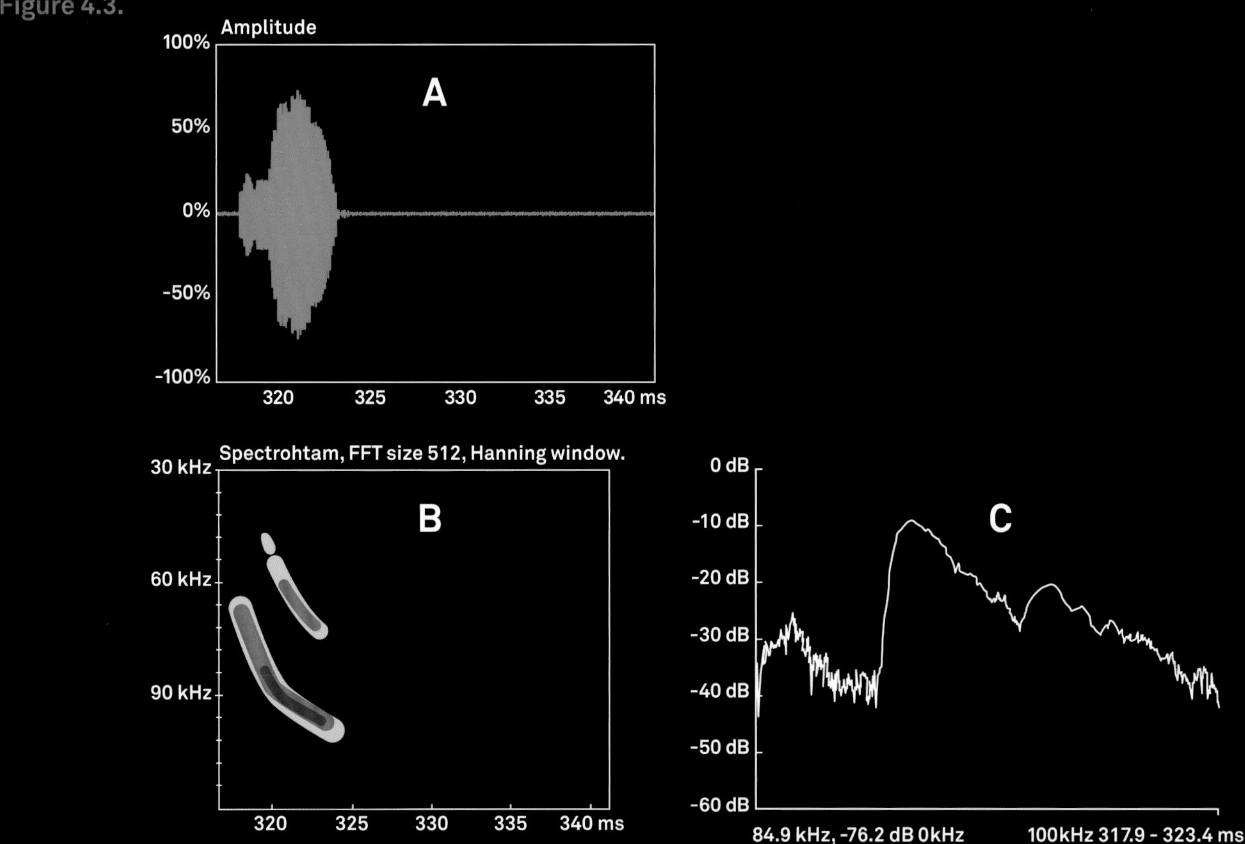

A
B
C
D
E
FREQUENCY
VOLUME
Pettersson
MODE
SETTINGS
REC
SPECTROGRAM
Back
110.0
99.9
89.9
79.8
69.7
59.6
49.5
39.5
0.000
0.011
0.022
0.033
0.044
0.055
0.066
0.077
0.088
0.099
0.110

what the bats were doing in the dark. To overcome this he suspended small bells on long ribbons from the ceiling. Tinkling bells now alerted him to collisions. Spallanzani found that in the dark flying bats did not collide with the ribbons, did not ring the bells, but owls did. If Spallanzani put a bat's head into a small bag, the bat would not fly. When he plugged one of a bat's ears it flew but rang bells, signaling its failure to negotiate as skillfully as bats with both ears unimpaired.

How did the bats do it? This question became "Spallanzani's bat problem", which remained unsolved for over a hundred years. Although several suggestions were made about how bats flew and maneuvered in the dark, it was biologist Donald Griffin and physicist George Pierce who provided the answer in 1938. Working with neuroscientist Robert Galambos, they used a "sonic detector" and showed that flying bats produced series of sound pulses that were "ultrasonic" or inaudible to people. (Figures 4.2 and 4.3) Careful analysis of the timing of the sounds revealed that bats were using echoes of sounds they produced to locate objects in their path. In 1944, Griffin coined the term "echolocation" to describe this behavior. No wonder Spallanzani had not realized that his bats were producing signals—being human, he could not hear them! Pierce's Sonic Detector allowed Griffin and Galambos to resolve Spallanzani's bat problem. Today we would refer to the sonic detector as a bat detector. (Box 4.2)

Box 4.2, Figure 1.
A sampling of bat detectors, including (**A**) a Batcorder, (**B**) a Song Meter, (**C**) a Pettersson detector and (**D**) an Avisoft Ultrasouind Gate 116e (with cable to connect to the USB port on a computer). A recent innovation, the Echo Meter Touch (**E**), turns an iPad into a bat detector that displays calls. The display on the iPad shows four echolocation calls. (See also Figure 4.6E.) The Batcorder, Songmeter and Avisoft detectors record sound files to flash cards; the Pettersson, which makes bat signals audible to people, to a computer. The Echo Meter Touch presents a visual image of the calls. The scale for **A**, **B** and **C** is 25 cm long. A, B and C courtesy of Toby Thorne; Figure E courtesy of Ray Crundwell.

Box 4.2

Bat Detectors

A bat detector is an instrument that converts an acoustic signal (such as a bat's echolocation call) into a display that humans can hear or see. Pierce's Sonic Detector, the first artificial bat detector, made possible the discovery of echolocation. This instrument consisted of a microphone and associated circuitry that converted high frequency sounds into sounds humans can hear. Today there are many commercially-available bat detectors that are used by scientists, educators and citizen scientists to monitor the behavior of echolocating bats. (Box 4.2, Figure 1) A bat detector can allow an observer to distinguish a commuting bat from a foraging one, and to determine which species of bats are active in an area. This fulfilled Griffin's expectation that echolocation would provide a window on the lives of bats. When Griffin was studying bats, he had to build his own bat detectors. The first commercially-available detectors appeared in the early 1960s, and today we have an embarrassment of riches including detectors, specialized recording systems and software for analyzing signals. In April 2014, a Goggle® search on "bat detector" returns nearly four million hits!

The output of bat detectors may be audible and/or visual, but many convert the original audible signal into a digital signal that can be stored, typically on a computer or flash card. Detectors such as Batcorders and Song Meters (Box 4.2, Figure 1 A and B) can be set and deployed to record bats all night every night for as long as the batteries last (or until the flash card fills with recordings.) Stored signals may be examinable only with appropriate software. There is now a bat detector that can be plugged into an iPhone or iPad with analysis software packed into a downloadable app. Nancy, Brock and their colleagues tested it in Belize in April 2014 and were very positively impressed.

Echolocation is an active system of orientation because animals use energy to produce the signals and analyze the echoes. By comparison, vision is more passive because light falling on photoreceptive cells generates neural stimuli. When you use a flashlight to see your way in the dark, you spend energy (from batteries) just as echolocating bats spend energy producing echolocation calls. By producing one echolocation call during each downstroke of their wings, bats appear to reduce and perhaps avoid a major cost of echolocation—the energy needed to produce loud (high-intensity) vocalizations. The details, however, are complicated because some bats produce echolocation calls that are much stronger than those of others, and the cadence of call production is not always linked to wingbeat.

Figure 4.4.
An echolocation call (**c**) of a Brazilian Free-tailed Bat is shown followed in 2 ms. by an echo (**e**) of its call. The bat was recorded as it flew in Falmouth, Jamaica.

Outgoing Signals and Returning Echoes

Echolocating bats use sounds (vibrations in air) to detect obstacles and objects, using information from both the original signals they produce and echoes of these sounds returning from nearby surroundings and potential prey. Echolocating bats register the outgoing signal in their brain for future comparison with returning echoes. The differences between what they say (their calls) and what they hear (the echoes) are the data that the bat uses in echolocation. Sound travels about 340 meters per second in air, so bat echolocation involves split-second timing. (Figure 4.4) By varying timing (call duration, time between calls) and frequencies in their signals, bats maximize the information they obtain from returning echoes. The extent to which an individual bat can vary its output is obvious in the call sequence associated with a "feeding buzz", an attack on a flying insect. (Figure 4.5) Changes in call frequencies, patterns of frequency change over time, duration of calls and time between calls affect a bat's ability to locate a target in space. (Figures 4.5, 4.6 and 4.7)

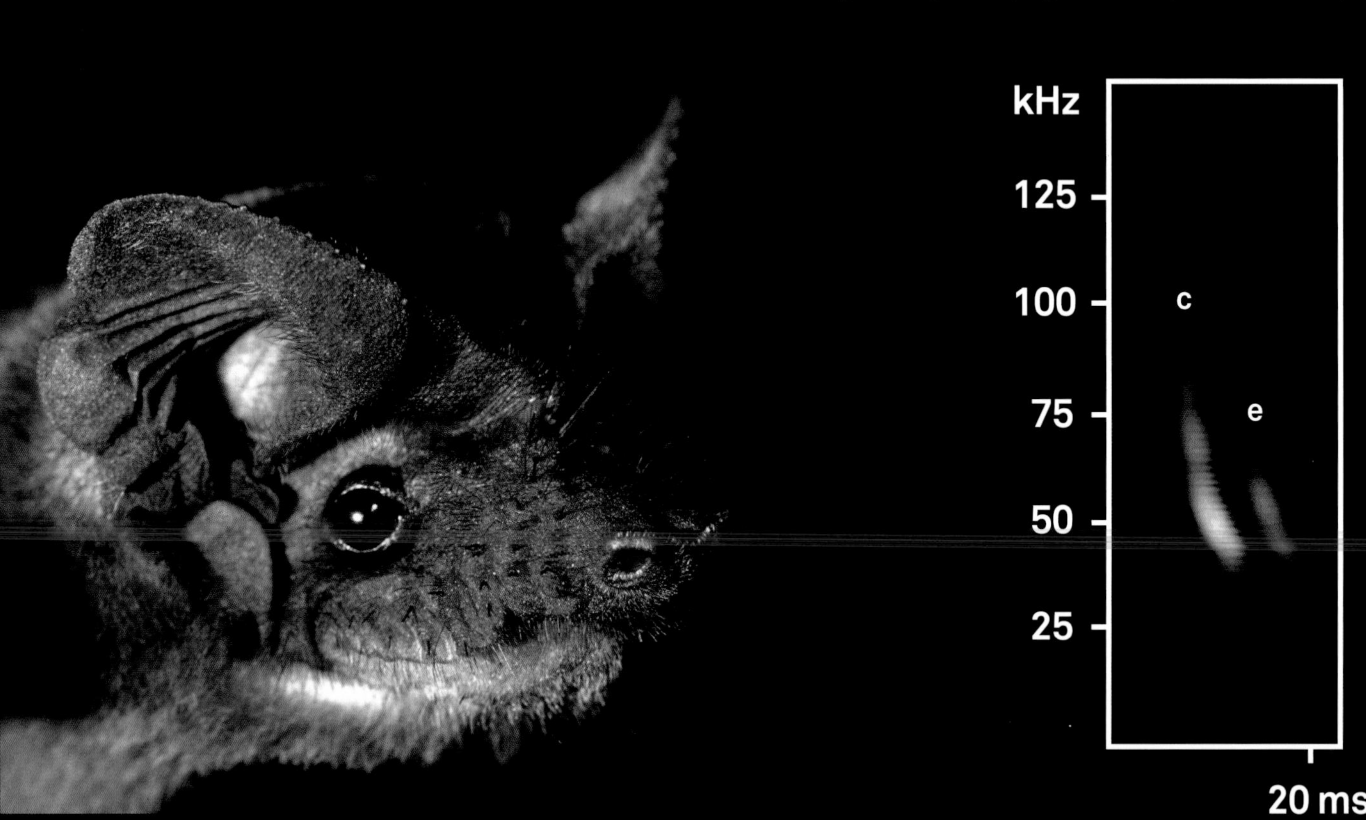

upward **FM** sweeps (**E**). Sometimes the calls have more linear elements (**F**), or are longer and more curvilinear (**G**). Some bats that separate pulse and echo in frequency produce calls like the one shown in **H**, while others use more gradual upward and then downward sweeps (**I**). See also the echolocation calls shown in Box 4.2, Figure 1E.

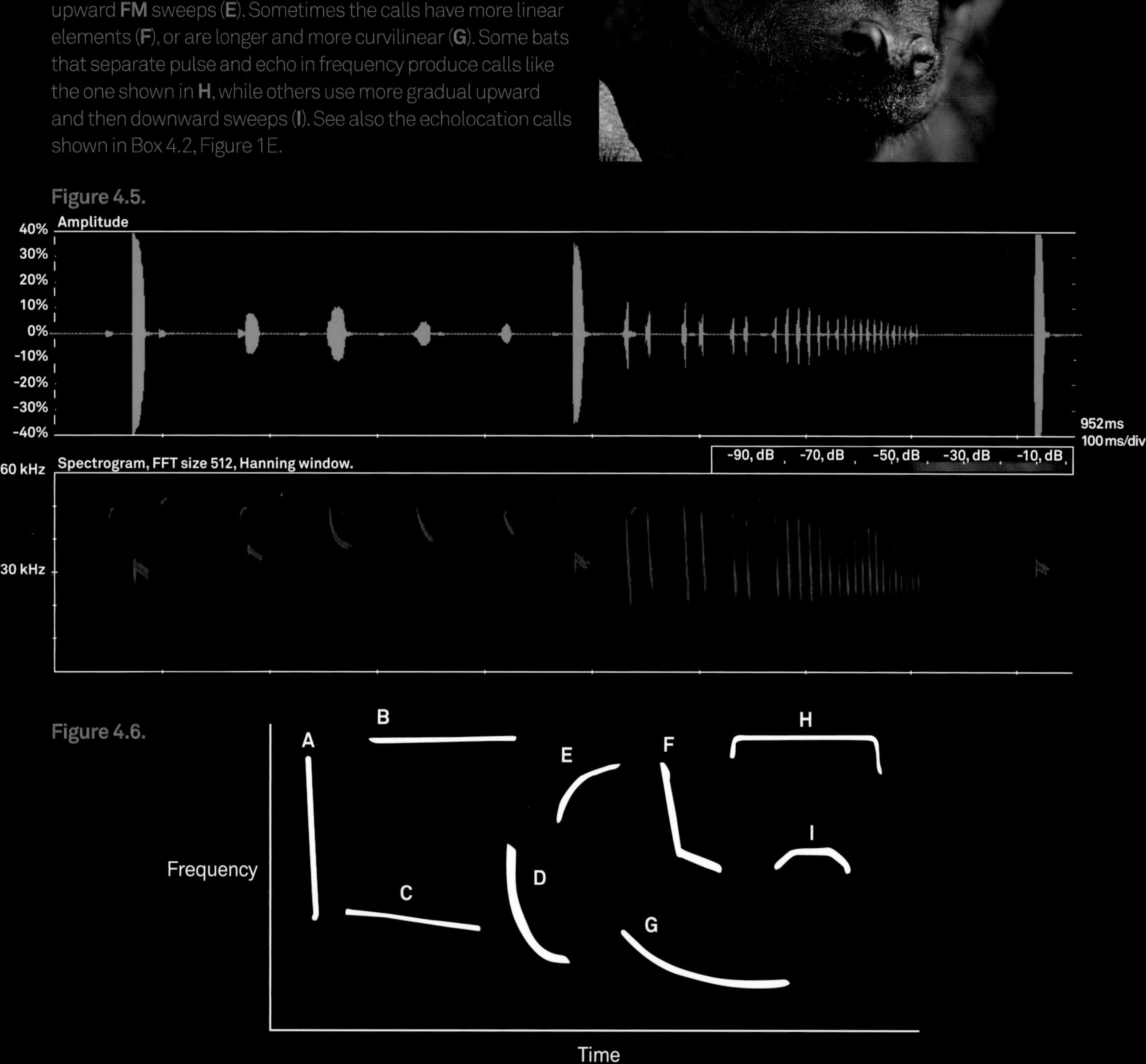

Bats tend to use echolocation signals dominated by high frequency sounds that have shorter wavelengths than lower frequency sounds. The wavelength of the sound affects the details the bat receives about its target. Shorter wavelengths (high frequencies) provide more detail than longer wavelengths (lower frequencies) due to the physics of how sound waves move through air. (Figure 4.7) Some humans can hear sounds up to around 20 kHz. (Hz. = Hertz, equivalent to cycles per second; k = thousand), but bats typically use echolocation calls above 20 kHz. (= ultrasonic). Echolocation signals in most species of bats cover a range of frequencies, affecting the information the bat can extract from the returning echoes. (Figures 4.6 and 4.7) Most bat species produce only one or a few types of calls when searching for prey, making it possible for scientists to identify many species by the frequency and structure of their echolocation calls alone. Closely related bat species often make very similar calls, so sometimes it is hard to distinguish among related species when several are present in the same area.

Most bats produce echolocation signals in the larynx or voice box. (But see Box 4.1 for a notable exception.) Although the size and arrangement of muscles in the larynx varies, the muscles employed by bats in producing echolocation calls are essentially the same ones that humans use to produce the sounds employed in speech. Air from the lungs passes through a narrow space in the larynx and causes muscles to vibrate—and produce sounds—when those muscles are contracted. Contraction of these muscles is under voluntary control, so bats can control the frequency and intensity of their calls.

A variety of different echolocation call designs is used by different bat species and under different circumstances. (Figure 4.8) In any call sequence a bat may combine different signal designs, reflecting the changes in challenges as the bat moves from obstacle detection to foraging, or from searching for prey to locking onto a particular target prey item. The information in Figure 4.7 makes it easier to appreciate the consequences of different signal designs. The importance of broader bandwidth (more frequencies) in precision of target location may partly explain the use the bats make of harmonics (or overtones). (Figure 4.7) Harmonics are identified by an "h" in Figures 4.8 and 4.9.

Figure 4.8.

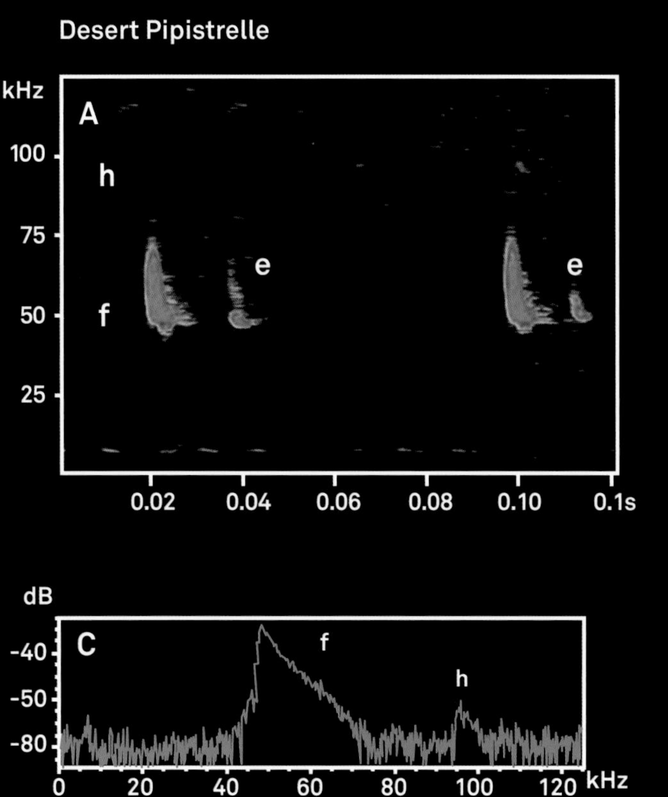

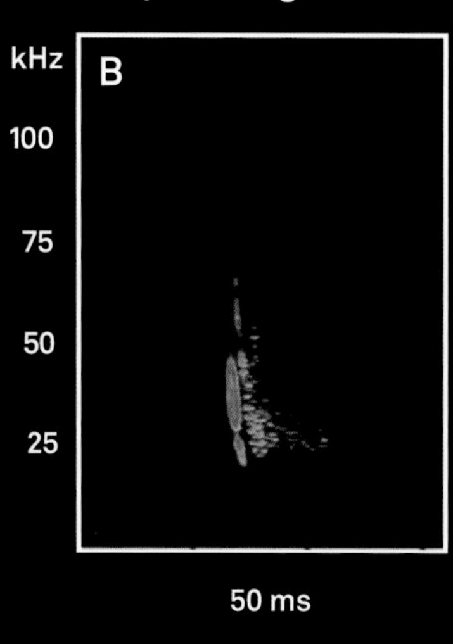

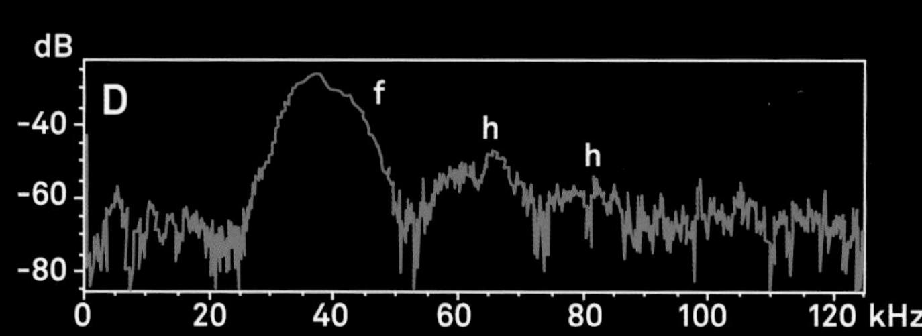

Figure 4.9.

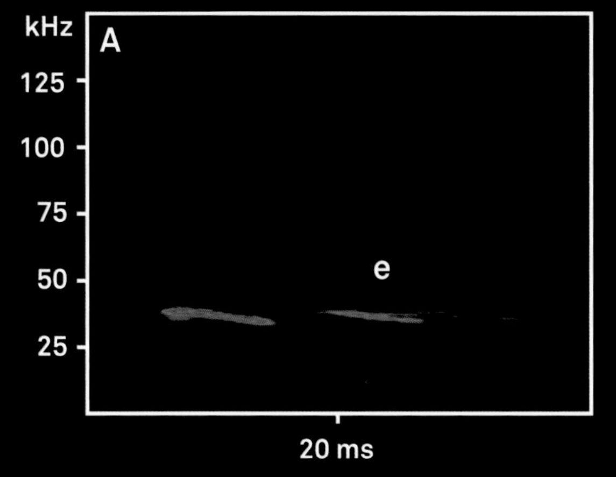

Figure 4.8.
Spectrogram (**A** and **B**) and power spectra plots (**C** and **D**) of two 6.8 ms. long calls of a Desert Pipistrelle (*Hypsugo ariel*) and a single 1.76 ms long call of a Hemprich's Big-eared Bat. Echoes of the pipistrelle (**e**) are obvious in the spectrogram plot (**A**). Harmonics (h) are more obvious in the power spectra plots (**C** and **D**). In the spectrograms (**A** and **B**) frequency is on the vertical axis and time on the horizontal axis. In the power spectra plots (**C** and **D**), there is no time information. Power is on the vertical axis, frequency on the horizontal axis.

Figure 4.9.
A 7 ms. long echolocation call of a Pallas' Mastiff Bat. An echo (**e**) is obvious in the spectrogram (**A**), and a harmonic (**h**) is evident in the power spectrum display (**B**). Note similarity of some calls shown in Figure 4.5, which shows a feeding buzz of this species.

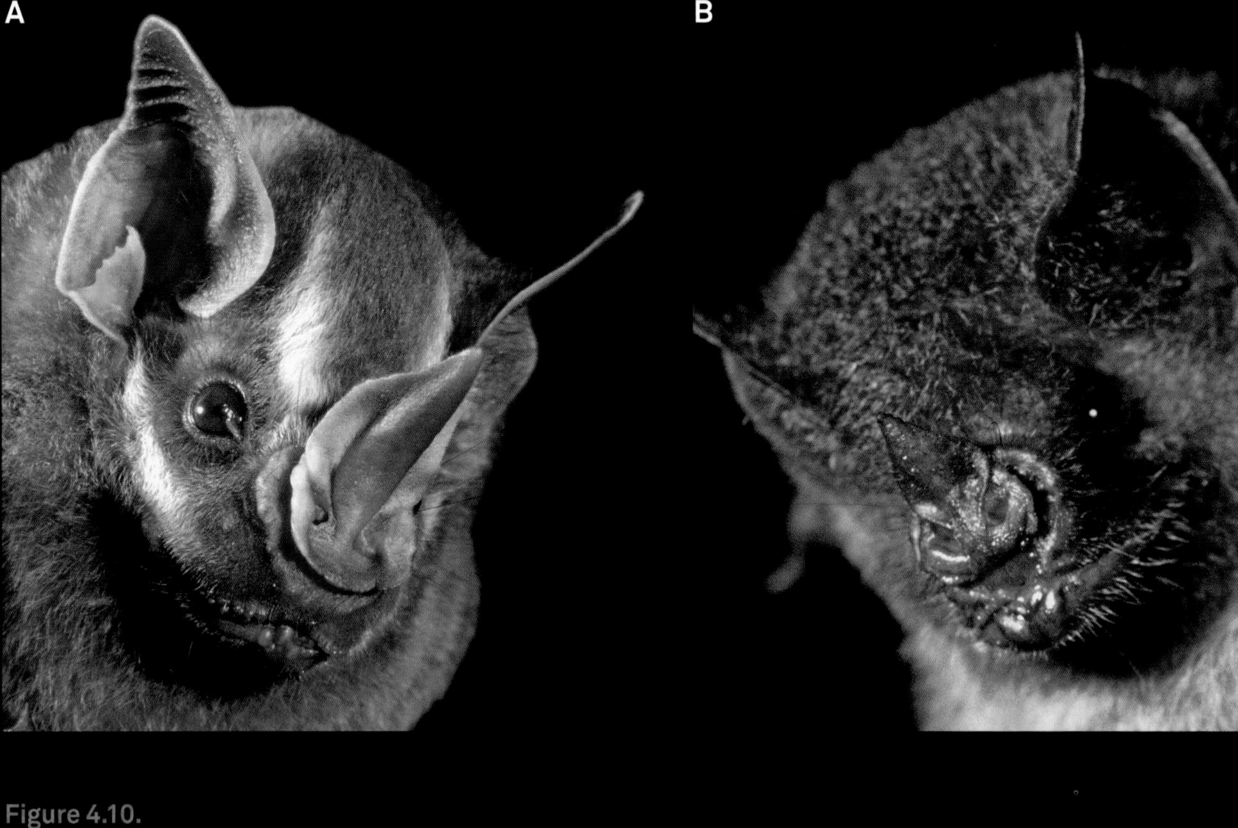

Figure 4.10.
New World Leaf-nosed Bats, such as (A) the Thomas' Fruit-eating Bat (*Dermanura watsoni*) and (B) the Yellow-throated Big-eared Bat (*Lapronycteris brachyotis*) produce low intensity echolocation calls.

Timing

Cadence—the timing of production of echolocation calls—is an important part of call design. A Little Brown Myotis searching for prey usually produces an echolocation call every 50 ms., which is twenty calls per second. At the same time and in the same place, a Hoary Bat similarly hunting will produce an echolocation call about every 200 ms., which is five calls per second. (Figure 4.2) Why different bat species make such different calls while doing the same thing in the same environment probably reflects their evolutionary histories. Different species may come from lineages that used different call types ancestrally. Clearly, for echolocating bats there is more than one way to detect, track and catch a bug. As is evident in Figure 4.5, the cadence of calls increases dramatically to 200 calls a second when a bat attacks a flying insect. This is usually true regardless of the cadence of the search-phase. For example, the twenty calls per second for a Little Brown Myotis and the five calls per second for a Hoary Bat in search phase are both increased to >100 (> = more than) calls per second during an attack on a flying insect target. Bats can vary the lengths (durations) of their echolocation calls from less than 1 ms. (one 1,000th of a second) to over 50 ms. Bats vary call duration depending upon the type of information the bat is seeking. During a single hunting sequence, a bat can go from calls of 10 ms. to those of 1 ms. duration as it closes on a potential prey and attempts to determine its exact location in three dimensional space.

Signal Strength

Bats that hunt flying insects often produce very intense echolocation signals, about 130 decibels (dB.) when measured 10 cm. in front of the bat's mouth. To put this in perspective, the smoke detector in your home produces a signal of about 108 dB. measured at the same distance. The difference does not sound like much until you remember that decibels are measured on a logarithmic scale, so a signal of 130 dB. is about twenty times more intense (stronger) that one of 108 dB. Many bat species produce strong signals to maximize the range at which they can detect insect prey because a more intense signal travels farther than the less intense one.

Some basic math makes the situation clear. A "typical" bat such as Pallas' Mastiff Bat (Figure 4.9) can detect a moth-sized target at 30 m. For a bat flying at 5 m. per second, there would be about six seconds from detection to contact, plus the distance the moth covers after detection. At 10 m. per second, there would be about three seconds from detection to contact. It is all about split-second timing.

Not all echolocating bats produce intense signals. (See Figure 1.10.) New World Leaf-nosed Bats, which typically glean insects from surfaces (such as leaves, tree trunks, or the ground), or eat fruit or visit flowers to obtain nectar and/or pollen, tend to produce much softer signals, perhaps about 100 dB. at 10 cm. Species in the family that eat insects apparently listen for prey-generated sounds, such as the rustle of insects landing on leaves. Most of these bats forage in vegetation where quiet (= less intense) signals minimize the cacophony of echoes rebounding from leaves and branches. In the 1950s, Donald Griffin described New World Leaf-nosed Bats as "whispering" because their echolocation calls were faint and difficult to hear. These bats may avoid hunting flying prey because their echolocation calls are too soft to detect them at distances effective for aerial capture. This trade-off, however, has been a successful strategy for species in this family as a whole. With the ability to maneuver and forage within vegetation and utilize numerous food sources (not just flying insects), New World Leaf-nosed Bats have become the most ecologically diverse family of mammals. As many as fifty species may coexist in some forest habitats in South America, a remarkable diversity that may be possible only because of their "whispering" echolocation behavior.

Nancy Spends a Night in a Rainforest

Spending hours alone in the rainforest waiting for bats has given me a great appreciation for the variety of flying insects around night in a rainforest. Humming, chirping, buzzing and whining in innumerable ways, flying insects are everywhere! My favorites are the click beetles, many of which have what I think of as a pair of "headlights"—patches that glow with bioluminescent light, much like the abdomens of fireflies. Watching them fly through forest often reminds me of taxicabs in New York City. It is, however, much more common to hear insects at night rather than to see them. Many of the larger beetles and katydids make a great deal of noise when they crash into vegetation (or into a waiting bat biologist). It is not hard to imagine how a bat listening for such sounds could easily zero in on potential prey.

Self-deafening

Echolocation works because the bat registers the details of outgoing calls it produces in its brain for future comparison with returning echoes. Almost invariably, however, the outgoing calls are stronger, sometimes much, much stronger than the returning echoes. Most bats cannot broadcast and receive signals at the same time because the outgoing calls deafens them to the faint returning echoes, so they separate pulse and echo in time. As an echolocating bat closes in on its target, the pulse of its calls must be shorter and shorter to minimize overlap between outgoing calls and returning echoes. The "feeding buzz" illustrates this tactic. (Figure 4.5) In this system, one of the critical challenges for bats is to avoid deafening themselves with the rapid, intense calls they are producing.

Neurological interactions between a bat's larynx and its middle ear play an important role in avoiding self-deafening. A bat's ear includes the pinna (outer ear), ear canal, ear drum, middle ear and inner ear. (Figure 4.11) About two milliseconds before a vocalization is produced in the larynx, muscles in the middle ear contract, disarticulating three bones in the middle ear (malleus, incus and stapes) that transmit sound from the ear drum to the inner ear. Immediately after the outgoing call has been produced, another nerve signals the muscles to relax, which puts the malleus, incus and stapes back in contact with one another. When articulated, these three tiny bones convey vibrations caused by sound waves hitting the eardrum (tympanum) to the oval window, and thence to the inner ear. Inside the inner ear, special nerve cells in the stiff structural basilar membrane are stimulated and signals sent to the brain, where they are perceived as sounds. The overall sensitivity of a bat's ear depends upon the eardrum and these auditory bones, as well as the structure of the inner ear. By interrupting the chain of transmission of sound vibrations from the outer ear to the inner ear, a bat can avoid deafening itself when it emits high-intensity echolocation calls.

This may seem like a complicated system, and you might wonder why the bat does not just turn off its hearing system while producing an echolocation signal. The answer is that the bat needs a "picture" of the original call in its brain for future comparison with returning echoes. Also, it is not clear that any mammal has the capability of ever turning off its hearing—in any way other than what bats do with their ear-ossicle chain.

Figure 4.11.
Path of echoes (white arrow in **A**) arriving at the outer ear (pinna) of (**A**) a Greater Bulldog Bat. The pinna directs sound waves down the ear canal towards the auditory meatus (the canal that carries nerves from inside the skull towards the middle and inner ear–white arrow in **C**), the eardrum, the middle ear (malleus, incus and stapes) and then to the inner ear (inside the cochlea). On a ventral view of the skull (**B**) the yellow line outlines the area enlarged in **C**. The manubrium of the malleus (**D**) sits inside the auditory meatus. Sound waves in air cause vibration of the eardrum (tympanum) and hence vibration of the manubrium, thus converting vibrations in air to vibrations in bone. The vibrations continue through the chain of bones in the middle ear—the malleus, the incus and the stapes. Where the stapes abuts the oval window (connection between the middle and inner ear), the vibrations are converted to waves in the fluid that fills the cochlea. The middle ear extends from the inner surface of the eardrum to the oval window. The inner ear begins on the inner side of the oval window. The diagram shows the relative positions of different labeled structures.

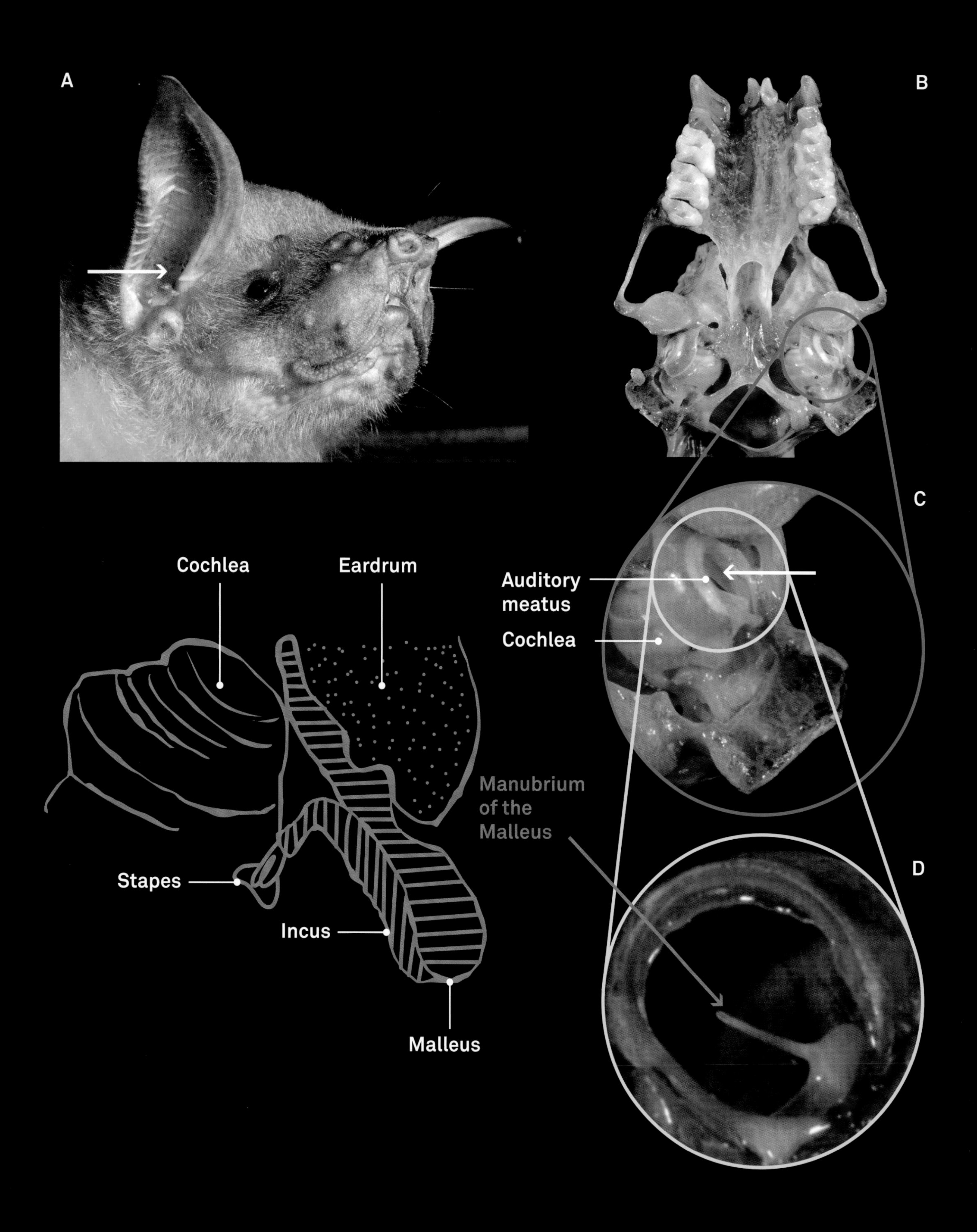
A
B
C
D
Cochlea
Eardrum
Auditory meatus
Cochlea
Manubrium of the Malleus
Stapes
Incus
Malleus

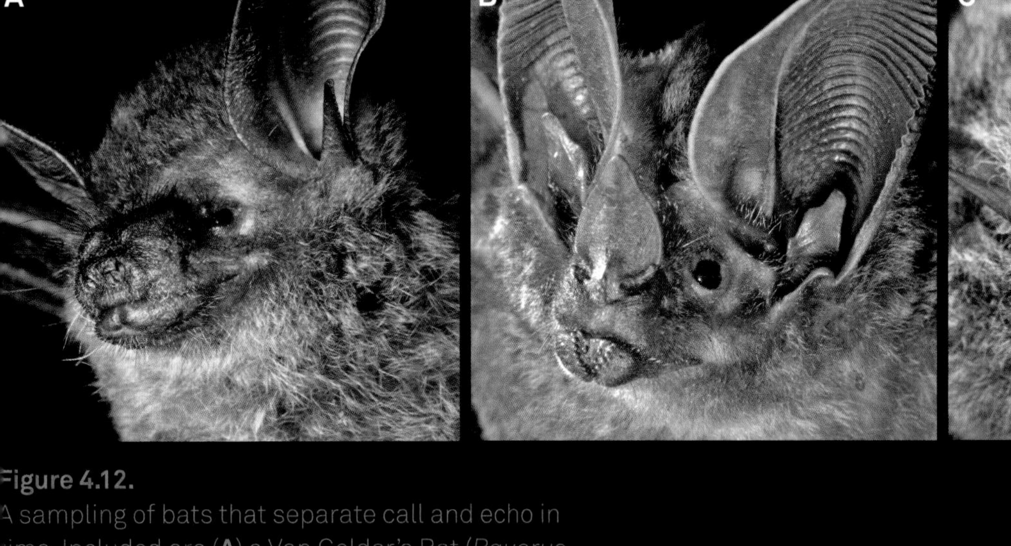

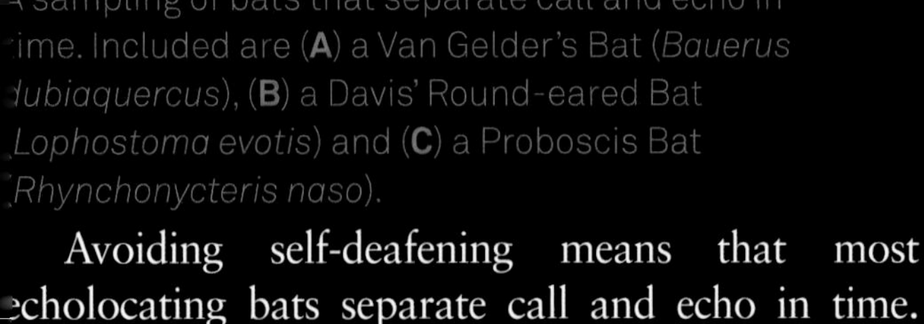

Figure 4.12.
A sampling of bats that separate call and echo in time. Included are (**A**) a Van Gelder's Bat (*Bauerus dubiaquercus*), (**B**) a Davis' Round-eared Bat (*Lophostoma evotis*) and (**C**) a Proboscis Bat (*Rhynchonycteris naso*).

Avoiding self-deafening means that most echolocating bats separate call and echo in time. (Figure 4.12) They cannot simultaneously broadcast and receive. Some bats, however, use an alternative way to avoid self-deafening. They separate pulse and echo in frequency. There are obvious differences in the patterns of production of echolocation calls between these two kinds of bats.

In the Old World tropics, approximately 160 species of Horseshoe Bats and Old World Leaf-nosed Bats separate call and echo in frequency. (Figure 4.13) In the New World, only a few very closely related species do this—Parnell's Moustached Bat, the Mesoamerican Moustached Bat (*Pteronotus parnellii*), and the Paraguanan Moustached Bat (*Pteronotus paraguanensis*). Bats that separate call and echo in frequency achieve this in four ways. First, they produce relatively long echolocation calls. Second, these calls are dominated by a single frequency. Third, in their inner ear there are many neurons each tuned to a different narrow band of frequencies that dominate returning echoes. Fourth, these bats adjust the frequencies of their echolocation calls to compensate for changes in the frequencies that dominate the returning echoes. The key to separating call and echo in frequency is the Doppler Effect. This effect is generated by changes in frequency caused by relative movements of sound source and receiver (such as when a fire truck drives towards and then away from a listener, the perceived frequency of the sound of the siren changes). Doppler shifts are responsible for the abilities of these bats to separate call and echo in frequency. (Figure 14.14)

Morphology and Echolocation

The faces of bats appear to influence both the pattern of sound radiation as an echolocation call is emitted by the bat and the bat's perception of the returning echoes. Many bats have a noseleaf or other structure around their mouth. (Figure 4.15) Differently shaped noseleaves may provide different acoustic views of the scene reflected by returning echoes. This has been shown for Pale Spear-nosed Bats (Figure 4.15A) and Common big-eared Bats (*Micronycteris microtis*) (Figure 4.15B) The noseleaves of Horseshoe Bats and Old World Leaf-nosed Bats may also affect the structure of outgoing signals, which are emitted through the nose.

Morphology of the ear pinnae, the visible part of the ear sitting outside the head, is also highly variable

Figure 4.13.

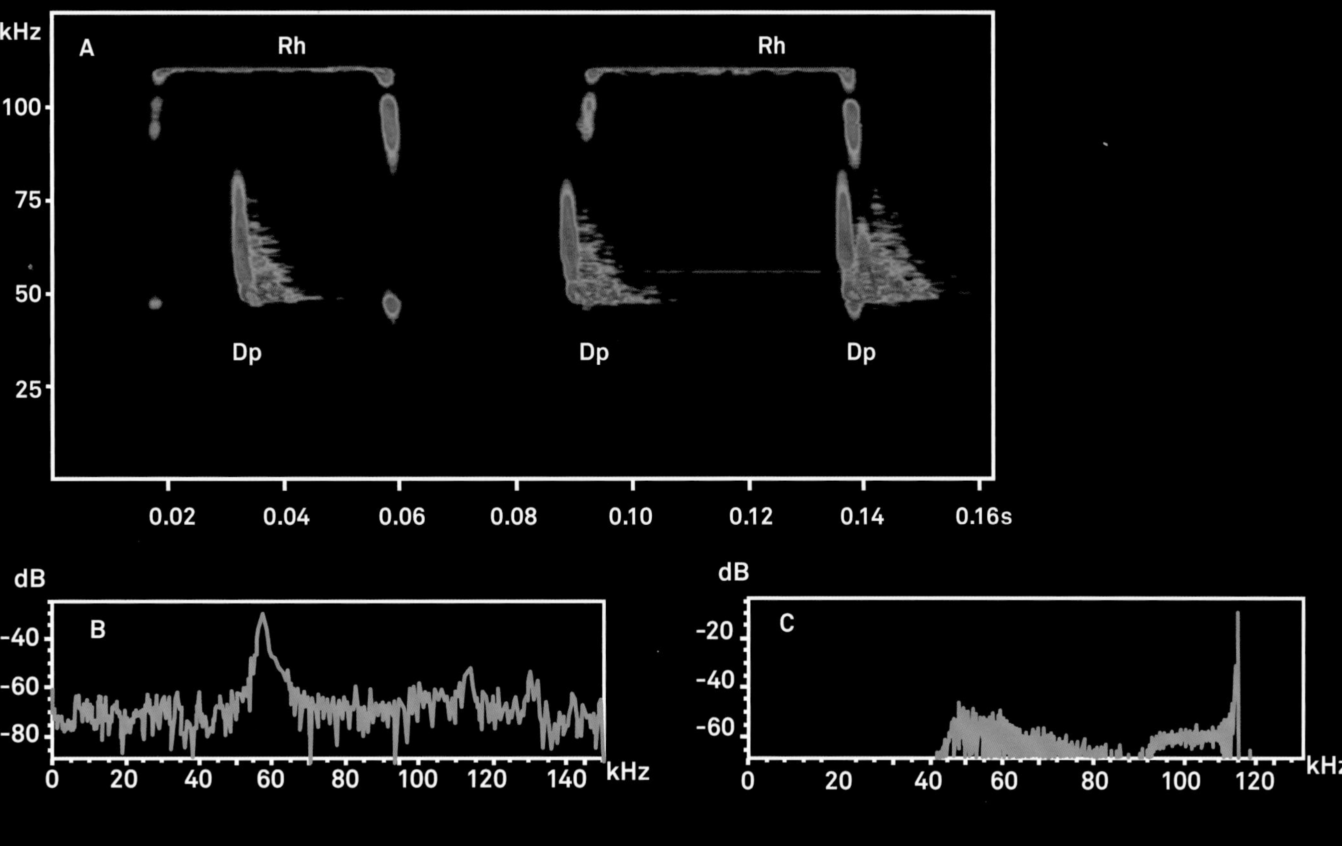

Figure 4.14.

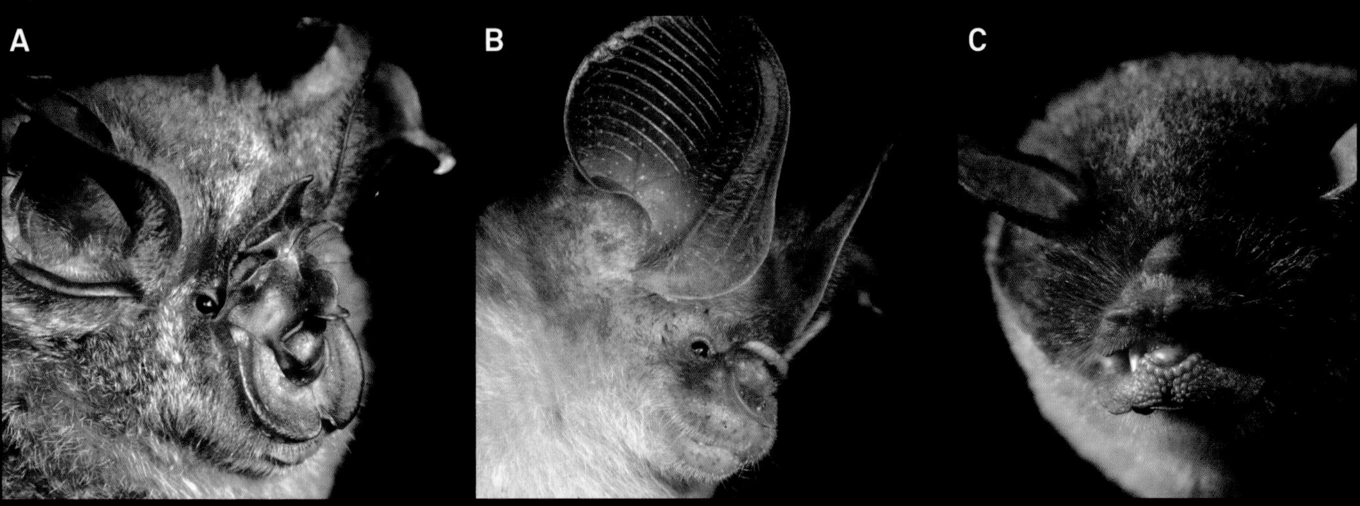

Figure 4.13.
Shown here are spectrograms (top) of two calls from a Lesser Horseshoe Bat (*Rhinolophus hipposideros*) (**Rh**) and three of a Desert Pipistrelle (**Dp**). Smearing to the right of the pipistrelle calls are echoes. The power spectrum plots (bottom) of a pipistrelle call (**B**) shows that they are broadband **FM** sweeps. The call (**C**) of the Horseshoe Bat is dominated by a single frequency.

Figure 4.14.
A sampling of bats that take advantage of the Doppler Effect and separate call and echo in frequency: (**A**) a Hildebrandt's Horseshoe Bat (*Rhinolophus hildebrandti*), (**B**) a Bicolored Round-leafed Bat (*Hipposideros bicolor*) and (**C**) a Parnell's Moustached Bat.

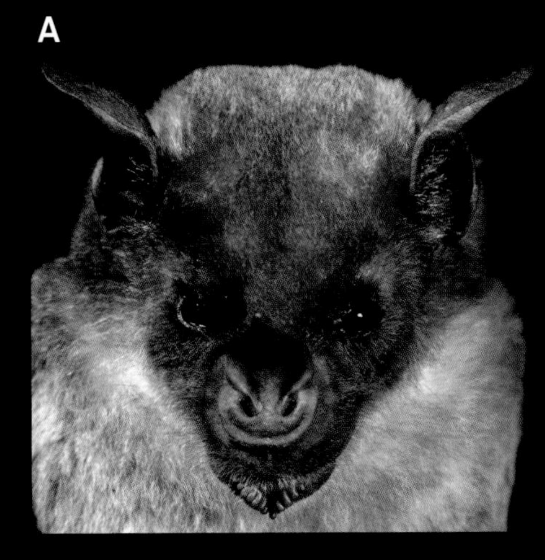

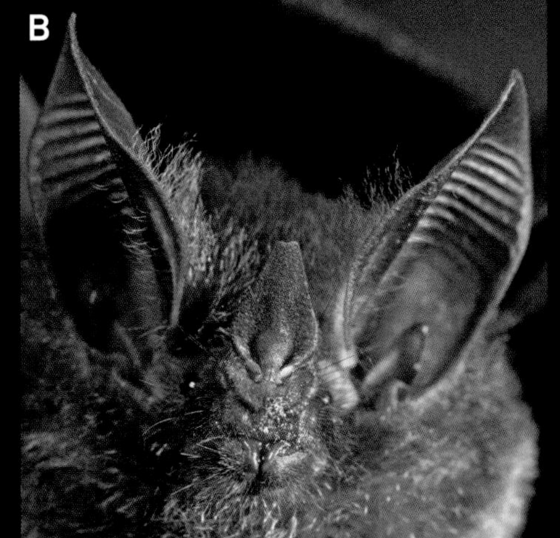

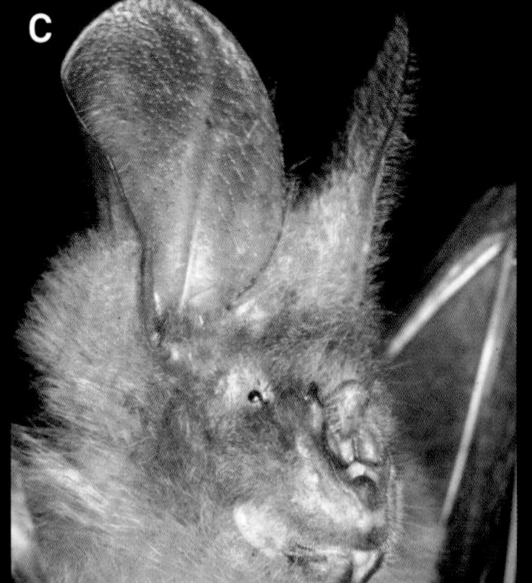

Figure 4.15.
The noseleaves of (**A**) a Pale Spear-nosed Bat (*Phyllostomus discolor*), (**B**) a Common Big-eared Bat and (**C**) a Large Slit-faced Bat play a role in echolocation. The large external ears (pinna) of some bats (**B**, **C**) apparently increases their ability to detect prey-generated sounds.

in bats. (Figure 4.16) The form of the pinnae affects how the bat hears echoes of their echolocation calls. While some ears may have evolved to facilitate echolocation, the really large ears of some species also increase their ability to detect faint prey-generated sounds. Bats such as Large Slit-faced Bats and Greater False Vampire Bats are gleaners that take their prey from surfaces such as the ground. They use echolocation to detect obstacles, but apparently also listen for the sounds that their prey make—for example, the croak of a frog or the rustle of an insect on a leaf—to locate their prey. Really large ears may make it easier to detect even faint sounds and movements of prey items.

Structures associated with the ear also can affect echolocation abilities of bats. In Big Brown Bats and Noctules (*Nyctalus noctula*) the tragus (both species) and the thickened lower margin of the ear (the noctule) influence the bats perception of returning echoes. (Figures 4.16A and B, respectively)

Bats Surprise Brock...Again

Most biologists do not expect to hear the echolocation signals of bats. In the 1970s, my colleagues and I could record bat echolocation sounds on a high-speed tape recorder and then, by slowing the tape speed, lower the frequencies of the sounds into a bandwidth we could hear. By 1976, I was very accustomed to listening to bat echolocation calls by recording them at one tape speed (76 centimeters per second) and slowing them down eight times. This changed the frequencies of the sounds and allowed me to hear what the bats said. But we learned that sometimes we could hear bat echolocation without the help of electronic equipment. In June 1977, working in the field in Zimbabwe, I vividly remember hearing a bat echolocating as it flew above me—it was very exciting to hear the entire echolocation sequence from the bat searching for a target all the way to the feeding buzz that culminated the attack. The bat was apparently a Large-eared Free-tailed Bat (*Otomops martiensseni*). (Figure 4.17B) At the time I did not catch the bat, nor did we record its calls, but the signals were utterly clear. They were not ultrasonic because I could

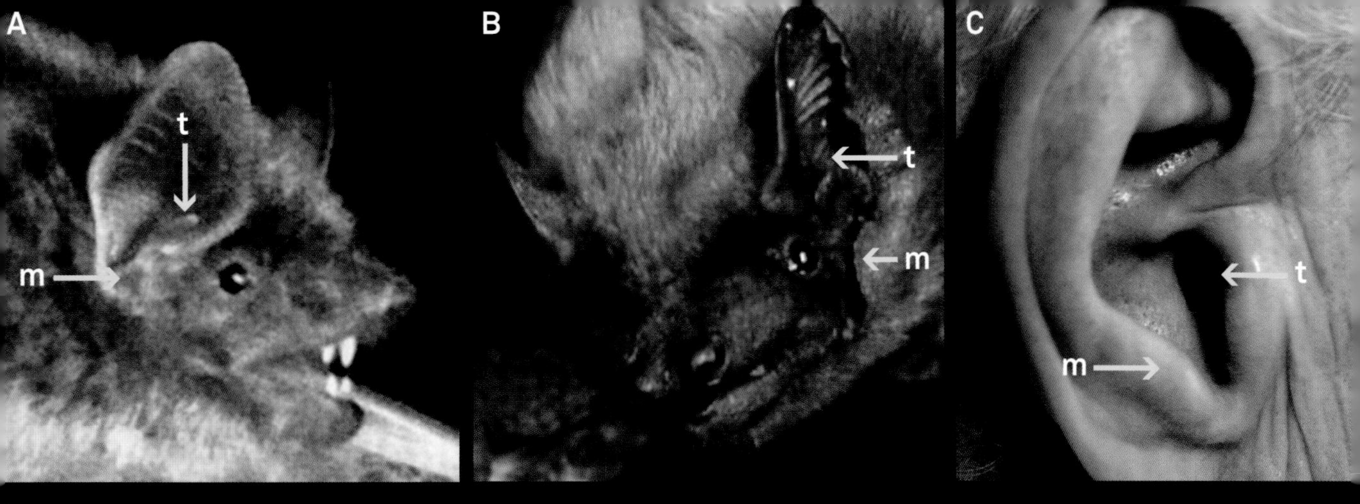

Figure 4.16.
The tragus and thickened lower margin of the ear (pinna) are shown in (**A**) a Big Brown Bat and (**B**) a Noctule. Yellow arrows identify the tragus (**t**) and the thickened lower margin (**m**) of the ear. The pinna of a human ear (**C**) is shown for comparison.

Then, in June 1979, I was studying bats in southern British Columbia where I saw a bat foraging in ponderosa pine woodland and could hear this bat's echolocation calls. Gary Bell, a student, and I did not succeed in catching our mystery bat and spent some time speculating about which species it might be. By listening to its calls, we knew that the bat foraged widely in the ponderosa pine woodland. Furthermore, some local naturalists were familiar with the calls, but were certain that they were produced by insects because they did not expect to hear bats. A year later, Greg Woodsworth, then a graduate student, went to the Okanagan Valley in Southern British Columbia to study the mystery bat. He shot one specimen, allowing us to identify it—it was a Spotted Bat (*Euderma maculatum*), a species not previously recorded from Canada. (Figure 4.17A) This was very exciting, and even the phone call from Greg at 3:00 am was a welcome way to receive the news.

I knew that in many animals large size equates to producing vocalizations of lower frequency, the difference between the woofs of a very large dog and the shrill yaps of a small one. But Spotted Bats are not "big", they weigh 10–15 grams, and even Large-eared Free-tailed Bats weigh only 25–30 grams. I can hear the echolocation calls of both species. At the same time, much larger bats, such as Commerson's Round-leafed Bats (*Hipposideros commersoni*) and Hairless Bats weigh >100 grams and their echolocation calls are ultrasonic. There is more to the story about echolocation calls than bat size.

Some bats with very big ears are likely candidates to produce echolocation calls audible to humans. But the degree to which a person can hear bats depends on how sensitive he or she is to hearing sounds at higher frequencies. Children can hear bats more often than adults because they have not yet suffered one of the common effects of human ageing—loss of hearing in the higher ranges. My favorite example is listening for bats along the Hudson River in New York. We were a team of three, me and two students, Aimee Macmillan and Yvonne Dzal. Aimee could hear the echolocation calls of Hoary Bats that were not audible to either Yvonne or me.

I think that far too many bat biologists refer to bats as using "ultrasonic echolocation." This idea is entertaining when one considers that what is ultrasonic to me and Yvonne was not to Aimee! The misconception about bats and ultrasonic echolocation supports an old saying "it ain't what you don't know that gets you into trouble, it's what you know for sure that ain't so." Why should some bats use lower frequency echolocation calls? See more about this on page 124.

A

B

5

What Bats Eat

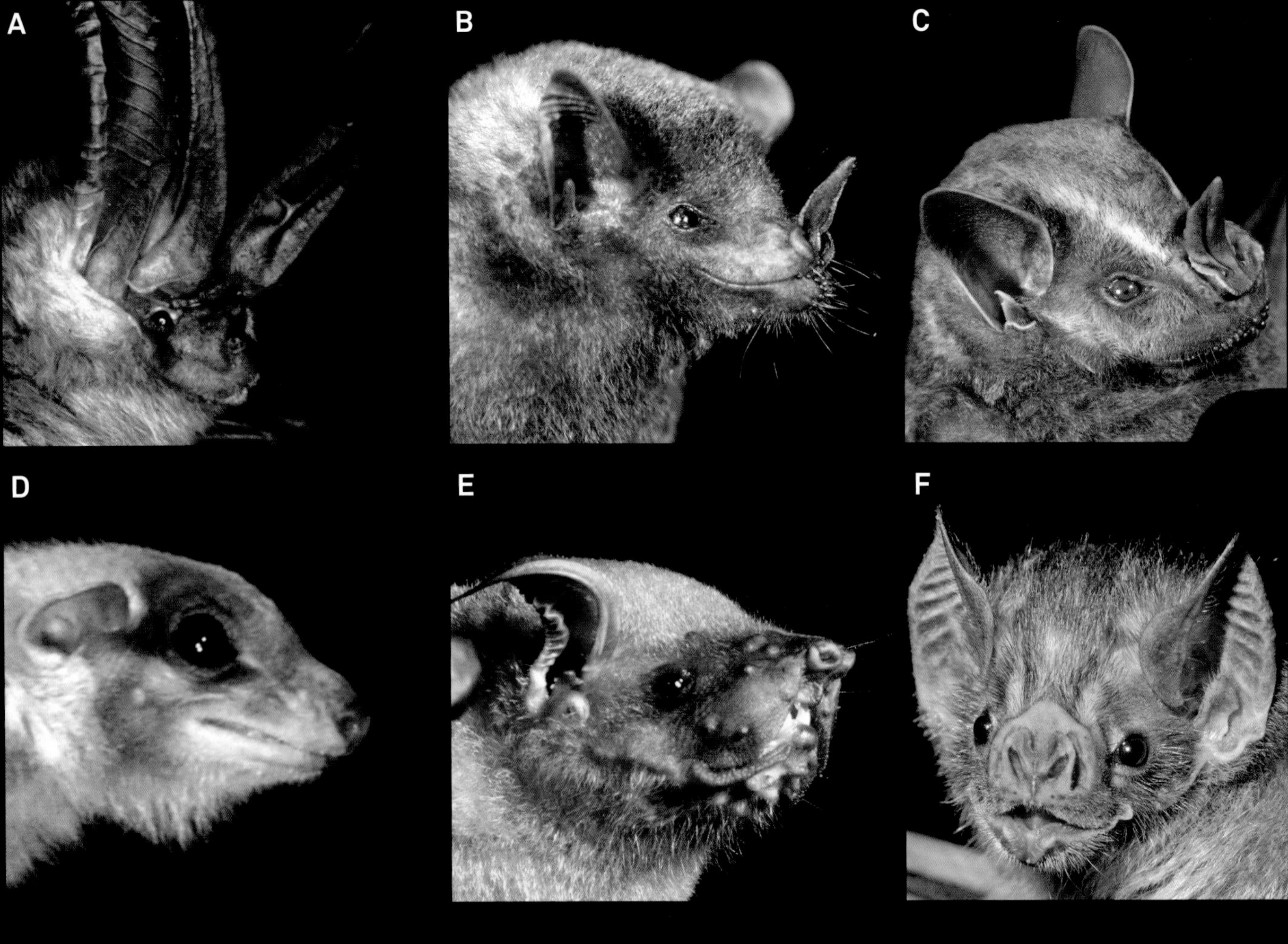

Figure 5.1.
A sampling of bats by diet: (**A**) an insectivore, Townsend's Big-eared Bat (*Corynorhinus townsendii*); (**B**) a nectarivore, Commissarisi's Long-tongued Bat (*Glossophaga commissarisi*); two frugivores, (**C**) Greater Fruit-eating Bat (*Artibeus lituratus*) and (**D**) Egyptian Rousette; (**E**) a piscivore (fish-eating), Large Bulldog Bat; and (**F**) a sanguinivore Common Vampire Bat.

Different Diets for Different Bats

Like most other mammals, milk is a young bat's food from birth until it is adult in size. Adult bats have different diets, and it has been traditional to recognize different general kinds of bats based on their diet. Most species of bats eat mainly insects. Some others eat a wider range of animals, extending to fish, frogs, birds, and other bats. Other species eat mainly fruit, while still others visit flowers to obtain nectar and pollen. The most notorious bats, the vampires, consume blood. By diet bats are often categorized as insectivorous, animalivorous (or carnivorous), frugivorous, nectarivorous, piscivorous and sanguinivorous. (Figure 5.1) When less detailed information about what bats eat was available, it was easier to believe that these largely non-overlapping diet groupings reflected the complete reality of bat diets. The more we learn about bats, however, the more difficult it is to be so categorical about diet.

Biologists have used four main approaches to determine what bats eat. First is direct observation of feeding bats, sometimes in a laboratory. Second is dissecting bats and examining the stomach contents. A third method is analyzing the droppings produced by bats. (Figure 5.2) A fourth way is collecting scraps of food dropped by feeding bats. (Figure 5.3) Occasionally bats carrying food fly into nets and drop their catch, further extending our knowledge about what bats eat. (Figure 5.4)

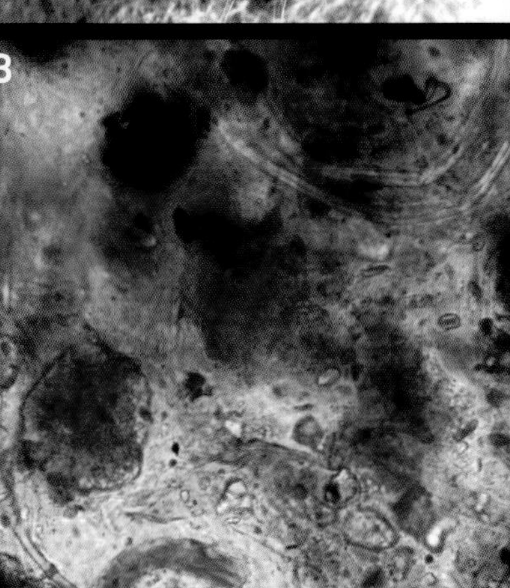

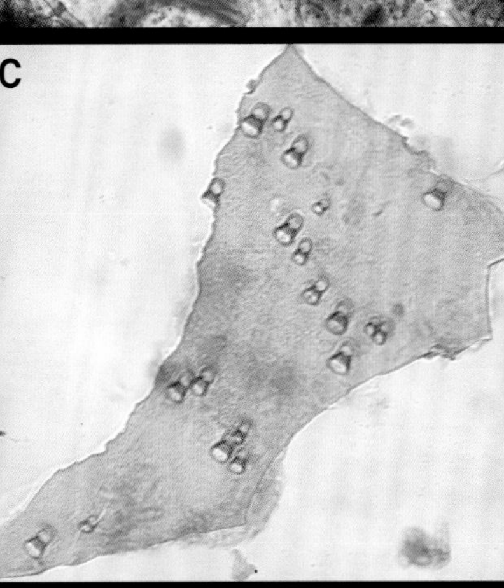

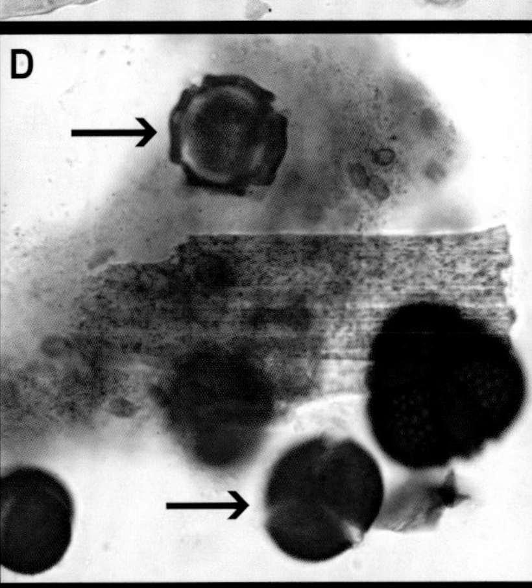

Figure 5.2.
(**A**) Leach's Single-leafed Bat (*Monophyllus redmani*) produces (**B**) droppings which under a microscope can be seen to include fragments of insects (**C**) and pieces of plants and pollen (**D**), including pollen for a Sausage Tree (*Kigelia pinnata*; black arrows).

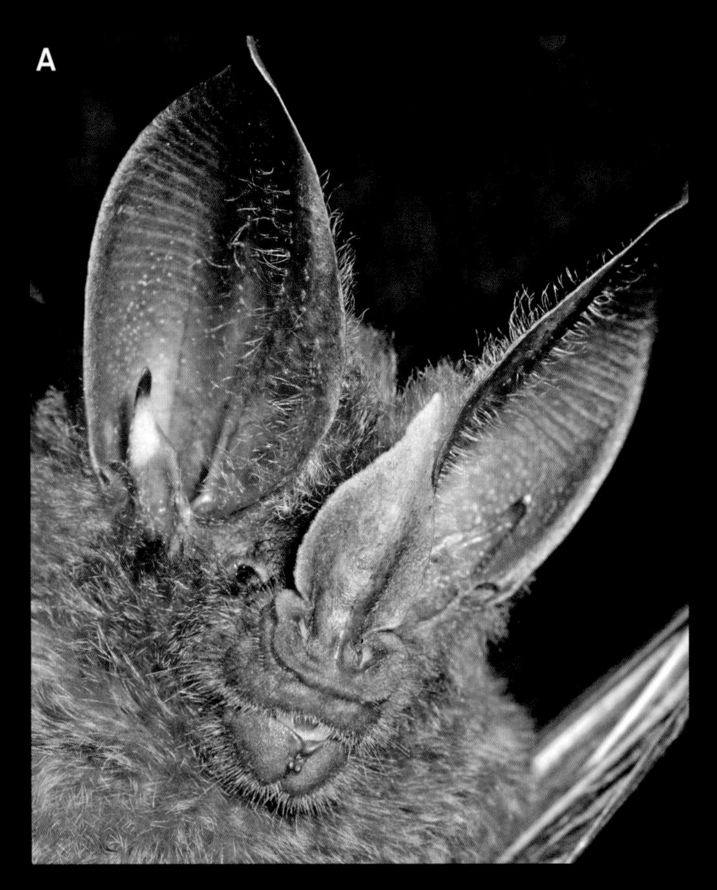

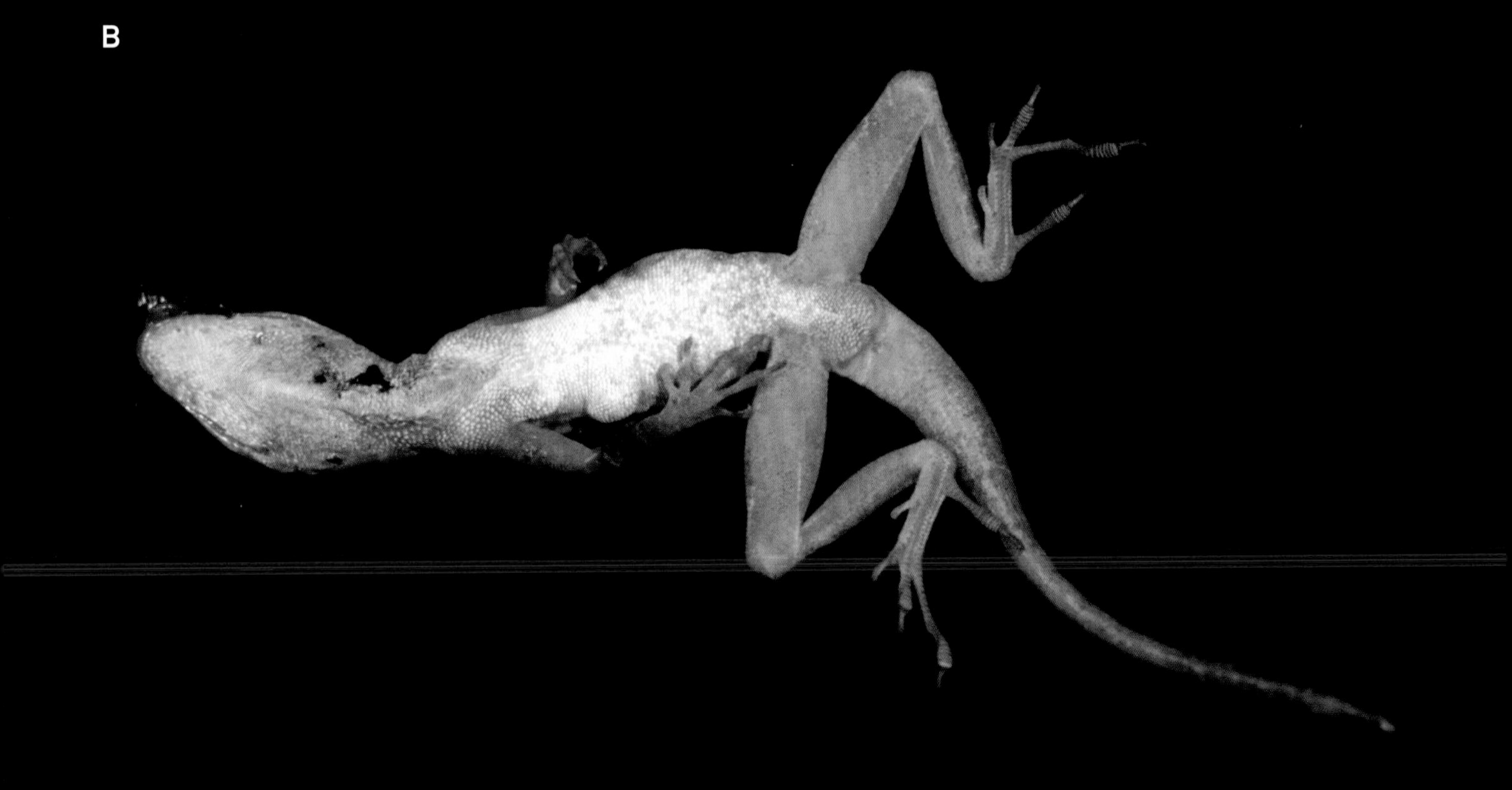

Figure 5.3.
A Golden Bat (*Mimon bennettii*) and a lizard that one of these bats was carrying when it flew into a mist net. Golden Bats weigh about 20 grams, the lizard 4.5 grams.

While everyone knows that bats eat insects, many people are surprised to hear that some bats eat larger prey. Several kinds of bats catch and eat birds. Spectral Bats (New World Leaf-nosed Bats) eat birds such as Groove-billed Anis (*Crotophaga sulcirostris*), presumably catching them when they are sitting on their nests. Three other bird-eating species hunt migrating birds. First described in 2007 from Spain and Portugal were Giant Noctules (*Nyctalus lasiopterus*). A second report that same year came from China and involved Great Evening Bats (*Ia io*). Then, in 2013 from Japan, Birdlike Noctules (*Nyctalus aviator*). In each case, the evidence of bird consumption is seasonal presence of feathers in bat droppings. Each of these bats weighs >50 grams and, on the wing, appears to detect, track, attack and eat the birds. Bird-eating bats probably use echolocation when hunting birds, but we lack details. There are no data about the bird-hunting behavior of other large carnivorous bat such as Large Slit-Faced Bats, Greater False Vampire Bats and Australian False Vampire Bats (*Macroderma gigas*).

Some bird-eating bats also eat other bats. The list includes bats from the Neotropics (New World Leaf-nosed Bats), from Africa (Slit-faced Bats) and from India to Southeast Asia and Australia (False Vampire Bats). The diet of Large Slit-faced Bats in Africa

Brock Explores the Diet of Large Slit-faced Bats

November 1979 was my first visit to Mana Pools National Park in Zimbabwe. Don Thomas and I went to collect information about bats eating frogs. Dolf Sasseen of Zimbabwe-Rhodesia Parks and Wildlife office had watched bats roosting in this disused water tower (Figure 5.4) as they ate frogs but had not witnessed how they caught them. Don and I visited the water tower and captured some bats, Large Slit-faced Bats, 30-gram super predators. We collected some of the remains of prey strewn beneath their roosts and established that these bats ate insects, sunspiders, fish, frogs, birds and bats. (Figure 5.4) This was the first record of a "carnivorous" bat in Africa. Although we obtained more details about the bats and their hunting behavior on subsequent trips, we still do not know how Large Slit-faced Bats hunt fish or frogs, or the details of their bat- and bird-hunting behavior, specifically how they search for, detect and then attack their prey.

I have watched a Large Slit-faced Bats catch, kill and then eat a 12 g. Egyptian Slit-faced Bat (Nycteris thebaica). These bats use a killing bite reminiscent of a cat, taking the victim's mouth and nose into its mouth and biting strongly. This method kills by suffocation. Large Slit-faced Bats discard the large ears and wings of the Egyptian Slit-faced Bats they catch and kill.

For people living in the world's temperate regions, most if not all local bats are insectivorous. Bats that eat plant products, as well as those taking larger animals, occur in tropics and subtropics around the world. Vampire bats occur only in parts of South and Central America. To an ecologist, bats as a group fill many trophic roles in ecosystems. Trophic distinctions are based on the food pyramid. Some bats are primary consumers, eating plants that convert solar to chemical energy. Other species are secondary consumers that eat animals that feed mainly on plants. Still other species are tertiary consumers (or top predators), species that eat animals that eat other animals.

DNA Barcode of Life

Today most shoppers are familiar with barcodes, identifying symbols that are read electronically to provide information about products, *e.g.*, what it is and how much it costs. Barcodes can quickly provide accurate details about a product, or, through implanted chips, information about your pet. Biologists at the University of Guelph in Canada have developed a "DNA Barcode of Life" technique that is based on new molecular techniques. DNA barcodes are short sequences of DNA that can be used to quickly identify organisms and that can sometimes provide a way to identify what bats have eaten.

The gene Cytochrome c oxidase subunit 1 (CO1) is species-specific for many animals, including bats and their animal prey. The CO1 gene is suitable as a Barcode of Life because its mutation rate among animals is often fast enough to distinguish one closely related species from another, while it is conservative enough that it rarely varies much within a species. This discovery, combined with polymerase chain reaction (PCR) technology—which makes it possible to determine DNA sequences from very tiny tissue samples—has allowed rapid processing of hundreds of specimens at low cost. By sampling known animal species identified by taxonomic experts, and then documenting the DNA barcode (*i.e.*, CO1 gene sequence) for each of these species, researchers have built a reference library of DNA barcodes against which unknown samples can be compared to obtain identifications. With these new methods, researchers can analyze pieces of insects from bat droppings and determine how many different and—with sufficient comparative insect data—exactly what species of prey a bat has eaten.

Barcode of Life analyses have revealed that Eastern Red Bats eat both Forest Tent Caterpillars (*Malacosoma disstria*) and Gypsy Moths (*Lymantria dispar*), as well as over 100 other species of insects. Little Brown Myotis sampled at sites from Alberta to Nova Scotia in Canada ate over 550 species of insects. Sampling at one location over two years

A disused water tower (**A**) along the Zambezi River in Mana Pools National Park was used by roosting Large Slit-faced Bats (**B**) that brought their prey there for consumption. The bats dropped pieces of their food as they ate (**C**). The discarded prey fragments included wings of insects (**i**), pieces of fish (**f**), a frog's leg (**fl**), bird wings (**b**), and a bat wing (**bw**). Scale in (**C**) provided by Donald Thomas.

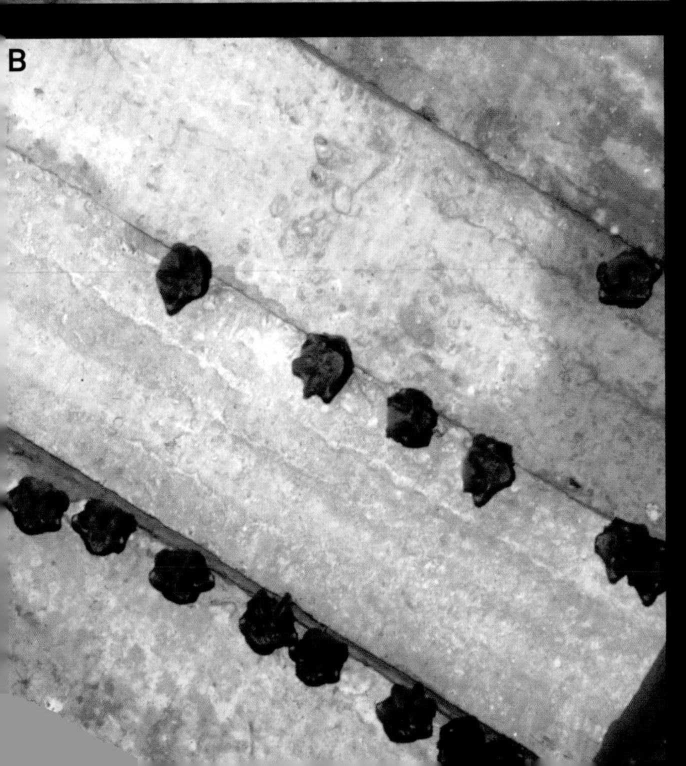

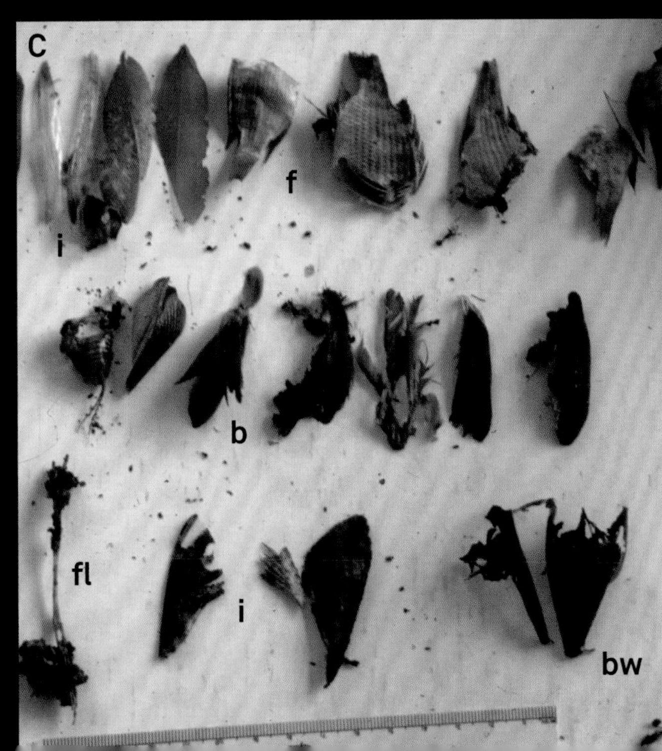

revealed that Big Brown Bats ate over 200 species of insects. Little Brown Myotis and Northern Long-eared Myotis (*Myotis septentrionalis*) ate several species of mosquitoes, and both Northern Long-eared Myotis and Big Brown bats occasionally ate blackflies (Simuliidae). This was surprising because blackflies are mainly diurnal.

Knowledge of the ecology of some species of insects consumed by Little Brown Myotis has revealed that these bats successfully forage over water bodies ranging from clear, clean streams to stagnant, polluted ponds. This was determined from the mix of insect species in their diets. For example, it is known that species of mayflies (Ephemeroptera) and caddisflies (Trichoptera) vary in their tolerance of pollutants in water, with different species of each favoring different types of streams/ponds in which their larvae develop. By looking at the mix of mayflies and caddiflies in the diets of the bats, researchers can tell where they were foraging (if there are different types of water bodies in the area). This also means that foraging bats could be used to monitor water pollution along watersheds.

The great diversity of insects taken by Little Brown Myotis and Big Brown Bats is not surprising for at least two reasons. First, insects are very diverse. Second, bats have huge appetites and, for energetic reasons, should take almost anything in the right size range that is available. In the evening, people fishing with artificial flies occasionally hook bats, for whom the lure designed for fish was appealing because of its size and motion. Experiments in British Columbia showed that foraging Little Brown Myotis and Yuma Myotis (*Myotis yumanensis*) attacked 12–15 mm. diameter objects moving in the airspace 2–10 cm. above the water. The bats attacked appropriately sized edible objects (insects) and inedible objects (pieces of leaf) indiscriminately but ignored larger ones, whether the latter were large pieces of duct tape or large insects.

Using the DNA barcode to identify species, whether of bats or insects, requires a library of barcodes based on samples from known species. But, even without a CO1 gene reference library to identify insects in a certain area to the species level, DNA Barcode analysis can reveal many details of the foraging ecology of bats since unknown prey can often at least be identified as to what large group they come from, *e.g.*, beetles, mayflies, *etc*. Each of the seven species of insectivorous bats living in the area of Windsor Cave in Jamaica was shown to eat different insects, even when the bats hunted in the same habitat. Included in the analysis were Antillean Ghost-faced Bats (*Mormoops blainvillii*), Waterhouse's Leaf-nosed Bats (*Macrotus waterhousii*), Parnell's Moustached Bats, MacLeay's Moustached Bats, Sooty Moustached Bats (*Pteronotus quadridens*), Brazilian Free-tailed Bats and Pallas' Mastiff Bats. Lack of overlap in the diets of species could reflect an abundance of prey combined with some inherent specializations.

Barcoding techniques do not work soley for identifying animals and prey—they can also be used for identifying plants. Ongoing work with fruit-eating bats in Jamaica using DNA barcoding has also provided details about what two species of fruit bats eat. Jamaican Fruit Bats and Jamaican Fig-eating Bats differ in size, with the former being nearly twice as large as the latter. Yet both eat many of the same fruits. The DNA Barcode of Life analysis revealed more details about the diets of these bats than did prior studies based on identification of seeds and fruit in bat feces and stomach contents. This may be because many of the identifiable parts of fruits eaten by fruit bats, *e.g.*, seeds, are discarded by the bats, with only the juice and a few squishy parts being swallowed. DNA barcoding can tease useful genetic data out of what is otherwise unidentifiable!

Classifying Bats by Diet

While many of the categorizations of bats by diet are broadly correct most of the time, in the last couple of decades data from detailed studies of fecal samples, stomach contents and DNA barcoding have revealed that many bat species do not conform to easy classification, or simply don't eat what we thought they did. More details about what bats eat reveal the flexibility of bats—whether indicated by behavior, diet or morphology. Pallas' Long-tongued Bat, a presumed nectar-feeder based on its long snout and tongue, is a case in point. Examining the droppings of these bats reveals that they often eat insects and fruit, as well as nectar and pollen. (See Figure 5.2.) This diversity of diet is seen whether one uses DNA barcode or traditional morphological analyses of what is in the bats' feces. There is still a great deal to learn about the diets of bats and the implications of this for their ecology, behavior and distribution. The discoveries that "insectivorous" bats sometimes eat fruit, "nectarivorous" bats often eat insects, and "frugivorous" bats sometimes eat leaves are changing the way scientists think about the ecological roles that various species play, and also changing perceptions about what resources may be necessary to support healthy populations.

Skulls, Teeth and Jaws

All bats have teeth, but the actual number varies from species to species. Like other mammals, bats have both milk teeth (baby teeth) and permanent (adult) teeth. (Figure 5.5) In bats the milk teeth help the young hang onto its mother's nipple. Perhaps as a result, the milk teeth are often shaped like Velcro© hooks—sharp and recurved. This tooth shape for the milk teeth is unique among mammals.

The permanent teeth of a variety of bats, shown in Figure 5.6, illustrate the diversity of tooth shape in bats that eat: (A) insects, (B and C) terrestrial vertebrates, (D) blood and (E, F and G) fruit. The names of permanent teeth are shown in Figures 5.5. and 5.6. The incisor teeth (upper front teeth) show considerable variation, from large and prominent (Figure 5.6 A, C, D, F and G) to small (B), sometimes with the outer incisors much smaller than the pair near the midline (E). Canine (eye) teeth are always present and range from blade-like in the vampires (Figure 5.7), to more spike-like in the other bats. There are small teeth (premolars) between the canine and the molars (cheek teeth) in all bats. In general, bats use incisor teeth for seizing food, canines for piercing and holding and the cheek teeth (premolars and molars) for cutting and/or crushing. The "W" or "V" shaped crests on the molars in Figure 5.6 A, B and C serve in cutting and puncturing, while the broader and flatter cheek teeth of E, F and G are more useful in crushing. The smaller bilobed (two cusped) lower incisor teeth are used to comb the bat's fur.

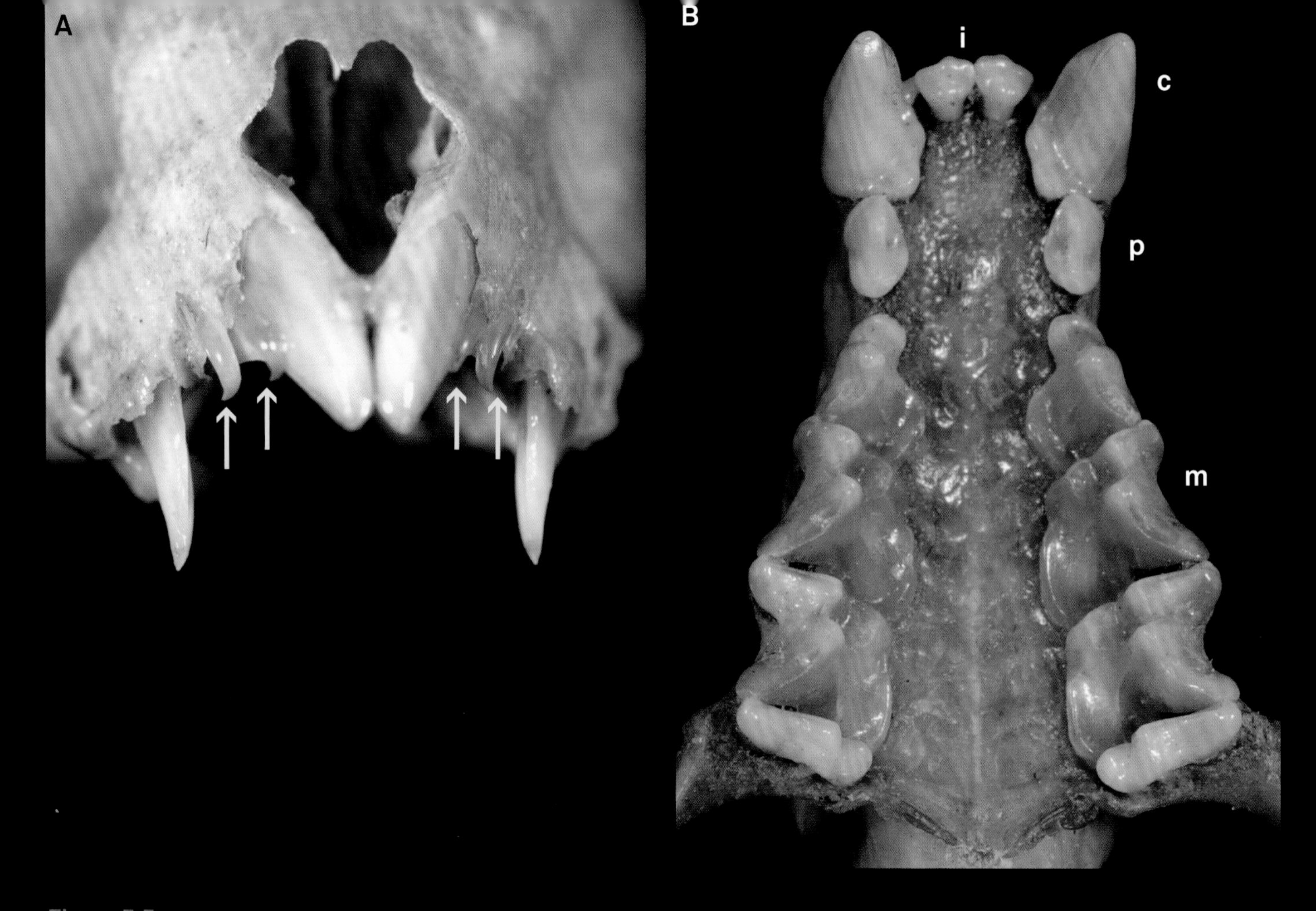

Figure 5.5.

Figure 5.5.
The Common Vampire Bat (**A**) shows both permanent (adult) and deciduous (milk) teeth, while the Spectral Bata (**B**) shows only adult teeth. (**A**) shows four milk teeth (yellow arrows) and four large adult teeth. The milk teeth of Common Vampire Bats are typical of bats in general—small pegs with hook-like structures on the cusp (biting end). (**B**) Permanent (adult) dentition of a carnivorous Spectral Bat showing the incisor (**i**), canine (**c**), premolar (**p**) and molar (**m**) teeth. There are two pairs of incisor teeth (note the tiny second incisors flanking the larger midline pair), one pair of canine teeth, three pairs of premolars and two pairs of molars.

Figure 5.6.
The diversity of bats' diet is reflected by the upper teeth. Shown here sorted by diet are (**A**) an insectivore Pel's Pouched bat (*Saccolaimus peli*); two carnivores (**B**) Large Slit-faced Bat; (**C**) Spectral Bat (*Vampyrum spectrum*); (**D**) a sanguinivore, Hairy-legged Vampire Bat (*Diphylla ecaudata*); and three fruit-eaters, (**E**) Little White-shouldered Bat (*Ametrida centurio*), (**F**) Antillean Fruit-eating Bat (*Brachyphylla cavernarum*) and (**G**) Golden-capped Fruit Bat (*Acerodon jubatus*). These bats differ greatly in size from Little White-shouldered Bats (10 g.) to Golden-capped Fruit Bats (1000 g.).

Figure 5.6.

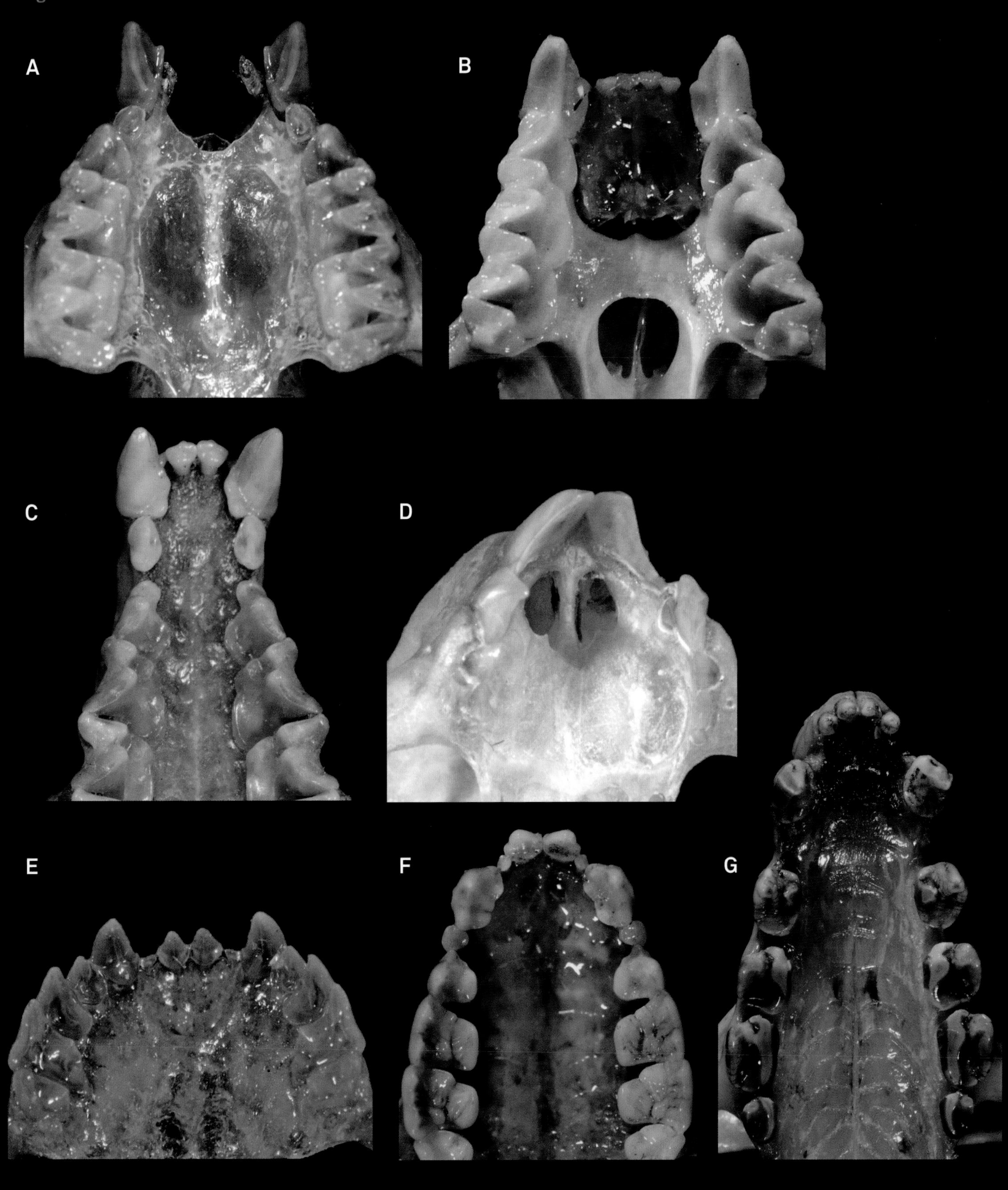

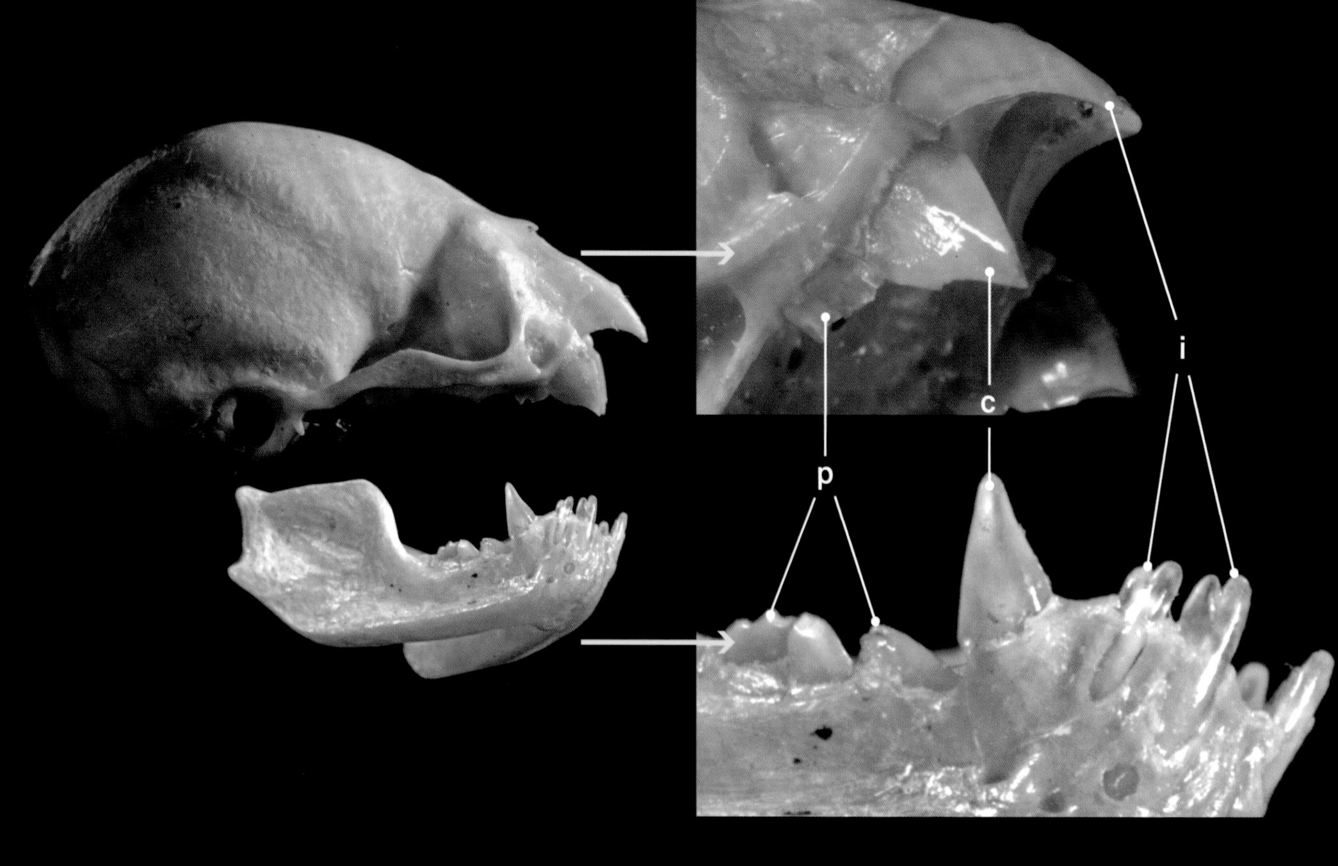

Figure 5.7.
Skull and lower jaw of a Common Vampire Bat showing enlarged views (yellow arrows) of teeth in upper and lower jaw. The different teeth (white arrows) are premolars (**p**), canines (**c**) and incisors (**i**). Also shown in a photograph by Anne Brigham is a view of the bat's face showing the prominent canines. See text for details of how the bat uses these teeth.

The teeth of vampire bats deserve special mention. (Figure 5.7) The upper incisors are greatly enlarged and are blade-like and specialized for cutting. The bat uses them to make its feeding bite. The canines (C) are used to deliver defensive bites on bat biologists and other predators, and are used conjunction with the premolars for clipping fur and feathers to expose bite sites (skin). In the absence of a need to chew their food, the cheek teeth of vampires have become reduced to small, blade-like spicules (spikes). Overall, vampires have one of the most highly modified dentitions seen in any mammal.

Nectarivorous bats often hover while feeding. (Figure 5.8) They have longer muzzles than their fruit-eating relatives as well as smaller teeth. (Figure 5.9) Nectar-feeding has evolved independently in bats in the Neotropics and in the Old World tropics. Note the similarities between the bats shown in Figure 5.9A and B and 5.9 C and D, and the differences between New World Leaf-nosed Bats (A and B) and Old World Fruit Bats (C and D).

Some of the flowers that bats visit are specialized to exploit the bats as pollinators. More than 500 species of angiosperm (flowering) plants (sixty-seven families, twenty orders) are pollinated by bats. Bat-pollinated flowers open at night and tend to be light colored. (Figure 5.10) Their nectar is often musty smelling, and the pollen is sticky. The syndrome of bat pollination (Chiropterophily) has evolved in plants in both the Neotropics and Old World tropics.

In the New World, the nectarivorous bats echolocate. Some chiropterophilous flowers in the Neotropics have ultrasonic nectar guides that make it easier for these bats to locate the flower and the nectar (and be covered with pollen in the process). At least one flowering vine, (*Marcgravia evenia*) from Cuba, has a specialized concave leaf above its flowers that reflects back echolocation calls, alerting bats to the presence of the flowers. In some cases, the petals of flowers shift slightly when a bat visits and extracts nectar, a change in flower form that might alert future bat visitors to the fact that the flower may have already been drained of some or all of its nectar.

Figure 5.8.
A migrating male Lesser Long-nosed Bat (*Leptonycteris yerbabuenae*) vists a hummingbird feeder in Tucson, Arizona. Photograph by Ted Fleming.

There may be an interesting dichotomy between chiropterophilous flowers in the New World and the Old World because different kinds of bats live in these two regions. Although the New World nectar-feeders echolocate (New World Leaf-nosed Bats), the Old World necatar-feeding bats (Old World Fruit Bats) do not. Chiropterophilous flowers in the Old World should lack ultrasonic nectar guides because the Old World nectar-feeding bats do not echolocate. However, this prediction has not yet been tested.

Energetic Realities and Food Consumption

As small mammals with high metabolic rates, bats burn large amounts of energy, especially when flying. This means that to survive bats must eat a great deal of food. For insectivorous bats, some estimates put this consumption at from 50 percent to over 120 percent of their body weight in food every day they are active. The same appears to be true of frugivorous bats. Peters' Dwarf Epauletted Fruit Bats (*Micropteropus pusillus*) and Beuttikofer's Epauletted Bat (*Epomops beuttikoferi*) handle 1.9 to 2.5 and 1.4 to 1.5 times their body weight in fruit every night. In this case, "handle" distinguishes between what the bat actually ingests as opposed to what it handles. You can appreciate this difference if you think of what you do not ingest when you eat an orange (the rind) or an avocado (the rind and the large seed). To put the amounts of food in perspective, think of your favorite food and then remember how much you weigh. Imagine trying to eat half your body weight in pizza (for example) in one night!

A bat's need for energy from food alerts us to some realities that face foraging bats. A bat's decision about what to eat will depend on what is available in its local habitat. There are several aspects to availability. There is the specific potential food type—insects, for example. Then there are the sizes of the insects and the distances between them (their density). Not surprisingly, bats often hunt where potential food occurs at high density because this reduces foraging time. For insectivores this can mean seeking out a swarm of mating insects or groups of insects that are active around lights or in the lee of a tree stand where they are protected from wind. For Natterer's Bat (*Myotis nattereri*) and perhaps others as well, a spider web can be a good place to find easy prey. Frugivorous and nectarivorous bats may find concentrations of food (fruit or flowers) on trees and bushes. This is true when there is synchronous flowering or fruiting of plants in an area. Bats use their spatial knowledge and memory to located good foraging areas within their home range, often checking out sites repeatedly over many days to monitor insect emergences, flower blooming patters or fruit ripeness depending on their diet.

For many bats, energy requirements are reflected by the range of food consumed. An early indication of problems with classification of bats by diet was the discovery that many species thought to eat mainly fruit also ate leaves. This makes sense because leaves usually more readily available than fruit. By 1995 bat biologists had demonstrated that that some species of Old World Fruit Bats and New World Leaf-nosed Bats that ate fruit also ate leaves. The 1999 discovery that New Zealand Lesser Short-tailed Bats (*Mystacina tuberculata*) regularly eat insects and fruit, but also visit flowers to obtain nectar and pollen illustrated that other bats crossed the boundaries proposed about the diets of bats. In 2011 researchers in Australia reported that Gray-headed Flying Foxes (*Pteropus poliocephalus*) thought to feed on pollen, nectar and fruit deliberately hunted, caught and then ate flying insects, specifically cicadas. That same year, Pallid Bats (*Antrozous pallidus*), which had be considered to be mainly insectivorous, were discovered visiting flowers of columnar cacti and agaves from which they obtain nectar and pollen.

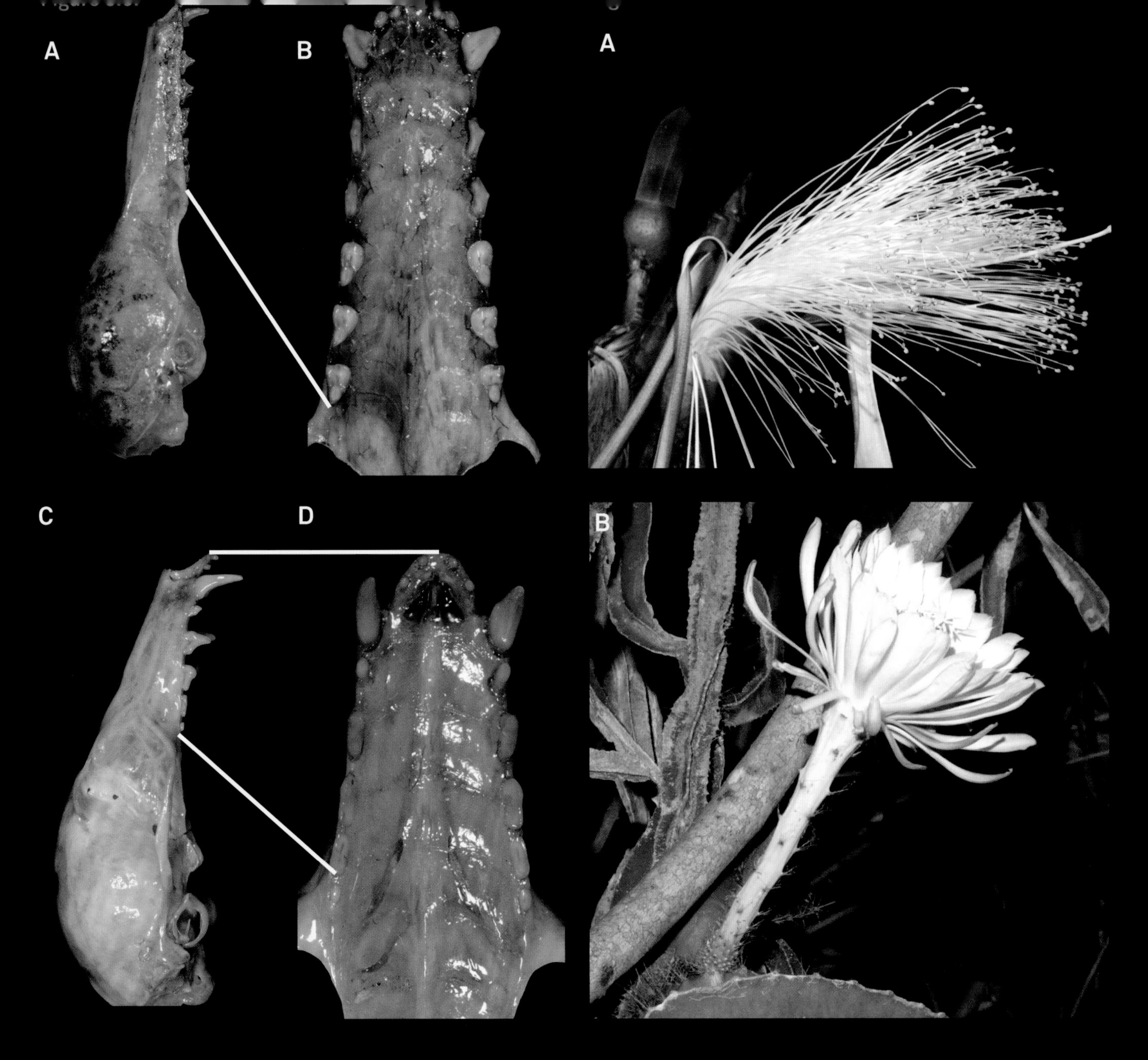

Figure 5.9.
Nectarivorous bats have elongated muzzles and smaller teeth. Compared here are (**A**, **B**) a New World Leaf-nosed nectar feeder, a Mexican Long-tongued (*Choeronycteris mexicana*); and (**C** and **D**) an Old World Fruit Bat nectar feeder, a Woermann's Bat (*Megaloglossus woermani*). Both species are shown from a side (**A** and **C**) and bottom (**B** and **D**) view. The white lines show how the rows of teeth are expanded between side and bottom views.

Figure 5.10.
Two New World chiropterophilous flowers: (**A**) a Shaving Bush Tree (*Pseudobombyx ellipticum*) and (**B**) an epiphytic Moonlight Cactus (*Selenicereus* spp.).

Figure 5.11.
A foraging MacLeay's Moustached Bat.

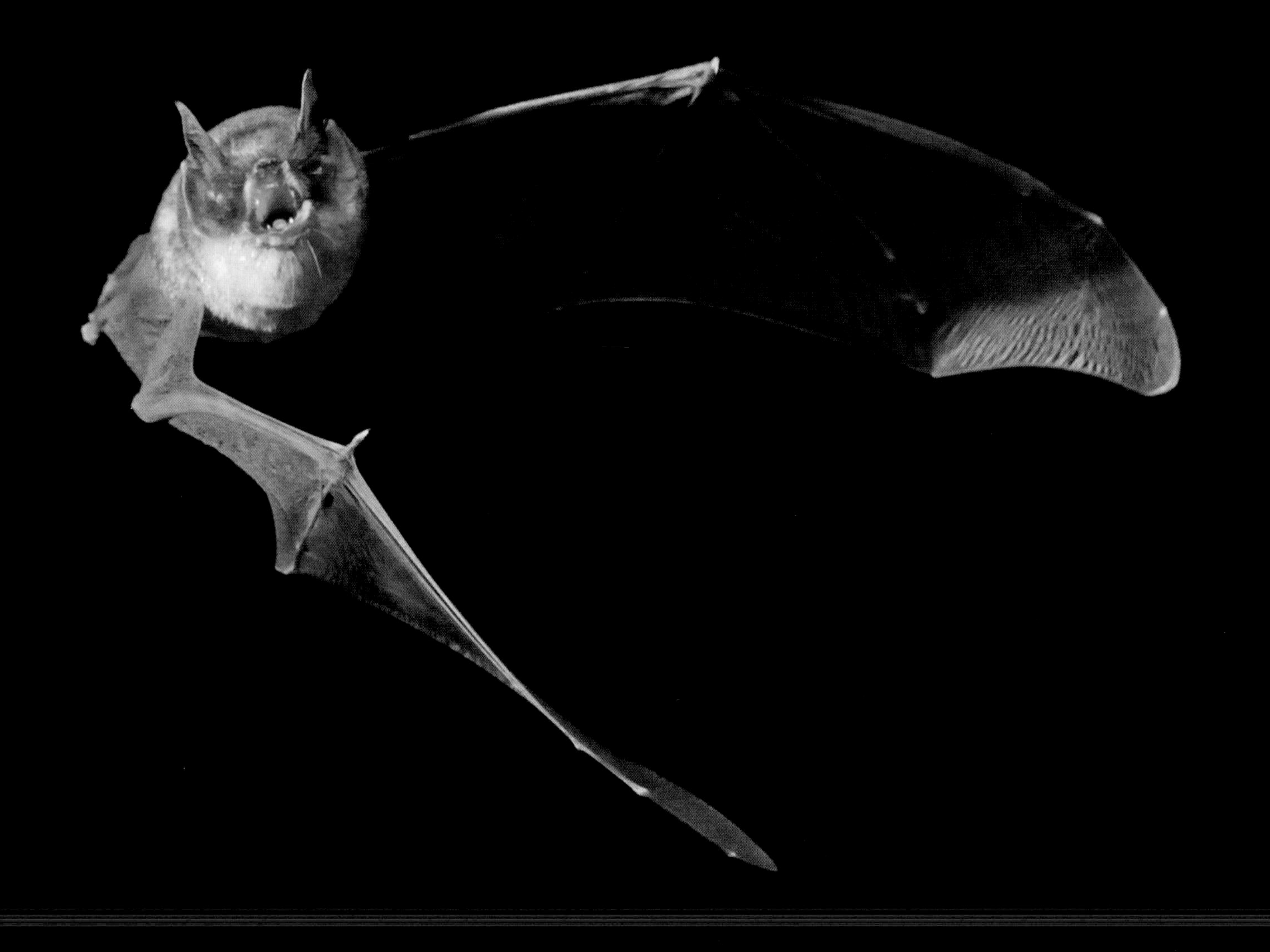

Brock and the Hoary Bats

My students and I worked with Hoary Bats at Pinery Provincial Park in Ontario for five years to gain a more complete view of the behavior of these bats. At 30 grams, Hoary Bats are the largest species in Canada. These bats roost in foliage and go south for the winter. There are occasional records of them from Iceland and the Shetland Islands, and there are populations on the Hawaiian and Galapagos Islands. Robert Barclay from the University of Calgary had studied Hoary Bats at the Delta Marsh Field Station located near the south end of Lake Manitoba. There he and his students routinely found and then caught roosting Hoary Bats.

Our first challenge was catching the bats that usually foraged well above the forest canopy. Two of my students, Brian Hickey and Lalita Acharya, discovered that Hoary Bats would chase and produce feeding buzzes as they pursued 1 cm. diameter pebbles tossed into the air. We used this technique to attract bats down closer to the ground and into a waiting mist net that two of us swung in unison. From captured bats we measured features of their wings and then either banded the bats or attached radio tags to them. The radio tags allowed us to find where the bats roosted and foraged. These tags also allowed us to measure just how long they spent foraging each night. Hoary Bats at Pinery Provincial Park often fed on the concentrations of insects around streetlights. We monitored their echolocation calls and counted feeding buzzes to measure how often they attacked insects. By direct observation, we determined that the bats succeeded in about half their attacks, and that they caught both medium sized (30 mg.) and larger (100 mg.) moths. (See Box 5.1, Figure 1) Using a modified police traffic radar unit and working with Horacio de la Cueva from the Centro de Investigación Científica y de Educación Superior de Ensenada, we measured the flight speeds of Hoary Bats. We used the flight speeds along with information about wing morphology we had gathered by measuring captured bats to estimate costs of flight. Combined with information from the literature about the bats' metabolic rates and the food value of the moths, we constructed a picture of the energy budget of a foraging Hoary Bat.

Hoary Bats at Pinery foraged for about 120 minutes a night, attacking an insect every 17.5 seconds. Time between attacks reflected the density of prey. At Delta Marsh, Robert Barclay had found that each foraging Hoary Bat attacked an insect about every sixty seconds and they foraged for 250 to 375 minutes each night. At Pinery, if the bats ate mainly 100 mg. moths and made an attack every 17.5 seconds, they could reduce their foraging time by half. In Manitoba, at one attack per minute, Robert's Hoary Bats would have had to be taking larger prey to cover their costs of foraging. This example illustrates the importance of prey density and prey size to a foraging bat. Another indication of the importance of density of prey to hunting bats was their tendency to hunt in concentrations of insects at lights.

Individual vampire bats appear to feed (procure blood) from one individual mammal or bird per night, so their prey need not be in a group. Vampire bats take blood from sleeping victims, and they are able to detect and recognize the deep breathing sounds of a sleeping mammal or bird. Each 30 gram Common Vampire Bat can consume 15–20 grams of blood in a night, about two tablespoons. In diet, vampire bats—Common Vampire Bat, Hairy-legged Vampire Bat and White-winged Vampire Bat (*Diaemus youngi*)—are the most specialized of bats because, as adults, they eat only blood. This raises an interesting question about whether the vampire bats are predators or parasites because they do not kill the animals whose blood they ingest.

A Variable Menu....

The diversity of species of food exploited by an individual bat of any species may vary considerably by season: spring, summer and early autumn in temperate zones and wet and dry seasons in the tropics and subtropics. Seasonal variability makes it unlikely that an insectivorous bat, for example, will specialize in feeding on a single insect species or small suite of species because those available tonight may be very different from those available in a month's time. But the reality of numbers remains. If an 8 gram Macleay's Moustached Bat needs to consume 50 percent of its body mass in insects every night, this means it must find and capture 133 medium-sized moths each weighing 30 mg, or 4,000 1 mg. midges—every night. (Figure 5.11) High energetic demands means that bats tend to feed in concentrations of insects and focus on locally abundant prey.

Biologists also have used analysis of isotopes to investigate the diets of bats. Comparing the isotopes of nitrogen (^{15}N) and carbon (^{13}C) in five species of Short-tailed Bats (genus *Carollia*) at seventeen sites in Central and South America revealed some interesting differences. Although generally considered to be fruit-eaters, the analysis of isotopes revealed that some species relied more on insects—enriched ^{15}N, Chestnut Short-tailed Bat (*Carollia castanea*)—or on fruit—enriched ^{13}C, Seba's Short-tailed Bat (*Carollia perspicillata*). Analysis of ^{13}C istopes using Common Vampire Bats revealed that they fed mainly on the blood of cattle. This additional cue provides yet another picture of the lives and diets of bats.

Foraging and Eating Behavior

Bats vary in their foraging behavior with some hunting and feeding on the wing (a behavior known as aerial hawking) while others take their catch to a perch to consume it. Typically, bats that hunt and feed on the wing eat flying insects. The majority of echolocating bats catch and consume their prey on the wing, stopping to roost and rest only occasionally. Some bats that eat fish such as the Fishing Bat (*Myotis vivesi*), the Rickett's Big-footed Bat (*Myotis ricketti*), and both Vesper Bats and the Greater Bulldog Bat also consume their prey while flying. (Figure 5.12) Fishing bats have enlarged hind feet with sharp claws. They gaff fish and other prey with their hind feet, quickly transferring them to their mouths. Enlarged hind feet are also found in a much smaller bat, the Long-legged Bat (*Macrophyllum macrophyllum*) of the Neotropics. This small bat, which weighs only 7 to 11 grams, apparently uses its large feet as gaffs just like the bigger fish-eating bats do, but goes after smaller prey like water striders.

Other bats usually land to feed. Vampire bats are an obvious example, as are some frugivorous bats. But this approach also occurs in bats that eat insects or other animals. Landing to handle and consume food appears to allow bats to eat larger individual food items. One advantage of consuming larger individual food items is reduced costs of travel and reduced time in food handling. A comparison of the diet of Hoary Bats and Large Slit-faced Bats (both weighing about 30 g.) illustrates a possible impact of foraging behavior on diet. (Figure 5.13) Hoary Bats fly continuously while hunting and appear to eat only flying insects that they catch and consume on the wing. Their diet in southern Ontario includes quite large Hawk Moths (Sphingidae). In contrast, Large Slit-faced Bats mainly hunt from perches, making short flights to catch insects and other prey. But they take their catch to a perch where they eat it, after culling the wings. (Figure 5.13) The menu of Large Slit-faced bats includes Hawk Moths more than twice the size of those taken by Hoary Bats. The upper size of prey accessible to a Hoary Bat is determined by its ability to process the prey in flight. Large Slit-faced Bats can handle larger prey because they do so from the comfort of a perch or roost. The same applies to humans and fast food. Some items can be eaten on the run, others require both hands and more attention.

Many bats that eat at a perch use their thumbs to manipulate the food. This is true whether bats are eating fruit or animals, especially large animals. Although Smoky Bats (Furipteridae) are sometimes referred to as "Thumbless Bats" members of this group in fact have thumbs although they are small and largely enclosed in the flight membrane. (Table 1.1) Ghost Bats (genus *Diclidurus*) also have small thumbs. Nobody knows why the thumbs have been reduced in these groups, but lack of any evidence that they use

Figure 5.12.

A

B

Figure 5.12.
The enlarged hind feet of a Greater Bulldog Bat carry prominent sharp claws and are used to gaff fish and other prey taken from surface of the water.

Figure 5.13.

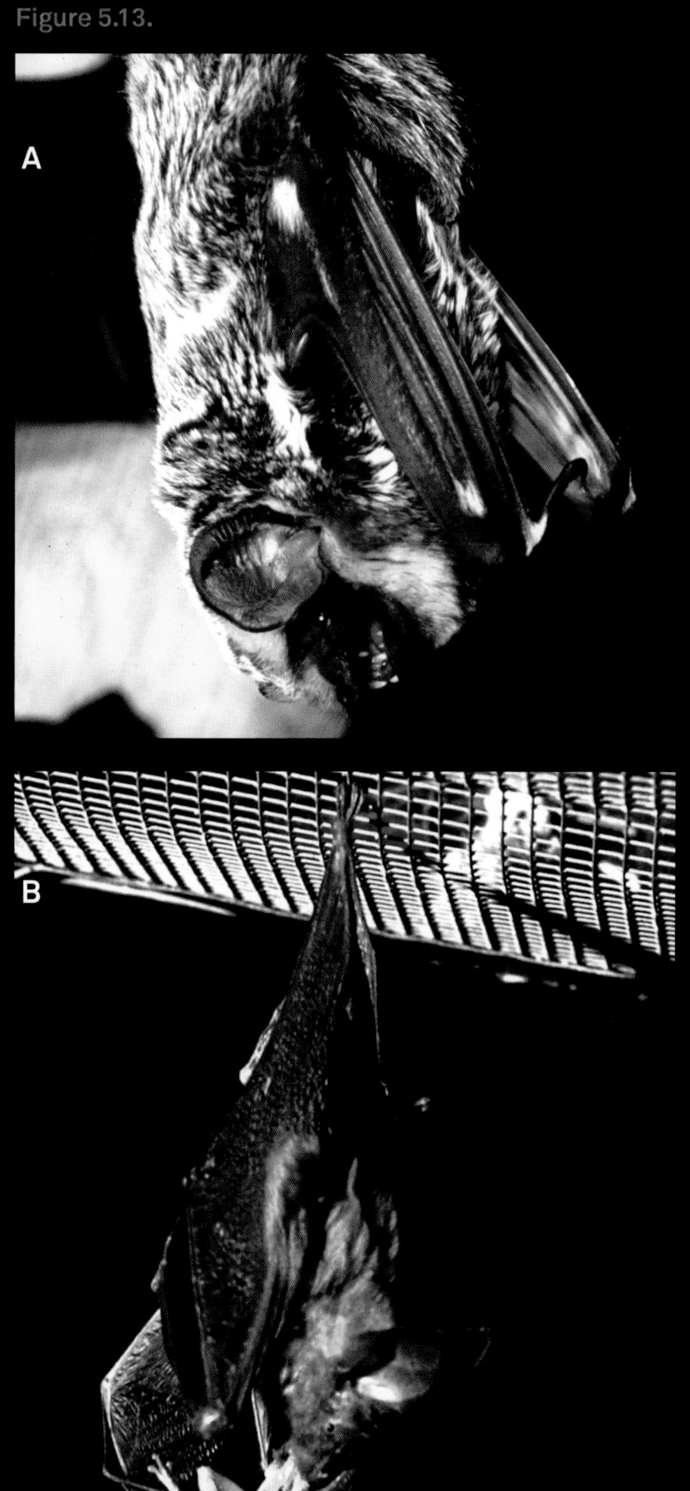

Figure 5.13.
A Hoary Bat (**A**) and a Large Slit-faced Bat, cupping the frog it has caught in its wing membrane as it begins to eat the prey (**B**).

Figure 5.14.

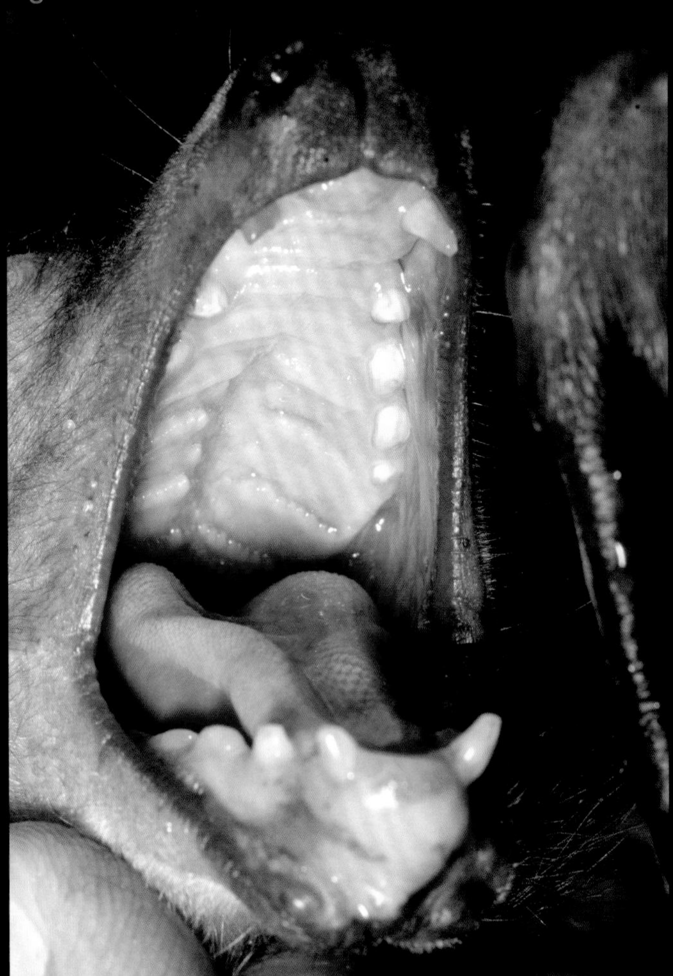

Figure 5.14.
Ridges on the palate of a Rousette Fruit Bat are used in conjunction with the rough tongue as a fruit juicer and pulp remover.

Figure 5.15.
Seeds that germinated in the darkness of a cave in Cuba. Jamaican Fruits Bats had brought the seeds into the cave.

Figure 5.15.

them in feeding suggests that they have evolved other ways to manage their prey. Most likely small-thumbed bats simply use their mouths and never "handle" their food at all.

A bat's hind feet seem to be used mainly for hanging onto a surface, branch, rock face, *etc.* One reason not to keep captive bats in a cage with metal screening is that the abrasion can quickly dull the bats claws. Even the Disk-winged Bats, which use the adhesive disks on their feet to hang on to their roosts, retain their toenails and, like other bats use their claws for cleaning their fur.

Bats appear to be very sensitive about adding extra weight to their bodies. When eating insects, bats typically discard most hard, indigestible parts, retaining and thoroughly chewing only the parts they can digest. Bats eating insects typically cull and drop the legs and wings of their prey. At sites where bats are feeding on the wing, *e.g.*, Pinery Provincial Park, it is common to see pieces of insects drifting down from above and littering the ground. (See page 110.) When bats eat fruit, they fill their mouths with the juicy pulp. These bats use their tongues to rub the fruit pulp against ridges on the roofs of their mouths while sucking up the juice and digestible material. (Figure 5.14) Afterward they often spit out any remaining pulp or seeds. In caves, seeds dropped or passed by fruit bats germinate in the darkness, producing a ghostly forest of doomed pale plants. (Figure 5.15)

BOX 5.1

Battlefields of the Night Skies

Insects from moths to lacewings, and beetles to praying mantises have ears with which they detect the echolocation calls of hunting bats. Typically, bat-detecting ears give the insects an advantage because they may hear the bats from 40 m. long before the bats have detected them. (Box 5.1, Figure 1) Moths with bat-detecting ears manage to hear and evade attacking bats 60 percent of the time (the other 40 percent of the time, the bat catches the moth). Detecting faint echolocation calls usually causes the moth to turn and fly away from the sound source. Loud echolocation calls stimulate the insect to fly erratically and/or dive to the ground. Deafening these moths by puncturing the tympanal membrane removes their ability to detect bats, leaving them more vulnerable to capture. Most insects that detect bat echolocation calls have two ears, allowing them to determine the direction from which the bat is approaching. Praying mantises have a single bat-detecting ear and are exceptions to the general rule about sense organs always coming in pairs. Their single, bat-detecting ear makes their hearing sensitivity less directional than that of moths and other insects with paired ears.

Tiger Moths (Arctiidae) have escalated the arms race with bats. Many Tiger Moths produce clicks when they hear intense bat echolocation calls. These moths often taste bad, so their clicks serve as a warning signal to the bats. Tiger Moth clicks also may startle the bats or even interfere with their echolocation. The varying ways Tiger Moths use clicks to stymie bats are not mutually exclusive partly because of the diversity of insectivorous bats and moths. The echolocation of some species of bats may be jammed by the clicks of some Tiger Moths, while the same clicks may startle another species of bat or alert it to a bad-tasting target.

To sneak up on their prey, some bats use sound frequencies that are not detected by some moths. This stealth approach to echolocation is another layer of complexity in the interactions between bats and insects. The fact that Eastern Red Bats and Hoary Bats sometimes capture moths that should have heard their calls suggests ever more mysteries about bats and the astonishing arms race between predators and their prey. (See page 110.)

Other bats, such as the Western Barbastelle (Barbastella barbastellus), use low intensity echolocation calls to foil the defensive behavior to moths. DNA barcode analysis confirms the success of the Western Barbastelle (*Barbastella barbastellus*) in hunting moths with bat-detecting ears.

Box 5.1, Figure 1.

Digestion and Food Passage

A factor that minimizes the weight that bats must carry during flight is rapid passage of food through the digestive tract. For example, Little Brown Myotis move food through their systems—from mouth to poop—in about twenty minutes. Comparable times of passage have been recorded from other species of bats, both insectivorous and frugivorous.

Blood is a specialized, liquid diet, and a large part of blood is plasma—basically water with some solutes. Vampire bats begin to urinate within two minutes of beginning to drink blood. Urination allows these bats to minimize the weight that they have to carry, while maximizing the payload of digestible components of the blood they have consumed. The plasma of the blood meal is the main component of the urine produced while the bats are feeding.

Digesting pollen to obtain proteins is important for nectar-feeding bats. And yet, pollen is extremely tough—each pollen grain is enclosed in a hard shell that must be opened or digested in some way to allow access to the nutrients inside. Lesser Long-nosed Bats drink their own urine, which creates an acidic medium in the stomach. This makes it possible for the bats to digest the pollen they consume. Flower-visiting bats often are coated with pollen and cleaning it off may involve mutual-grooming sessions by individuals roosting together. Mutual grooming is most common between mother and offspring in some species such as Antillean Fruit-eating Bats. Such grooming sessions may allow bats ready access to pollen collected by others, spreading out the nutritional benefit of the pollen load.

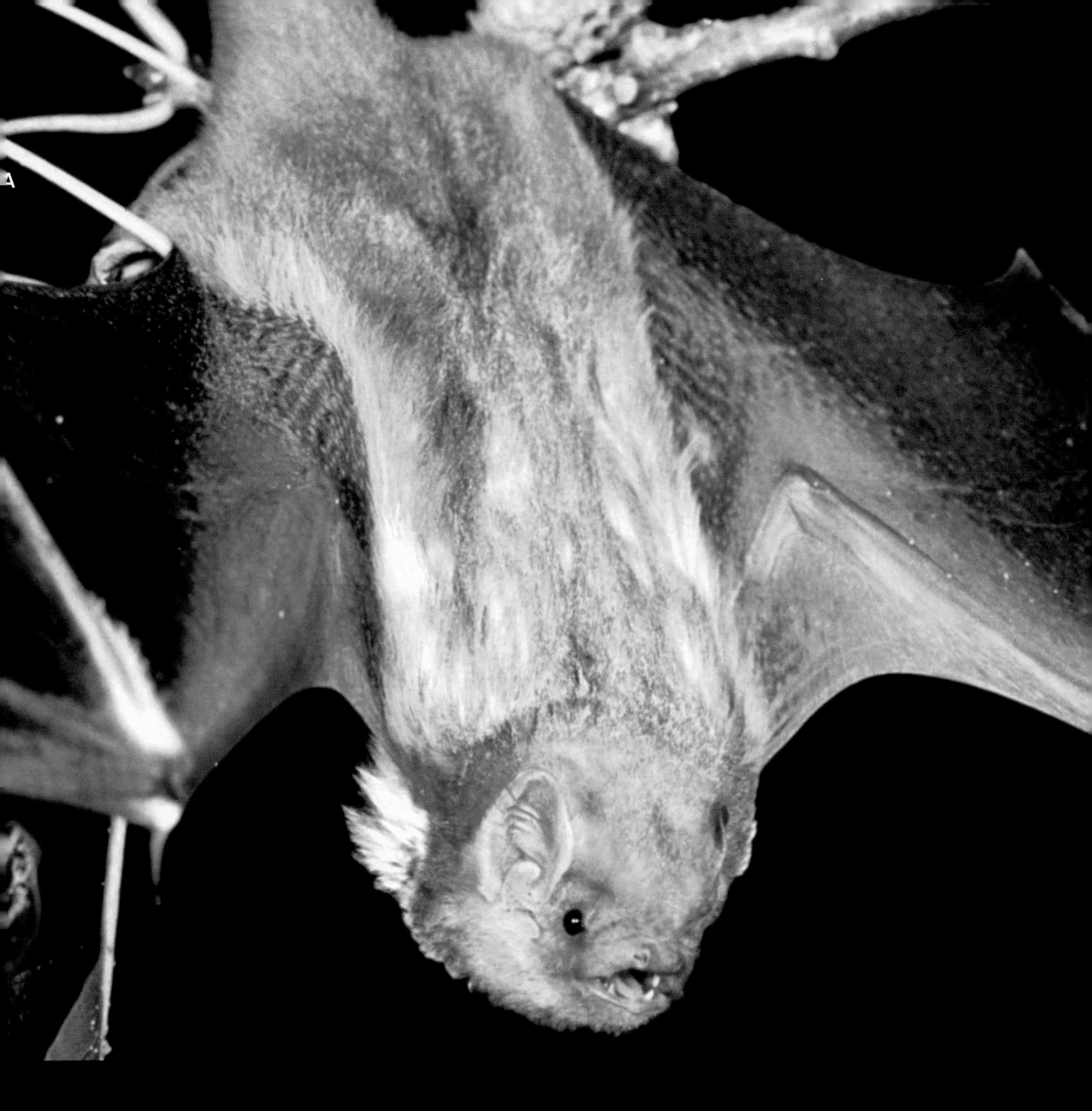

Bats and Pest Control

People often believe that because bats eat huge quantities of insects, bats must be important agents of biological control of insect pests. Brazilian Free-tailed Bats are an obvious example of bats as control agents. Over ten million of these bats roost in some caves in Texas, and over one million of them under the Congress Avenue Bridge in Austin, Texas. These bats average about 12 grams body mass, and if each of them eats 50 percent of its body weight in insects each night, ten million bats would consume sixty million grams of insects in one night. That's the equivalent of the weight of sixteen elephants! In Texas, Brazilian Free-tailed Bats also eat Corn Earworm Moths (*Helcoparva zea*) that plague corn and cotton. In 2006, the estimated economic contribution of Brazilian Free-tailed Bats was US$741,000 in an eight-county region in southern Texas.

What about the contribution to pest control of other species of bats? In Pinery Provincial Park in southern Ontario, Eastern Red Bats and Hoary Bats occasionally attack, capture and eat medium-sized Gypsy Moths and Forest Tent Caterpillar Moths

Figure 5.16.
DNA barcode analysis revealed that (**A**) Eastern Red Bats eat both (**B**) the Forest Tent Caterpillar Moths (the silken net containing molting larvae) and (**C**) Gypsy Moths.

each weighing about 30 mg. (Figure 5.16) Both species of moths are notorious pests because their caterpillars regularly defoliate large tracts of forest. The bats eat the moths, so do not be misled by the first part of the name "Forest Tent Caterpillar Moth." Each bat needs about 154 of these moths a night to meet its energy needs. Assuming these bats eat only Forest Tent Caterpillars Moths, how many bats would it take to control an outbreak of these moths? When there are <1000 (< = less than) moths per hectare, they are not readily detected by entomologists. At 1000 moths per hectare, 3.2 Eastern Red Bats eating only Forest Tent Caterpillar Moths might consume the entire population in three nights. But an outbreak of Forest Tent Caterpillar Moths can involve 10,000 or even 10,000,000 moths per hectare, requiring between 32.5 and 32,500 Eastern Red Bats per hectare to control the populations of moths. Years of work at Pinery Provincial Park documented a local population of only thirty-eight Eastern Red Bats in the whole park (2,532 hectares), suggesting there are not enough these bats to control a major infestation of Forest Tent Caterpillar Moths there.

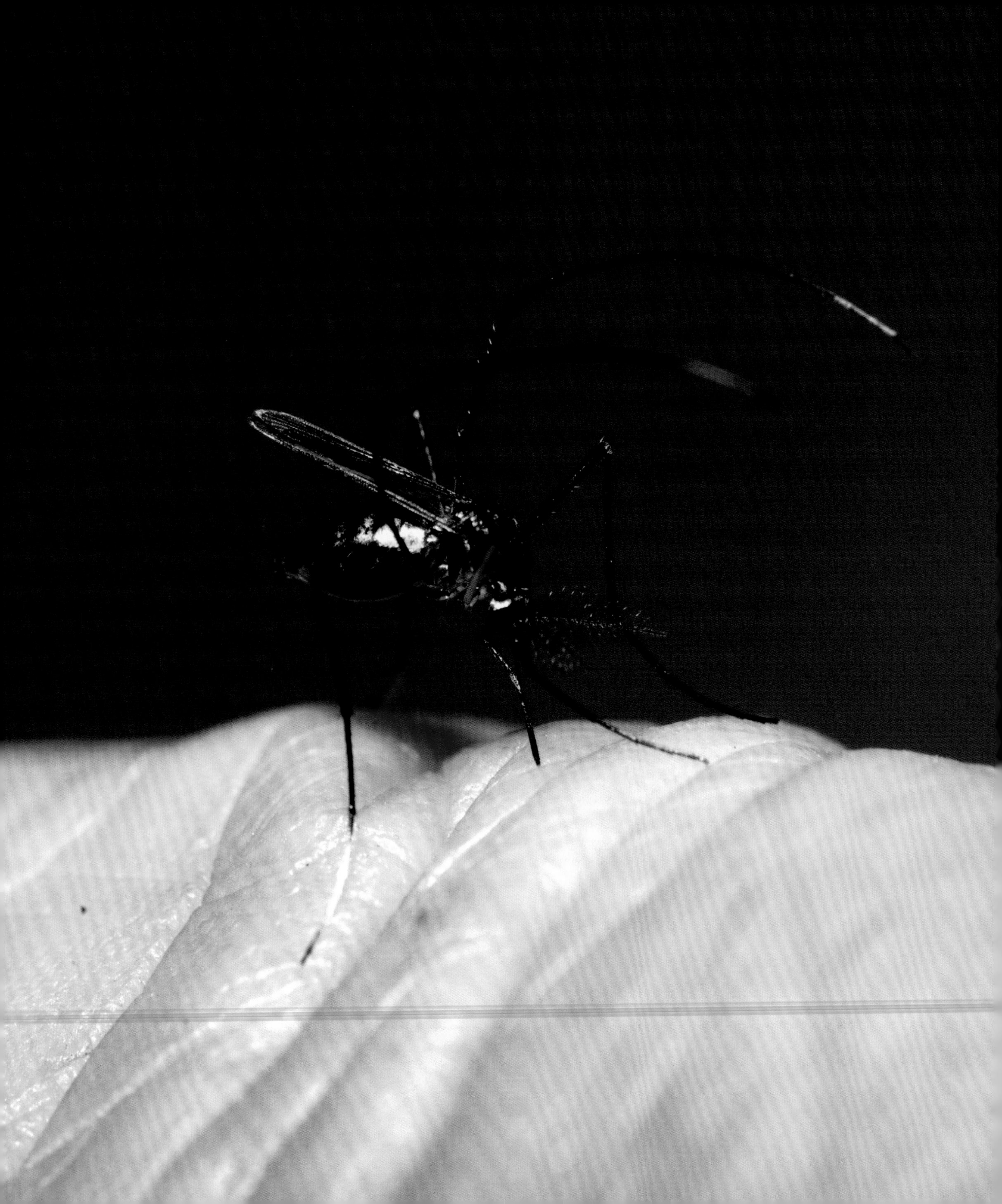

Bats and Mosquitoes

The idea that bats eat and thus control the mosquito population is firmly entrenched in the public's image of bats. In the late 1950s, Donald Griffin and colleagues were studying the echolocation behavior of bats in a laboratory, and they often used mosquitoes and fruit flies as prey for the bats. The flying bats regularly detected, tracked and caught them. Some Little Brown Myotis caught five to nine mosquitoes (each weighing 2.2 mg.) a minute, the equivalent of 300 and 540 per hour. The idea that bats eat mosquitoes that feed on people is attractive, but to what extent is this really true? (Figure 5.17)

What happens in the wild? From DNA barcode analysis of insect remains in bat droppings, we know that Little Brown Myotis and Northern Long-eared Myotis eat some mosquitoes, while Big Brown Bats and Eastern Red Bats apparently do not eat mosquitoes at all. All of these bats apparently prefer other insect prey. Do the bats that sometimes consume mosquitoes eat enough to control populations of mosquitoes? This question remains unanswered.

Limitations on Bat Diets

Bats are equipped with teeth and their teeth show some degree of specialization according to diet, but their teeth also pose some limitations. Specifically, bats lack carnassial teeth, which are the scissor-like specialized cutting and slicing teeth of carnivorous mammals like dogs and cats. This means that when a Large Slit-faced Bat, for example, eats a smaller bat that it has caught, the chewing process can take well over an hour.

Bats do not have digestive tracts specialized for fermenting large molecules such as chitin and cellulose. Other animals that eat plant material (cellulose) use fermentation and help from symbionts that live in the gut including fungi, bacteria and ciliates. This is true of kangaroos, cattle, rodents and rabbits (among mammals), and also of termites, which are renowned for their ability to digest cellulose and lignin. The inability of bats to digest these plant materials means that like humans they miss out on some of the energy stored in plants that they eat.

Interestingly, a bacterium (*Aeromonas hydrophila*) that lives in the digestive tract of some leeches has also been found to inhabit the guts of Common Vampire Bats. In both hosts, leech and Vampire Bat, the bacteria live in a specialized organ called the bacteriome. The bacteria appear to provide these bats with a variety of amino acids and vitamins not normally found in the blood of their prey.

Where do bats obtain their vitamins? Measuring the amounts of vitamin D (serum 25-hydroxyvitamin D) in the serum of free ranging bats suggested yet another area where bats may harbor some surprises. Bats that roost mainly in dark hollows, including Antillean Fruit-eating Bats, Leach's Single-leafed Bat and Jamaican Fruit Bats, apparently have very low levels of vitamin D. If they were human, they would be considered deficient in vitamin D. What effects does this lack of vitamin D have on their physiology? Nobody knows for sure, but they seem to do just fine with low vitamin D levels. Interestingly, some other bats, such as Greater Bulldog Bats and Common Vampire Bat, have higher levels of vitamin D. Perhaps they acquire their vitamins from their prey or perhaps they do not rely upon them as much as we do.

Figure 5.17.
In Belize, a mosquito (*Wyeomyia* spp.) sucks blood from Brock, even as several species of insectivorous bats hunted in the immediate area.

6

Where Bats Hang Out

Most people are surprised to see bats roosting in plain sight because they imagine bats as secretive, nocturnal animals that hide by day. (Figure 6.1) But sometimes the best place to hide is in plain sight because your potential predators do not expect to find you there. Bats are small animals, and their ancestors probably roosted or hid out of sight and as do most living species of bats and other small mammals. In most parts of the world, going for a walk during the day and planning to see roosting bats is not a realistic expectation. In fact, we do not know where roughly half of the living species of bats roost. Finding a bat roost usually means catching a bat, putting a transmitter on it and following it home.

It is easy to generalize about the types of roosts that bats use: hollows, crevices, foliage or unexpected specialized "tents." But some and perhaps many species of bats are flexible in their choice of roosts, often using man-made structures when natural roosts are not plentiful or available. The roosts preferred by a particular species may vary with region, habitat and time of year. By tracking the movements of individual bats, biologists often learn that they also regularly switch among a number of roosts, sometimes on a daily basis. Roost switching may be one way to minimize the risk of predation because your predator will not know which address you are at on any given day. Greater White-lined Bats (*Vampyrodes carricioli*) and Great Fruit-eating Bats roost in foliage. In Costa Rica individuals chose roosts in the same general area (zip code) but switch actual roosts (addresses) from day to day. It is not just bats that roost in foliage that regularly switch roosts. By following eight radio-tagged White-bellied Yellow Bats (*Scotophilus leucogaster*) in Kruger National Park, Brock and several colleagues found that these insectivorous bats roosted in the hollows of Mopane Trees (*Colophospermum mopane*). Their roosts were about 2 km. from the Luvuvhu River, and every night the bats foraged along it for about an hour. The bats regularly switched between roost trees, and the researchers never saw them in their roosts, although they could locate them to a tree via their radio tags. By tracking signals from their radio-tags, they determined these bats regularly switched among roost trees. They did not switch to be closer to their feeding areas because all of the roost trees were about the same distance from the river.

One of our very favorite accounts of opportunism by roosting bats came to us from Roy Horst, a friend and colleague from the State University of New York at Potsdam. At the Mission at San Xavier del Bac a few kilometers south of Tucson, Arizona, many of the statues are clothed in fine costumes made by ladies of the congregation. According to Roy, Bernardo Villa-Ramirez, a well known Mexican bat biologist and his wife Clemencia visited the Mission. While praying at the foot of the statue of Mary, Bernardo noticed a sprinkling of bat droppings around the statue's feet. He offered to catch and remove the bats to minimize any possible damage to the statue or her clothing. The local priest accepted the offer, but insisted that Clemencia be the one reach under the Virgin's skirts and catch the bats. The Yuma Myotis were carefully removed and added to a museum collection, with a label that read "...taken from under the skirts of the statue of the Holy Mother in the Mission of San Xavier del Bac, Tucson, Arizona, 29 Nov 1970." Roosting bats are where you find them!

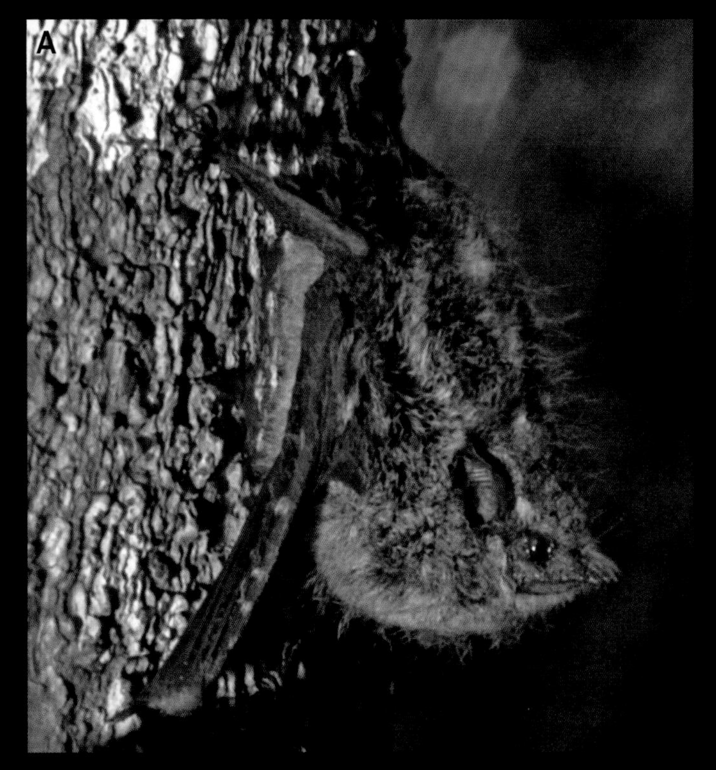

Figure 6.1.
Roosting Proboscis Bats (*Rhynchonycteris naso*), one close-up (**A**) and five roosting on the trunk of a tree (**B**). The third view is a thermal image (**C**). These small (5 g.) bats are widespread in the New World tropics and usually roost on tree trunks or branches hanging over water. The bats show up clearly in the thermal image because the bats are warmer than the trunk of the tree on which they are roosting. Photographs (**A**) and (**B**) by Mark Skowronski; thermal photograph (**C**) by Greg McIntosh.

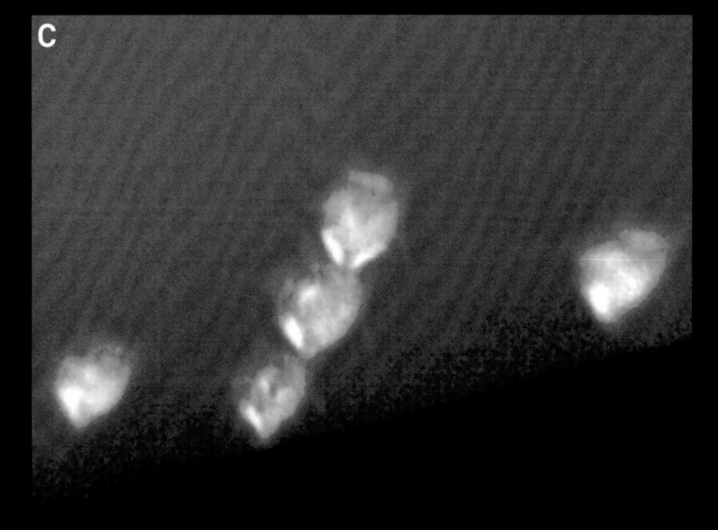

BOX 6.1

Bat Boxes and Bat Houses

Many bats exploit man-made structures for roosting sites. In addition to houses and other buildings, bats are regular residents of military bunkers, mines, culverts, bridges, football stadiums and clothed statues. The rapidity with which many species of bats exploit artificial structures as roosts suggests that people should be able to provide housing for bats. The variety of bat houses and bat boxes is considerable ranging from wooden, plastic and stucco boxes to do-it-yourself kits and construction projects. In some areas, very large bat houses—the size of small sheds—have been erected to try to attract bats for mosquito control.

The secret to a successful bat house (which attracts bats) is known only to the bats themselves. Some bat houses that appear perfectly placed never attract bats, while others installed in seemingly identical circumstances are soon overflowing with residents. Nobody knows why some bat houses are used and others rejected by bats. Two factors, however, seem to be particularly important to bats: the height and placement of a bat house relative to obstacles such as trees, brush and buildings, and the internal temperature of the bat house during the day. Painting bat houses different colors that absorb different amounts of heat and putting them on different sides of buildings in warm versus colder areas may help provide bats with optimal living conditions. [See http://www.batcon.org/pdfs/bathouses/bathousecriteria.pdf for Criteria for Successful Bat Houses.]

Bat biologists were puzzled by the persistent absence of bats from a pine forest and surrounding lake at a site in Sweden. So they erected bat boxes in the pine plantation and within two years at least five species of bats were using the boxes, some raising their young in them. This illustrated that the bat fauna of that area had been limited by the availability of roosts. In this situation, providing roosts for bats can be important in their conservation. (See page 227.)

Figure 6.2.
Victorian houses at Chautauqua, New York which were once routinely used as roosts by Little Brown Myotis are seen in (**A**). The chest waders in which Little Brown Myotis roosted are seen in (**B**).

Brock Offers Bat Boxes (And Is Rejected)

Bats are generally thought to roost in natural hollows such as those in trees or in crevices under bark. Some species of bats are opportunistic in their choice of roosts. In the early 1990s at the Chautauqua Institution in New York (Figure 6.2A), Brock and Universidade de Coimbra biologist Alison Neilson tried to find a bat house or bat box (Box 6.1) that would be used by Little Brown Myotis there. At that time, Little Brown Myotis were common around the Institution, and some of them roosted in large numbers in the older cottages and other buildings. The bats at the institution consistently preferred their "normal" roosts to the bat boxes we offered. Then a local resident told us about bats roosting in his garage. These Little Brown Myotis roosted in the chest waders that hung, suspended from the garage ceiling between duck-hunting seasons. (Figure 6.2B)

Another way to find roosting bats is to set a camera trap, arranged so that emerging bats take their own photographs. (Figure 6.3) This approach revealed that Heart-nosed Bat (*Cardioderma cor*) roosted in a well in Kenya, while Common Vampire Bats roosted in a hollow tree in Belize. Particularly startling to those "sitting on the throne" are the bats that fly out of the darkness of pit toilets, passing between the legs of the seated person and off into the night. In Africa, Slit-faced Bats are the usual residents of such roosts.

When looking for roosting bats, some people expect that because birds and bats have wings, both also lay eggs. The cup-shaped nests of birds contain eggs that otherwise would roll out. But bats are mammals, and bear live young, which at birth have adult-sized hind feet for holding on to things. From birth young bats can hang (See Figure 7.2); they do not "roll away" so bats have no need to build a nest. Some bats, however, make tents, and others excavate roosting spaces within termite nests. (See pages 150–153.)

The Roost Needs of Bats

As mainly nocturnal animals, bats need secure places to spend the day where they are out of sight and reach of would-be predators. (See page 176.) The specific needs vary according to species, size, season and environment. Pregnant bats and those with nursing young ensure better conditions for growth of their young by roosting in places that are warm to hot. At other times of the year, the same bats may choose cooler roosts to minimize energy expenditure needed to maintain an even body temperature. Many bats also use night roosts—places to pause during feeding where they can rest and digest prey, again out of reach of potential predators.

To appreciate a bat's roost choices often means remembering that some of them have a variable internal thermostat (= are heterothermic; see page 9). This is particularly true of species living in temperate regions or at high altitudes. The variable thermostat allows bats to let their body temperatures follow the ambient within a certain range. (Figure 6.4) This manner of body temperature management allows bats to save the energy costs associated with being warm-blooded on a cool day.

In general, bats face perpetual problems about heat. Sometimes they can be too hot because of the heat generated by their bodies during flight (like a track athlete after a long run). At other times they may be forced to expend a great deal of energy to stay warm. This difference can be critical because bats are small animals with high metabolic rates. Their small size and relatively large wings (which are essentially uninsulated sheets of skin) mean that they have a very large surface area relative to their volume. This combination makes it expensive to stay warm (let alone hot) in cool or cold conditions because their bodies readily lose heat. This probably explains why bats often roost in groups. During the summer, a Big Brown Bat roosting in contact with another bat saves about 9 percent of the cost of staying warm. When forty-five bats roost together in a cluster, the savings jump to 53 percent of the heating costs they would incur had they been roosting alone. The cost of staying warm depends upon the difference between air temperature and body temperature. In tropical and subtropical settings, even small bats can remain warm in roosts without expending too much energy. (See Figure 6.1C.)

Pregnant and nursing Gray Myotis (*Myotis grisescens*) congregate in large numbers in domed chambers in caves mainly in the Southeastern United States. The collective body heat of all these bats raises the temperature in the dome enough to foster rapid growth of young. Some species of Bent-winged Bats (*Miniopterus*) use the same strategy. This is an extension of the strategy of energy conservation by clustering illustrated by Big Brown Bats.

At other times of the year or in other environments, some species of bats conserve energy by

A

Figure 6.3.
A Heart-nosed Bat (**A**) photographed as it emerged from a well in Kenya, and the Common Vampire Bat (**B**) emerging from within a tree hollow in which it had roosted in Belize. Note that the bat had to fold its wings to fly out through the opening in the tree. Photographs by Jens Rydell (**A**) and Brock Fenton (**B**).

B

Figure 6.4.

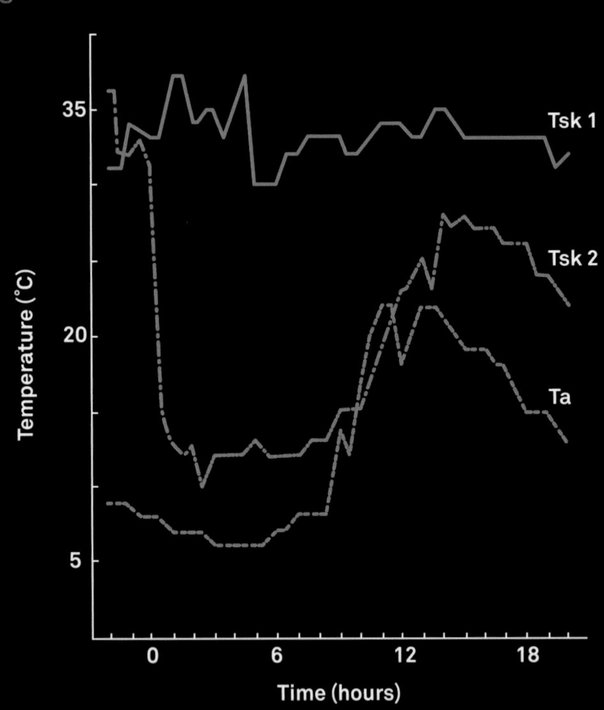

Figure 6.4.
Skin temperatures (**T_{sk}**) from two Big Brown Bats that roosted inside the attic of a house near Ottawa, Canada. Also shown is the ambient temperature (**T_a**) in the roost. While one bat (**T_{sk1}**) returned from foraging (time 0) and maintained its body temperature near 35° C, the other (**T_{sk2}**) allowed its body temperature to follow the ambient and entered torpor.

Figure 6.5.
Looking out of the entrance to St. Clair Cave in Jamaica. At least seven species of bats roost in this cave. The entrance is a busy place at dusk and again at dawn.

Figure 6.5.

lowering their core body temperatures, effectively going into a torpor-like sleep, lasting only a few hours, at most a day or two. This is not the same thing as hibernation, lasting weeks or even months, but uses the same principle: reducing its metabolism to keep body temperature higher than ambient temperature helps the bat to save energy. The use of daily short-term torpor as an aspect of thermoregulation appears typical of post-reproductive females as well as males and young bats. Pregnant females that enter torpor prolong their gestation period, with the result that their young are born later in the season and have less time (for temperate species) to prepare for winter. Bats that roost alone often take advantage of sites that are warmed by late afternoon sun. In such settings, bats can rely on passive warming and do not expend metabolic energy to arouse from their daily torpor.

During times of food shortage (such as winter), insectivorous bats in temperate latitudes and/or high altitudes areas may go into more prolonged torpor, the sleep of hibernation. When hibernating, the body temperatures of bats approach those of their surroundings, but typically stay above freezing. As far as we know, hibernating bats cannot withstand freezing. When temperatures in their hibernacula (hibernation sites) go below freezing, bats must arouse and move, or spend energy by metabolically raising their body temperature. Temperature, combined with relative humidity, are fundamentally important features of sites where bats can hibernate. Relative humidity is important because some hibernating bats are vulnerable to dehydration by water loss through breathing and evaporation across their skin (= pulmocutaneous).

Sites Used As Roosts by Bats

Hollows

As we have seen, it is easy to overlook roosting bats. People often associate bats with caves, and many of the most famous bat roosts are caves (Figure 6.5) Frio and Bracken Caves in Texas each have over ten million Brazilian Free-tailed Bats roosting inside them during the summer, placing them among the world's largest known concentrations of mammals. But Brazilian Free-tailed Bats also roost in crevices, for example, the one million or so Brazilian Free-tailed Bats that roost in crevices under the Congress Avenue Bridge in Austin, Texas, which have become a major tourist attraction. Elsewhere in the world there are many other examples of large numbers of bats roosting in caves. Caves and other hollows are well-suited as roost sites for animals that echolocate because the darkness makes them relatively inaccessible to other animals, such as predators that orient by vision. Echolocating birds (Oilbirds and Cave Swiftlets) also roost in dark hollows to provide some protection from predators.

Not surprisingly, bats that roost in caves also use other kinds of hollows, from those in trees to those inside buildings. The attics of homes are common bat roosts in some areas. Other hollows used as roosts include bird houses, large rolled leaves of banana and *Heliconia* plants and the pitchers of some pitcher plants. (See page 148) The essential feature from the bat's point of view may be protection from daytime predators, which usually—but not always—means darkness. Other important considerations include microclimate conditions such as temperature and humidity. The roost conditions, however, essential at one time of the year may be entirely inappropriate at other times. Thus the roosts used as nurseries by pregnant and nursing bats are typically not suitable for hibernating bats, and vice versa.

Although large caves can accommodate hundreds, thousands or even millions of bats, most colonies are much smaller. Little Brown

Myotis usually roost in hollows, and in this species a nursery may consist of twenty to thirty females with population doubling after birth of the young. Little Brown Myotis sometimes form much larger colonies in the attics of buildings, ranging up to 3,000 individuals. Large numbers of bats in a single roost face at least three dis-advantages. First is the congestion that ensues when many individuals try to leave (or return to) the roost at the same time. The degree of congestion is determined by the size of the population and the size of the opening(s). Second, a group of emerging bats often attracts predators, perhaps reducing any anti-predator advantage associated with roosting in dark hollows. (See page 132) Third, ectoparasites such as blood-feeding bat flies, fleas, bed bugs, and mites often thrive in large aggregations of bats.

Nancy Finds a Cave Inside a Tree

While working in French Guiana, I often wondered how bats that were supposed to be cave or hollow-roosting species found places to live in the local forest, which grew on sandy soils. There were no rocks anywhere in the area, let alone caves. What did this mean for bats? Where did they live? Part of this mystery was solved for me one day when my research team found a huge buttressed rainforest tree that was hollow at the base, forming a vertical chimney where bats roosted. Further exploration of the forest led us to other ancient trees that were similarly rotted out at the bottom. Inside the cave-like centers of these trees we caught roosting groups of Common Big-eared Bats, Seba's Short-tailed Bat, Hairy Big-eared Bat (*Micronycteris hirsutus*), Pallas's Long-tongued Bat, and Lesser Spear-nosed Bat (*Phyllostomus elongatus*). Still other bat species roosted inside smaller hollows in other trees. It turned out that caves were not absent at our work site—they were just inside the trees!

BOX 6.2

Ectoparasites of Bats

Bats are an important resource for many blood-feeding animals. (Box 6.2, Figure 1) Ectoparasites, animals that mainly live on the outsides of their hosts, are abundant on some species of bats. There are both long- and short-legged species, some that live on the bats' wings or in their fur and others that may cling to the ears. These include fleas, ticks, mites, bed bugs and various bat flies, *e.g.*, Streblidae and Nycteribiidae. Fortunately, most of these ectoparasites are very bat-specific and rarely move to humans (such as bat biologists). Hairless Bats of Malaysia are known to host an earwig (*Arixenia esau*). Although initially thought to be an ectoparasite, further research revealed that these earwigs feed on skin and other secretions of the bats, not on the bats' blood. Technically, this means that the earwigs were not parasites of the bats. These earwigs are very large relative to the size of the bats, the equivalent of a human with a dinner plate-sized animal living on its skin.

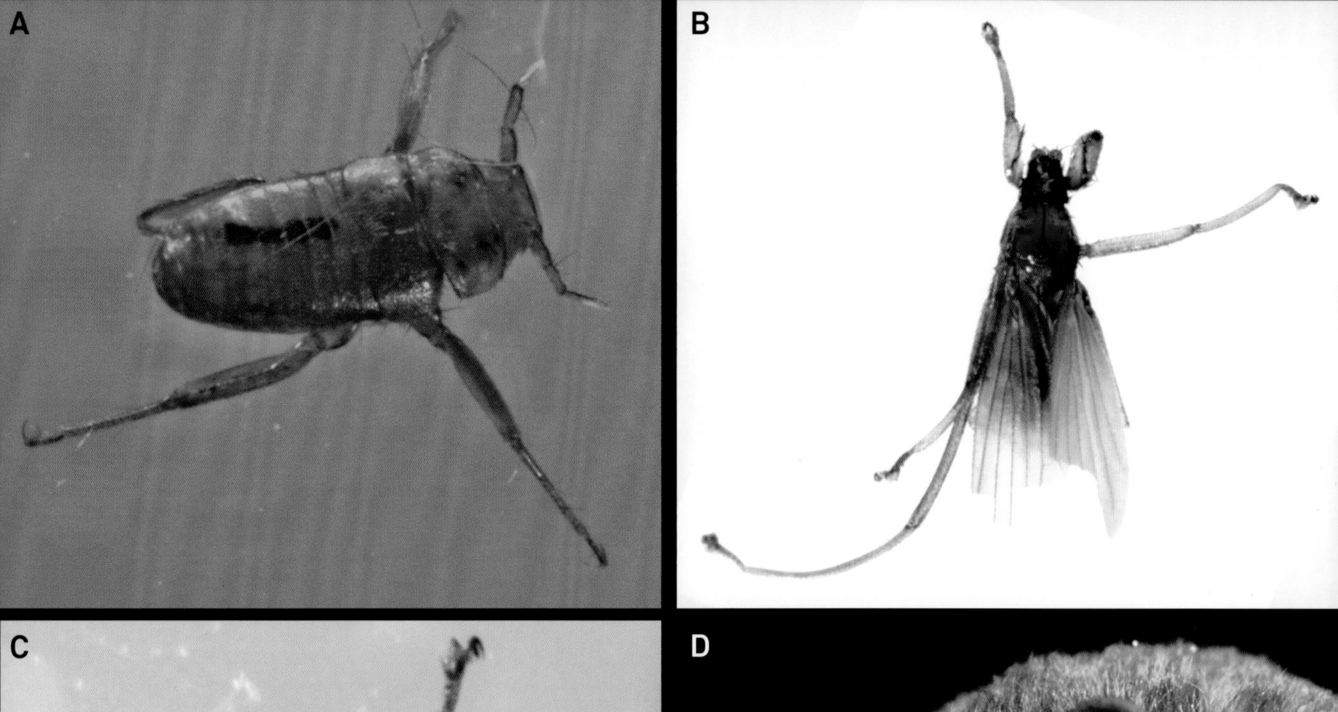

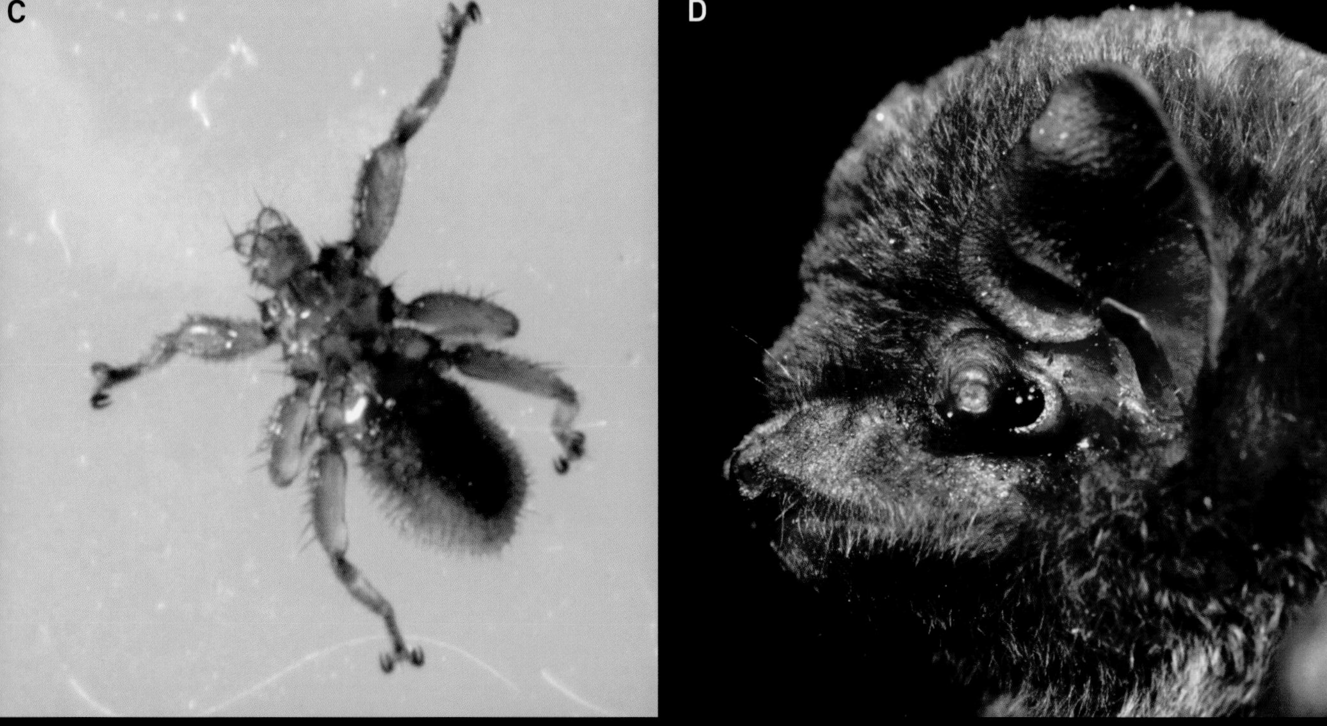

Box 6.2, Figure 1.

Bats host a variety of blood-sucking ectoparasites. From Belize (**A**) a polyctenid bat bug, (**B**) a bat fly (*Streblidae*) and (**C**) a nycteribiid bat fly. From South Africa, (**D**) a

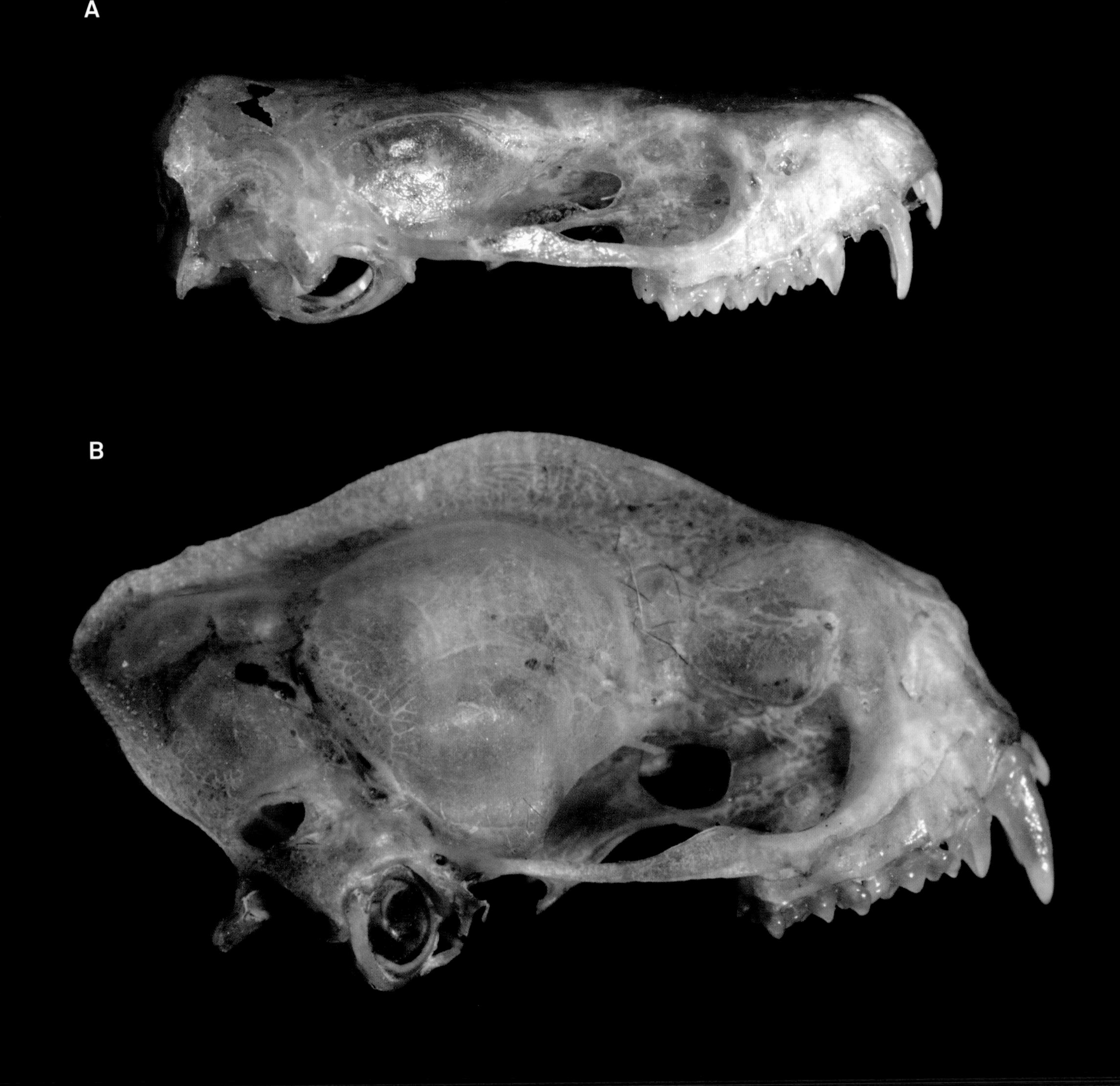

Figure 6.6.
Skulls of two species of Free-tailed Bats (Molossidae), including one (**A**) Robert's Flat-headed Bat (*Mormopterus petrophilus*) flattened for roosting in crevices; the other not so specialized (**B**) a Black Mastiff Bat (*Molossus rufus*).

Crevices

The arrangement of flight muscles and the skeletal anatomy of bats makes them relatively thin from back to front (their rib cage is shaped much like that of a human; see page 38.) These features allow bats to squeeze into narrow spaces and through very narrow openings whether in rock, tree bark or spaces between boards of buildings out of the reach of predators. Relatively flat skulls are another feature allowing bats access to crevices or hollows served by narrow openings. Some species of bats have flattened skulls that seem to be an adaptation for living in crevices. (Figure 6.6)

Two intriguing examples of crevice specialists are Bamboo Bats (Vesper Bats) and Flat-headed Bats (Free-tailed Bats). In Southeast Asia, Bamboo Bats (*Tylonycteris pachypus* and *Tylonycteris robustula*, Lesser and Greater, respectively) roost in the hollow spaces in bamboo stems. These bats have flattened heads allowing them access to the openings into the bamboo made by Leaf Beetles (chrysomelid). In South America, Mato Grosso Dog-faced Bats (*Molossops mattogrossensis*) roost under flat stones in the open savannah. In Africa, two other species of Free-Tailed Bats roost under flat stones, Peters' Flat-headed Bat and Roberts' Flat-headed Bat (*Mormopterus petrophilus*). (Figure 6.5) Among the bats that roost under flat stones, Peters' Flat-headed Bats and Mato Grosso Dog-faced Bats have wart-like protuberances on their forearms, perhaps protecting them from abrasion. The first specimens of Roberts's Flat-headed Bats known to science were collected in 1917 by South African zoologist Austin Roberts turning over rocks looking for scorpions. The name "petrophilus" means rock-loving.

Foliage

A number of bat species habitually roost in foliage, for instance, among the leaves of trees, vines and shrubs. (Figure 6.7) Being small, quiet and blending into the background makes foliage-roosting bats difficult to spot. Indeed, most foliage-roosting bats depend for safety on staying inconspicuous. However, some species of large bats, including many species of flying foxes, roost in foliage, often hanging from branches. Smaller foliage roosting bats, such as Tricolored Bats (*Perimyotis subflavus*) or species in the genus *Lasiurus* (*e.g.*, Eastern Red Bats or Hoary Bats), hang directly on leaves. Groups of flying foxes can be both noisy and conspicuous. Their roosting spots are called "camps", and they are often well-known to local people. Two species of Flying Foxes (genus *Pteropus*) were once common in American Samoa in the South Pacific. Samoan Flying Foxes (*Pteropus samoensis*) each weigh about 300 g. and they typically roost alone. Pacific Flying Foxes (*Pteropus tonganus*) are larger (550 g.) and roost in camps. Hunters equipped with shotguns have hunted Pacific Flying Foxes to the edge of extinction. The less conspicuous Samoan Flying Foxes are less often taken by hunters.

Little Yellow-shouldered Bats (*Sturnira lilium*) are much smaller (15–20 g.) and are widespread in the New World tropics. Although commonly caught in mist nets during surveys for bats, it was not clear where they roosted until recently. At Paracou in French Guiana, Nancy and her colleagues caught over eighty of these bats but did not find a single roost site despite considerable searching. In Belize, Brock and his colleagues caught a similarly large number Little Yellow-shouldered Bats in mist nets and attached small (0.6 g.) radio tags to seventeen of them. By following the radio signal, they quickly located some of our bats in their day roosts. These Little Yellow-shouldered Bats roosted in hollows in trees, in tangles of vines, in foliage or at the base of palm fronds. One radio-tagged bat roosted in a relatively small shrub (3 m. in diameter) so we made a special effort to spot it. After two hours of careful searching all around and under the bush, none of the six biologists saw the bat. Then Brock lightly tapped the bush with the blunt side of a machete blade and at least six bats flew out of it, one with the radio tag.

Along the Luvuvhu River in Kruger National Park in South Africa, Brock and colleagues had a similar experience with the much larger Wahlberg's Epauletted Fruit Bats (*Epomophorus wahlbergi*). They fitted ten of these bats with radio tags and quickly learned that while some spent the day in foliage along and usually overhanging the river, another flew 4 km. south and spent the day in a rock shelter with about eight others. The bats in the rock shelter were easy to see, once the researchers realized they were there, but we could not ever see any that were roosting in foliage. In the rock shelter, the roosting bats that could be seen were not in physical contact with one another. (Figure 6.8) Bats may roost with spaces between roost mates for a variety of reasons ranging from thermoregulation to social factors.

Figure 6.7. (below)
Three Butterfly Bats (*Glauconycteris variegata*) roosting among the leaves of a Natal Mahagony Tree (*Trichelia emetica*) are visible in this photograph. The bats' vocalizations drew attention to them. By catching them in a bucket trap, it was determined there were ten in the group, including four nursing young.

Figure 6.8. (opposite)
Wahlberg's Epauletted Fruit Bats were thought to roost in foliage (**A**), but some roost in shelter caves (**B**), and still others around buildings, under the eaves or in the thatch.

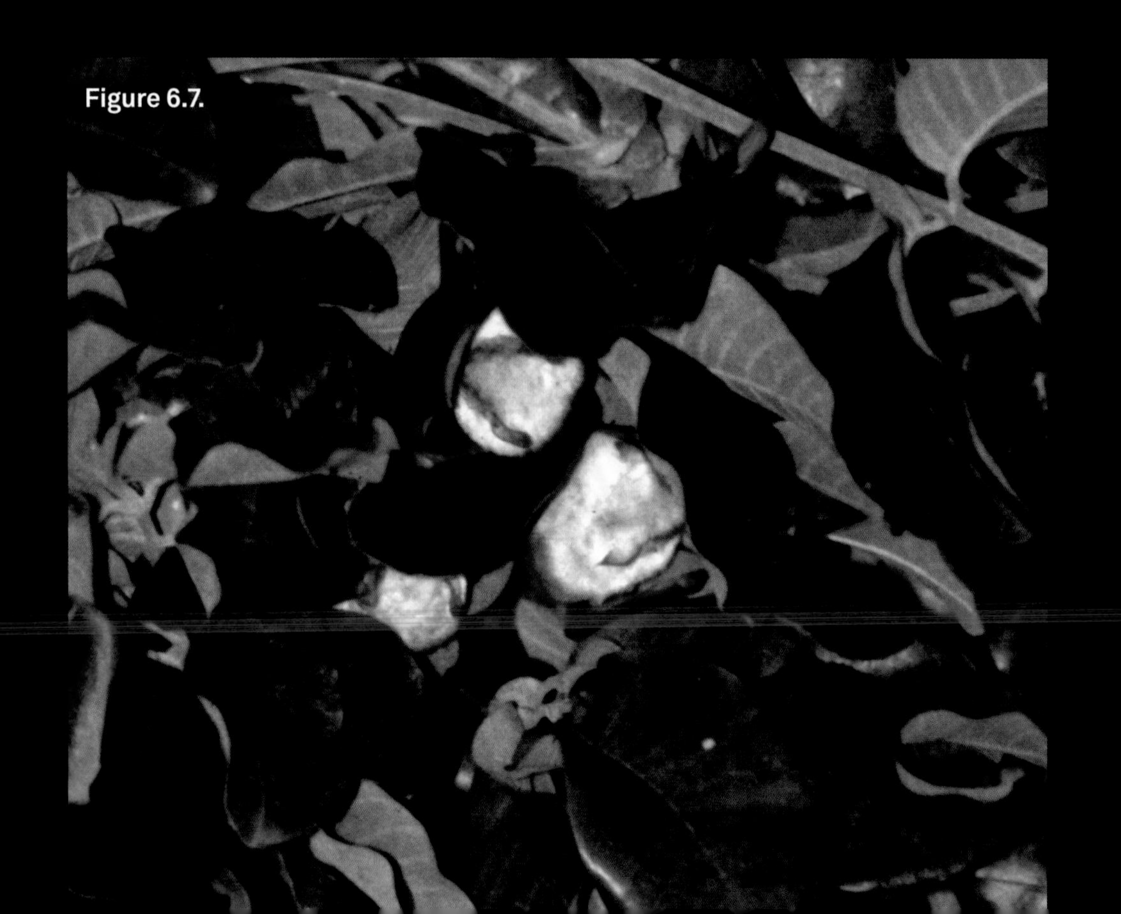

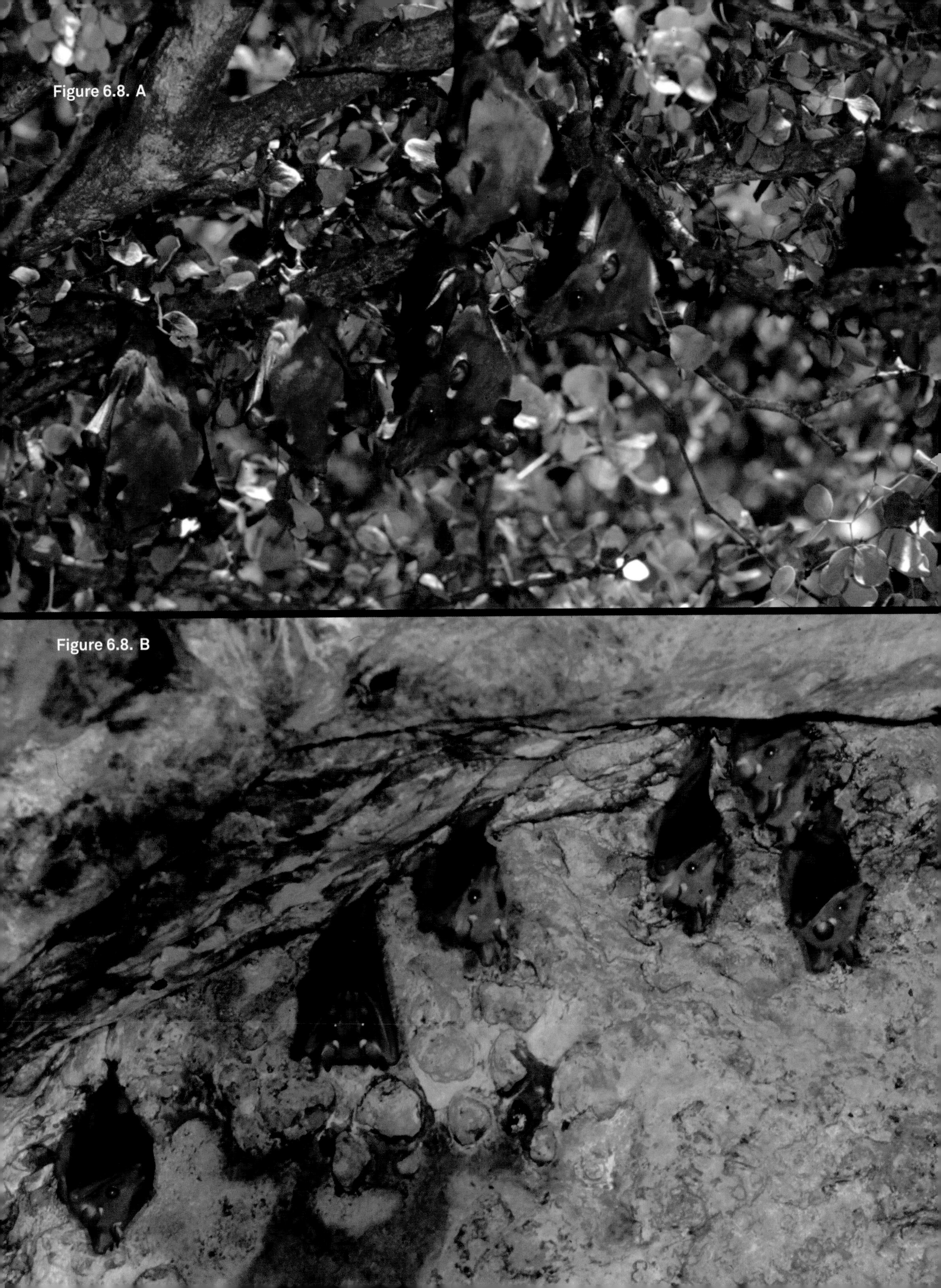

Figure 6.8. A
Figure 6.8. B

Do bats randomly select foliage roosts? Female Hoary bats with young definitely prefer foliage roosts that offer shelter from prevailing winds, exposure to sun and a south-facing opening in the foliage. While this is easy to say, the work of biologists Robert Barclay, University of Calgary, and Brandon Klug, University of Regina, demonstrated this in the Delta Marsh area of Manitoba, Canada. They are exceptionally sharp-eyed obser-vers experienced at spotting roosting bats.

Specializations and Remarkable Roosts

A few species of bats exhibit extremely specialized roosting behavior. In the New World tropics, Disk-winged Bats (Thyropteridae) have sucker disks on both their forelimbs (wings) and hind legs (feet) that they use to assist with roosting. (Figure 6.9) Details of the roosting behavior of Spix's Disk-winged Bat (*Thyroptera tricolor*) and Peters' Disk-winged Bat (*Thyroptera discifera*) are quite well known. Both species roost in the tube-like unfurled leaves of plants such as False Bird-of-Paradise (*Heliconia*) and bananas. These bats have circular suction cups on the soles of their feet and at the base of the thumb, allowing them to cling to and move about on the slippery surfaces of the leaves.

In one study area in Costa Rica, bats occupied 10–15 percent of the available unfurled leaves, most of which were suitable as roosts for only one day. The bats were very particular about the degree of rolling of the leaves and apparently rejected those that are rolled too tightly or unrolled too much. The bats roosted in groups of three to fifteen, usually adult females and their young, sometimes with an adult male. Genetic analysis indicated that the adult males with females and young were rarely the fathers of the young. This leaves a mystery for bat biologists. Where does mating take place and where do most of the adult males roost?

Figure 6.9.
The silhouette of a Spix's Disk-winged Bats roosting head up in unfurled leaves (**A**). These bats hang on by suction that occurs at specialized disks (arrows) on wrists and ankles (**B**).

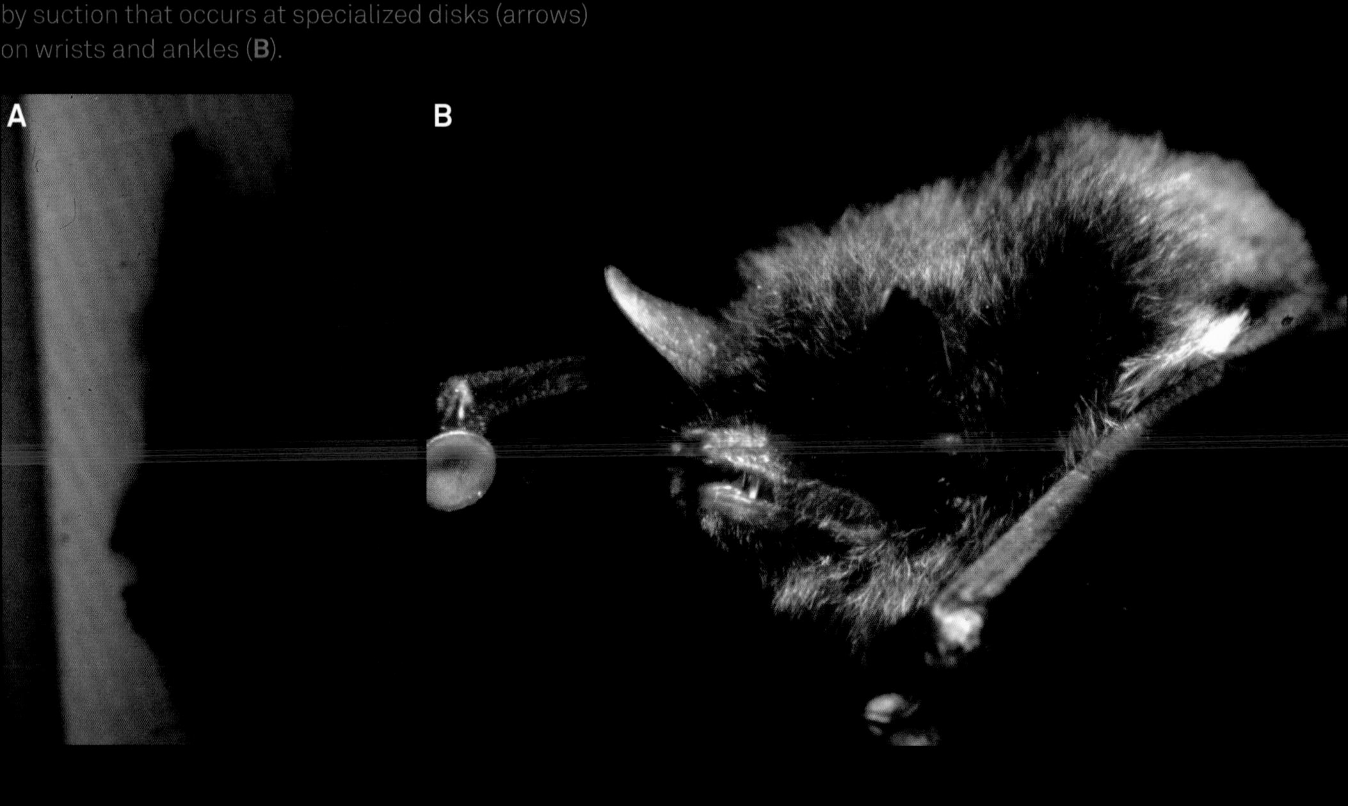

A

In Madagascar, two species in another family of bats, Old World Disk-winged Bats (Myzopodidae), have specialized pads on their wrists and ankles that secrete a wet adhesive used to anchor themselves to smooth leaves, usually of Travelers' Palms (*Ravenala madagascarensis*). (Figure 6.10) Other bats, for example Rufous Mouse-eared Bats (*Myotis bocagei*), Banana Pipistrelles (*Neoromica nanus*) and Thick-thumbed Bats (*Glischropus* spp.) have also been reported roosting in unfurled leaves. None of these species has the specializations of either Disk-winged or Sucker-footed Bats. Although the discs in Sucker-footed Bats look superficially similar to those of Disk-winged Bats, their internal structure and the way they function (suction versus wet adhesion) are different. This has lead researchers to conclude that they evolved independently in the two groups.

One of the most bizarre bat roosts described to date is used by Hardwicke's Woolly Bat (*Kerivoula hardwickei*). In some parts of Borneo, these bats roost in specialized leaves of pitcher plants. The Raffles' Pitcher-Plant (*Nepenthes rafflesiana*) has two kinds of pitchers, ones that catch insects and ones used by roosting bats. Each modified leaf can accommodate one bat. A comparison of nitrogen stable isotopes reveals that the plants derive much of their nitrogen from bat feces and urine. Hardwicke's Woolly Bat occurs from Sri Lanka to Philippines and Borneo as well as other islands in the East Indies. The bats probably use other roosts, and it is not clear if they roost in pitchers throughout their range.

Bats Making Roosts

Tents

In 1932, the American zoologist Thomas Barbour reported finding bats roosting in a "tent" along the Panama Canal. He reported that the bats appeared to have made their tent by biting along the underside of banana and palm leaves causing them to collapse and fold over. The resulting tent provides shade, shelter from the rain and some protection from predators. The bats Barbour observed subsequently became known as Tent-making Bats (*Uroderma bilobatum*).

By 2005, nineteen species of bats had been reported to roost in tents, most of them (fifteen) members of the New World Leaf-nosed Bat family. (Figure 6.11; see also Figure 8.1) The other tent-roosting species include three Old World Fruit Bats—Spotted-winged Fruit Bat (*Balionycteris maculata*), Lesser Short-nosed Fruit Bats and Greater Short-nosed Fruit Bats (*Cynopterus sphinx*)—as well as one Vesper Bat, the Greater Asiatic Yellow Bat (*Scotophilus heathi*). The details of tent construction vary with the leaf and at least eight basic types of tents have been described. Bat tents have been found in the leaves of species from eighteen families of plants.

The leaf side of tents stories may be as interesting as the bat side. To a botanist, it is a surprise to learn that leaves modified as tents may remain green for from 45 to 365 days. This is surprising because of the extent of damage done to the leaf's veins by bats in making the tent seems like it should interrupt the leaf's functioning. It appears, however, that in leaves modified as tents, small xylem vessels (10 µm. diameter) take over the delivery of water and nutrients when the larger veins are damaged. If parts of the leaf beyond the cuts made by the bats turned brown, they would be conspicuous, defeating any

Figure 6.11.
Two Little Yellow-eared Bats in a tent in Costa Rica.

A

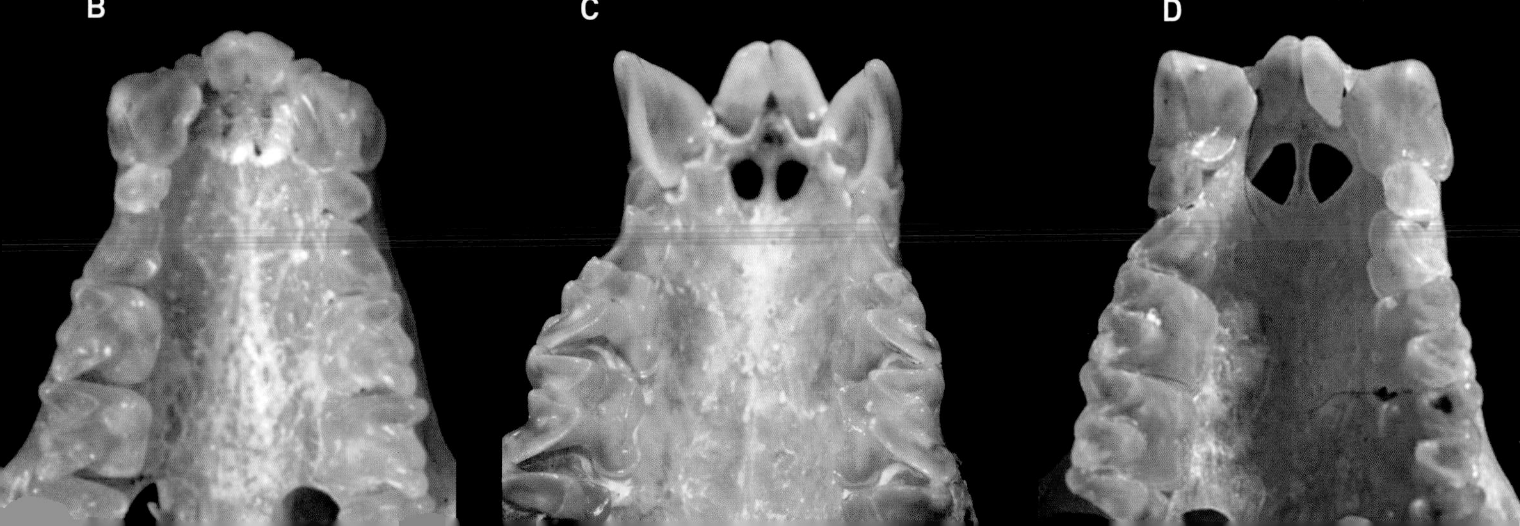
B
C
D

Excavations

Just as intriguing as the discovery that bats make tens out of leaves was the 1961 discovery that some bats excavate small cavities in the nests of arboreal termites and ants, which the bats then use as a roost. (Figure 6.12) Among these are New World Leaf-nosed Bats, including White-throated Round-eared Bats (*Lophostoma silvicola*), Carriker's Round-eared Bats (*Lophostoma carrikeri*), Pygmy Round-eared Bats (*Lophostoma brasiliense*) and Greater Spear-nosed Bats (*Phyllostomus hastatus*); an Old World Fruit Bat, the Lesser Short-nosed Fruit Bat; and a Vesper Bat the Flores Tube-nosed Bat (*Murina florium*). The roosting cavities are walled off from the active parts of nest by the insects themselves, thus creating tiny caves for the bats to roost in. It is an amazing experience to stick your hand inside one of these bat roosts and feel all the little warm, squirming bodies of the bats roosting within!

Some of these bats (the Round-eared Bats) are known to use their upper incisor teeth to excavate a roost hollow in arboreal termite nests—chewing their roost cavities into what may be nearly rock-hard nests. Males are the excavators, and apparently use the roosts they construct to attract females. Groups of bats found in excavations in termite nests are typically an adult male with several females and their dependent young. We expect that the male will be the father of the young, but this remains to be confirmed. In some cases when more than one adult male is present, there are reports of young bats being killed by one of the males.

Figure 6.12.
In **A**, three Pygmy Round-eared Bats looking out of a roost hole the bats excavated in a termite mound. Bats such as Pygmy Round-eared Bats use their upper incisor teeth to excavate the roost. In **B**, the upper teeth of another species that excavates roosts (White-throated Round-eared Bats) are show for comparison with two species (**C** and **D**) not known to excavate roosts. **C** shows the upper teeth of a Fringe-lipped Bat *(Trachops cirrhosus)*, and **D** the upper teeth of a Striped-headed

Physiological Adaptations of Bats for Roosting

There are records from Tanzania of large numbers of bats, probably Persian Trident Bats (*Triaenops persicus*), roosting in abandoned Koalin mines. Lack of air flow, combined with high temperatures and large concentrations of bats, means that the concentration of oxygen in the mines is low—too low even to allow human visitors to enter these mines. It seems likely that bats face low oxygen levels in many roosting situations, and the details of how they deal with the physiology of this are not known.

High concentrations of ammonia in some bat roosts are another consequence of large congregations of bats. Humans can detect ammonia by its odor at concentrations of 17 parts per million (ppm.), and regular exposure to 25 ppm. of ammonia is not considered dangerous. However, exposure for thirty minutes to air with 500 ppm. will cause irritation to both eyes and nasal passages. Concentrations above 5000 ppm. can be lethal to humans. In contrast, at least some bat species have adaptations, which allow them to deal with higher levels of ammonia exposure. Brazilian Free-tail Bats show an astonishing tolerance for ammonia and can withstand four days exposure to concentrations above 5000 ppm. without apparent ill effects. Secretion of mucous along the respiratory tract appears central to protecting bats and other animals from high concentrations of ammonia.

This combination of circumstances means that humans, including bat biologists, are often well advised to stay out of roosts housing many bats. Both of us have found that the concentrations of ammonia in underground roosts of Common Vampire Bats are high enough to clear your nasal passages and bring tears to your eyes.

Hibernation

Traditionally, temperate species of Horseshoe Bats and Vesper Bats were considered to be the only hibernators in the Chiroptera. These bats were the ones most often found hibernating during winter months in underground sites such as caves and abandoned mines. A few species also hibernate in buildings and in hollows in trees, but finding hibernating bats in their roosts is at least as challenging as finding active bats in their roosts.

Bats hibernate by going into a deep torpor-like sleep during which their metabolic and breathing rates decrease and their core body temperature is lower than normal. At sites in southern Ontario (Canada), hibernating Little Brown Myotis have body temperatures just above freezing. These bats breathe about once an hour, usually in short bursts. Hibernating Little Brown Myotis periodically arouse from torpor, raising their core body temperatures to about 36°C for about three hours. They use this time to urinate and groom. Each arousal cycle costs a bat about 108 mg. of body fat. That may not sound like much, but when your body weight is only a few grams, it can be a lot.

Little Brown Myotis in good condition enter hibernation with about 1800 mg. of body fat. Others, often young of the year, have <500 mg. of body fat. Each arousal from torpor costs the bat the fuel that would support it for sixty days of hibernation Information about the cost of an arousal cycle and the total fat store make it clear that bats like the Little Brown Myotis only survive hibernation by minimizing the number of times they arouse. As we will see below (page 208), this has important repercussions for conservation and invasive diseases such as White Nose Syndrome (WNS).

Brock Has a Cool Time in the Tropics

In 1996, I and a group of colleagues studied Black Mastiff Bats in Akumal, south of Cancun in the Mexican Yucatan. A group of about thirty of these bats roosted in the cinder block wall of one of the houses in Akumal. We caught the bats as they emerged and put radio tags on five of them. Then, for two days the weather was cool (<13°C), windy and rainy, and the transmitters never moved. We thought that the bats had shed the transmitters, but on the third night, they all emerged went out to forage and returned. Then we decided that these bats were just staying the roost and conserving energy, but could not prove that they had been torpid. If they were torpor, it was a variation on more extensive torpor typical of hibernating bats.

It is likely that other bats faced with inclement weather also use a torpor-like strategy to conserve energy. For example, in 2012, colleagues demonstrated that in Taiwan, roosting Formosan Round-leaf Bats (*Hipposideros terasensis*) drop their body temperatures to 10°C during winter, down from a normal >30°C. When these 60 g. insectivorous bats roosted at 150°C, they reduced their cost of thermoregulation by 94.6 percent (compared to the cost of being warm blooded at that temperature). These bats would require 4.9 g. of fat to survive in hibernation for seventy days.

Inclement conditions for a bat may be times when food is scarce, often reflecting some combination of prevailing temperatures and rainfall. In very hot deserts, bats such as Lesser Mouse-tailed Bats accumulate large bodies of fat at the base of the tail and endure periods of extreme heat by remaining in their roosts. This strategy may also help them conserve water and effectively thermoregulate. Although we tend to think of winter as the time of most limited resources for bats, this may not always be the case.

Figure 6.13.
Four views of hibernating bats. In three (**A**, **B** and **C**), the bats are Little Brown Myotis, in **D** a Horseshoe Bat. In **A** and **D** the bats are alone, in **B** and **C** there is a cluster of bats. In **A**, condensation is visible on the bat's fur and on each of its whiskers. The thermal image (**C**) shows four individuals (the bright spots) arousing from torpor and the rest (the dark mass in between) still in torpor. Thermal image courtesy of Greg McIntosh.

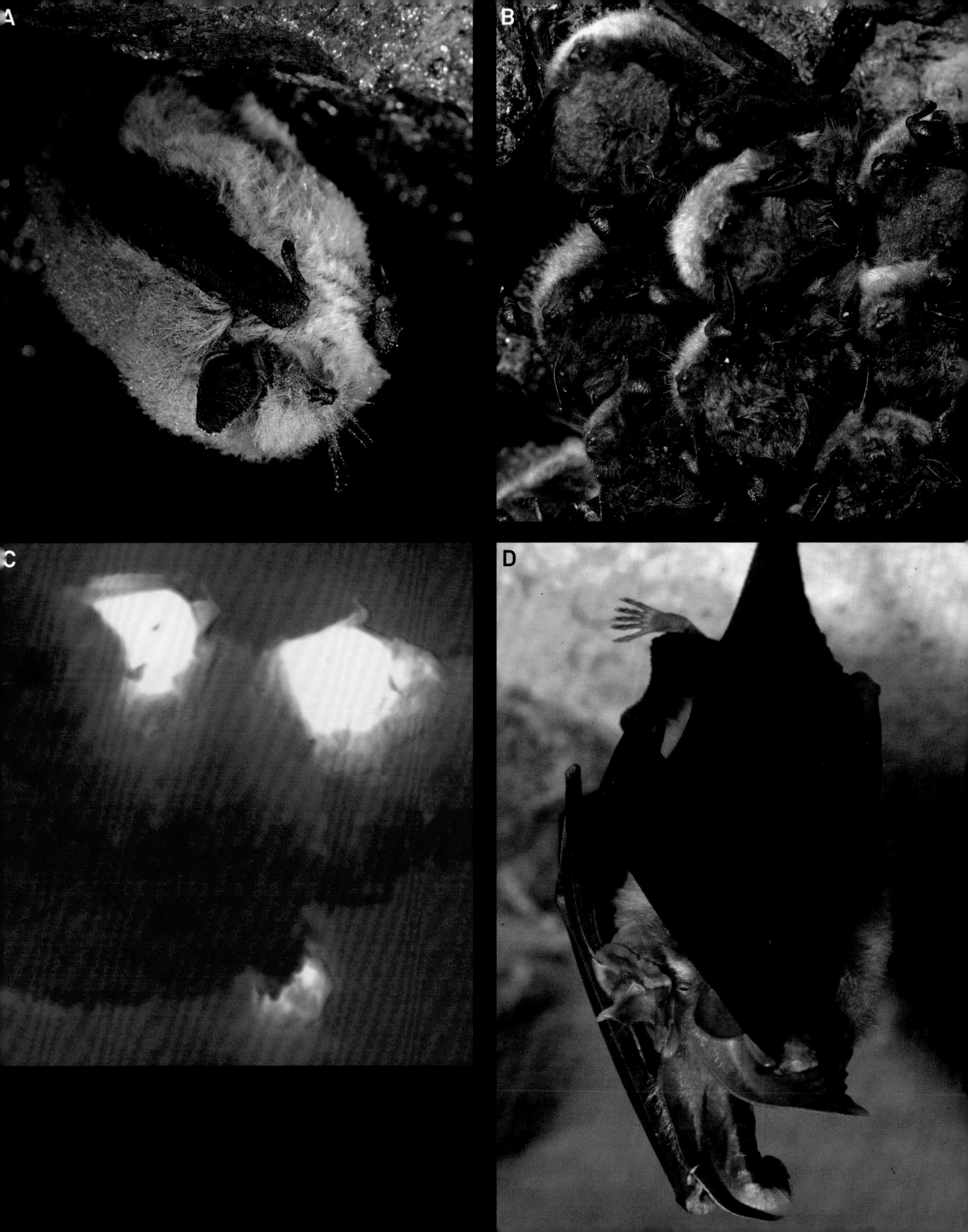
A
B
C
D

7

Life Histories of Bats

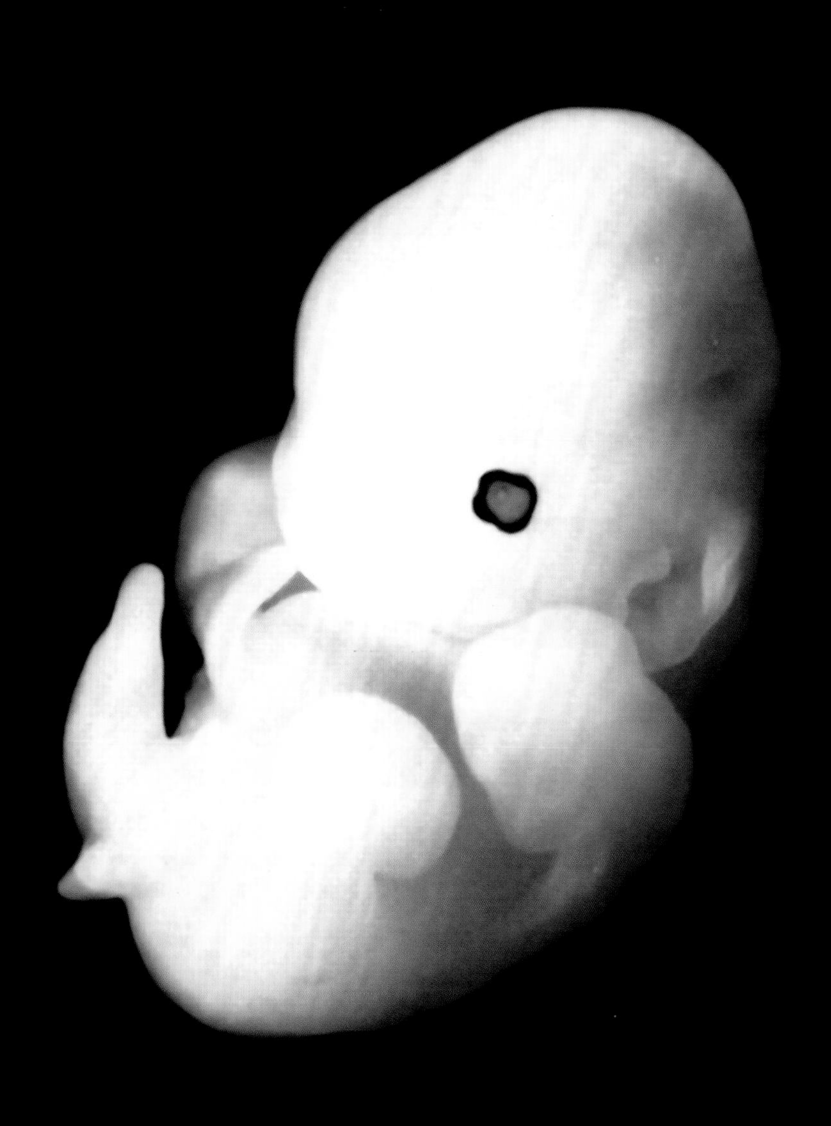

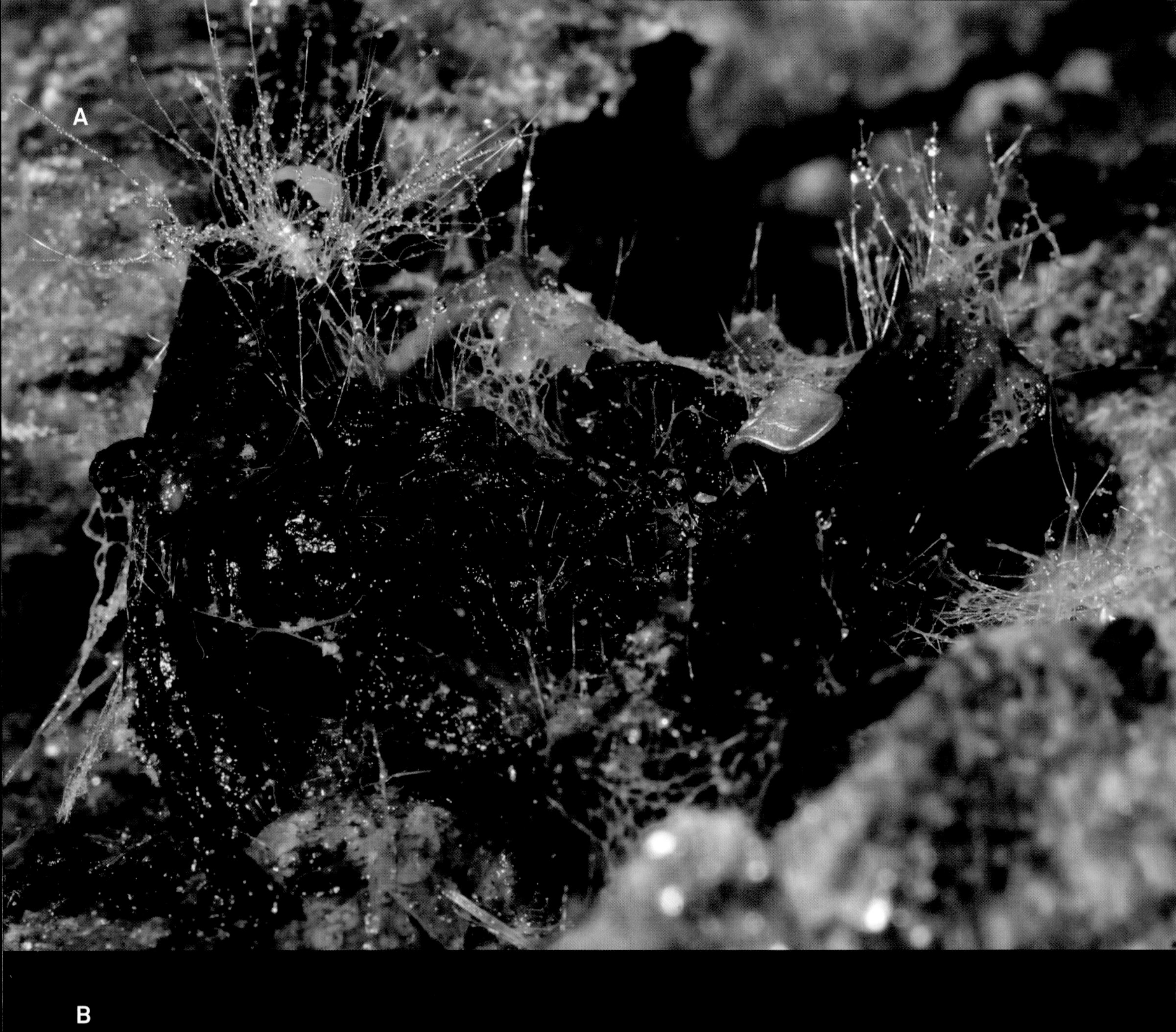
A

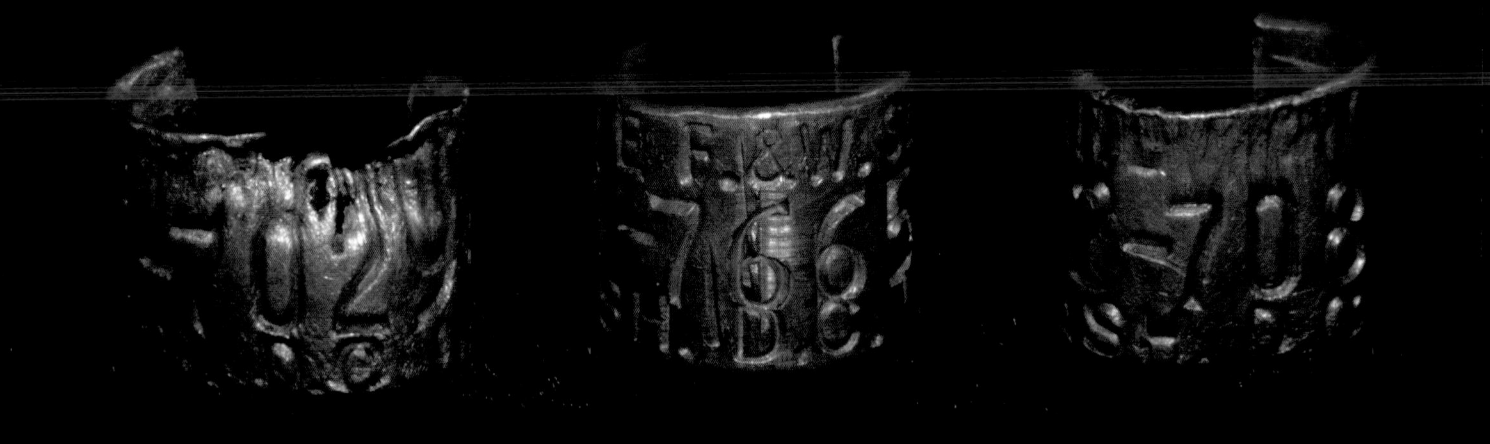
B

Three Statistics

The life histories of typical bats are characterized by three astonishing statistics. First, most bats have litter sizes of one (a single offspring per birth), although a few bear twins. Second, newborn bats are huge relative to the sizes of their mothers—25 to 30 percent of the weight of their mother. This is the equivalent of a woman who weighs 45 kilograms (kg.) giving birth to a baby weighing 12 kg. Third, some bats live to over the age of thirty in the wild. These statistics suggest that the life histories of bats are much more like those of larger mammals than smaller ones.

Old Bats

Biologists have studied the longevity of bats by banding (ringing) them, which involves placing numbered bands on the forearm just above the wrist. (Figure 7.1) Provided that the bander has kept accurate records, and there is an appropriate (regional, national) registry of banding records, it is possible to document the survival of banded bats.

Biologists do not yet have a reliable, non-invasive method of determining the age of a bat after it has been weaned and is independent of its mother. Newborn and very young bats are often easy to recognize because they lack fur and their eyes are closed. Although some bats are born with their eyes open, *e.g.*, Moluccan Naked-backed Fruit Bats, Sharp-nosed Tomb Bats (*Taphozous georgianus)* and New World Leaf-nosed Bats, most bat pups are born with their eyes sealed shut, their eyes opening one to ten days after birth. Similarly, many bats are born naked but some, such as New World-Leaf-nosed Bats and Slit-faced Bats, are born with fur. During the first months after birth, the fur of young bats (known as the natal coat) tends to be much darker or greyer than that of the adults, and their finger joints retain obvious growth plates. (Figure 7.2) But when young bats molt for the first time, the dark colored fur is replaced with fur of normal adult coloration. Soon their finger bones are fully grown and ossified, making it impossible to distinguish young bats from older ones. Timing of the first molt varies by species, ranging from a few weeks to a few months after birth.

Most available data about the ages to which bats live in the wild come from the temperate zone. In 1980, an examination of banding records of Little Brown Myotis from Southeastern Ontario indicated that at least one individual in the wild was still alive when it was at least thirty-five years of age. In 2004, the oldest Brandt's Myotis (*Myotis brandti*) found in the wild was at least forty-four years old. The "at least" qualifications reflect the fact that at time of banding these bats might already have been several years old.

At least two species from the tropics and subtropics also may live a long time in the wild. Banded Black Myotis (*Myotis nigricans*) have survived at least seven years in the wild in Panama, and Common Vampire Bats may live to at least nineteen years of age in the wild in Costa Rica. At least five species are known to live more than thirty years in the wild, including Little Brown Myotis, Brandt's Myotis, Lesser Mouse-eared Bat (*Myotis blythii*), Brown Long-eared Bat (*Plecotus auritus*) and Greater Horseshoe Bat (*Rhinolophus ferrumequinum*). An analysis of survival information from sixty-four species suggested that hibernation may be positively associated with longevity—that is, bat species that hibernate may live longer than those that do not but for now, the

Figure 7.1.
Sometime before it died, this Little Brown Myotis was banded (**A**). A colored plastic band was placed on the left wing and a numbered aluminum band was placed on the right wing. The banding was done as part of a study of the impact of disturbance on hibernating bats exposed to White-Nose Syndrome. (See page 207.) Inset shows more detail of bands. Also shown (**B**) are three number-2 Bird Bands used on bats. Each band is 6 mm. long. Note that one band had been chewed by the bat that carried it.

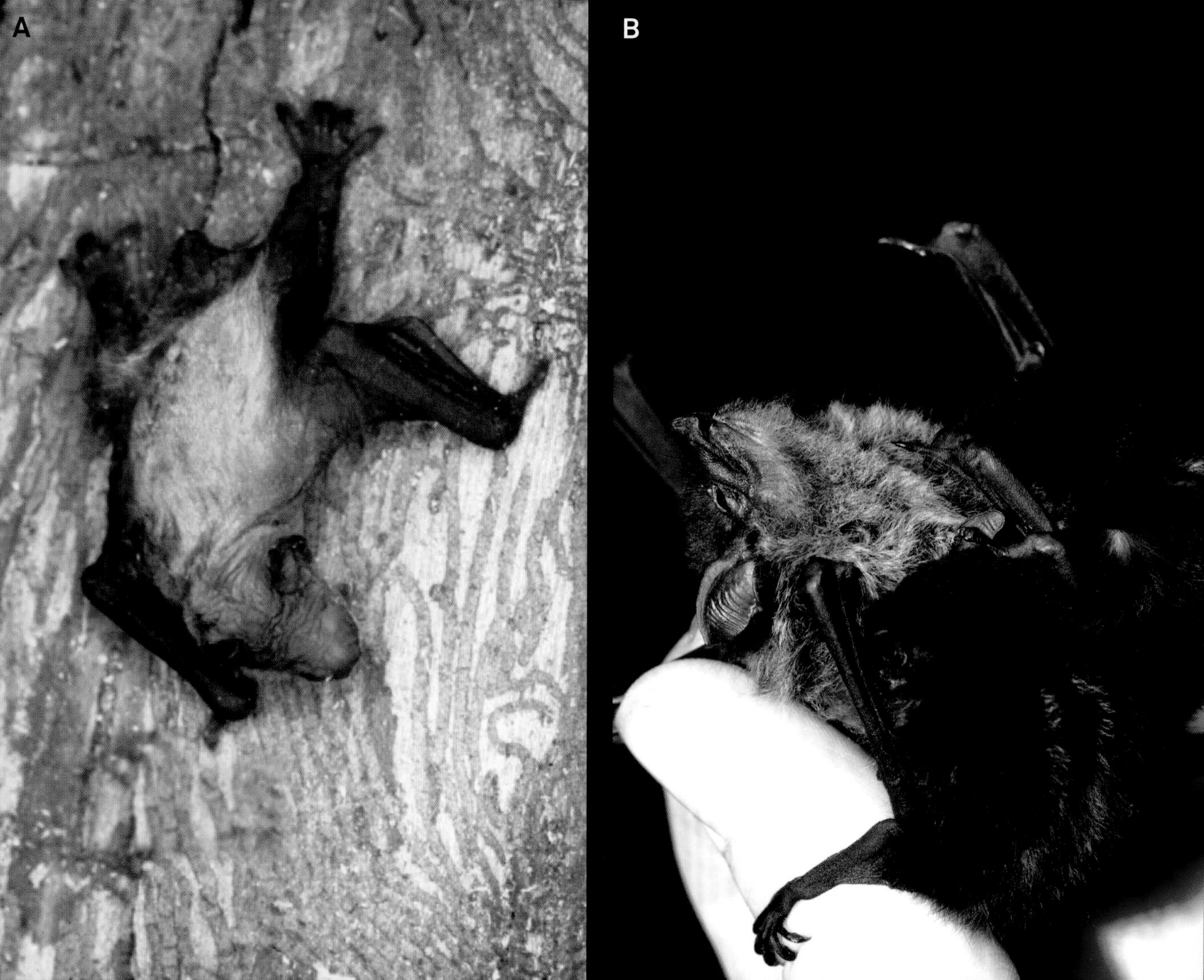
A
B

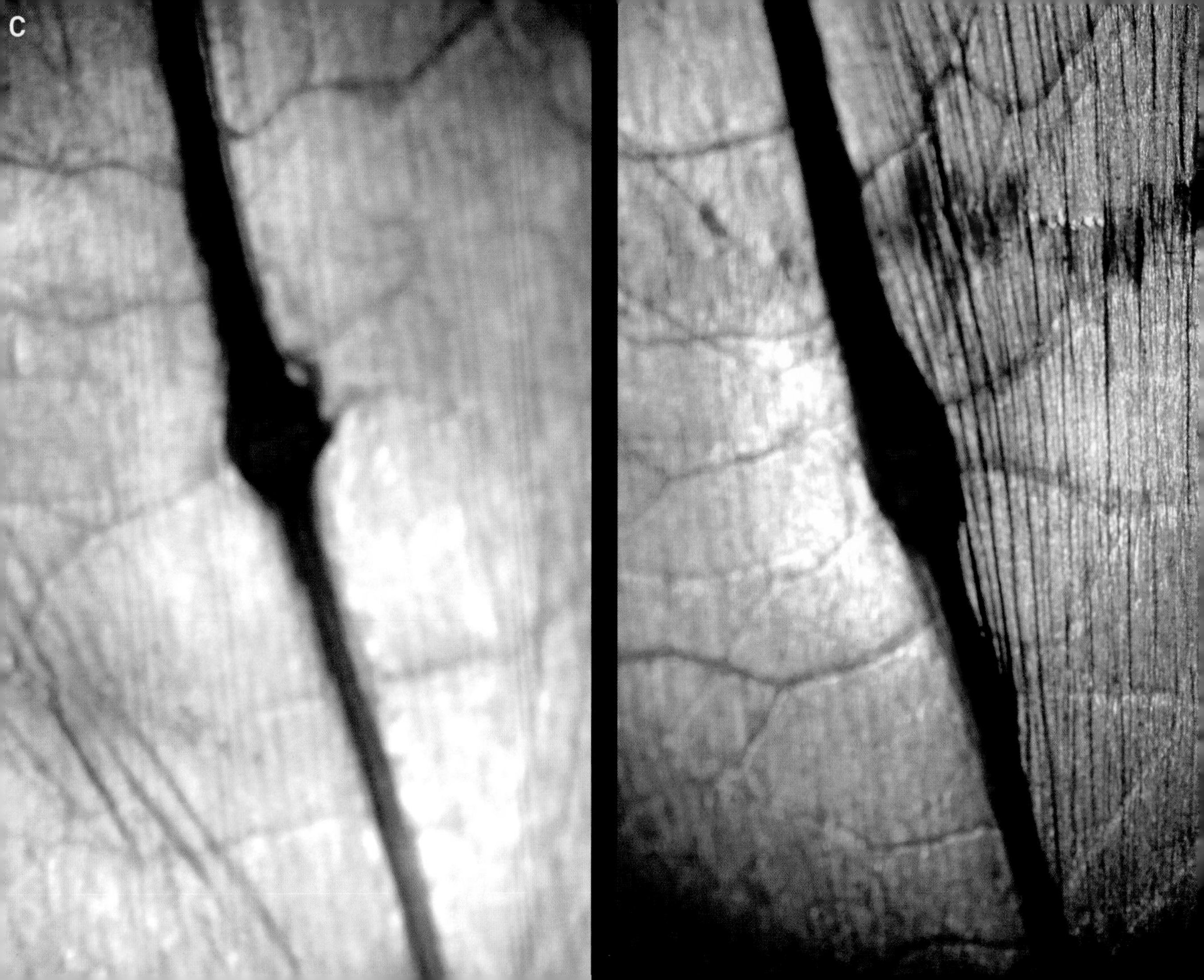
c

Figure 7.3.
A pair of copulating Indian Flying Foxes (*Pteropus giganteus*) photographed by Skanda de Saram in Sri Lanka.

reason(s) for the difference remain elusive. Increased reproductive output (larger litters) has a negative effect on survival. Diet and colony size had no influence on bat longevity. Not surprisingly, with >1300 species of bats, for most there are no data on longevity.

Researchers working on humans and other organisms have documented that mechanisms at the cellular level play a role in longevity. Mitochondria are centers of oxidation that function as the powerhouses of cells in animals, including bats. Active mitochondria produce reactive molecules known a ROS (reactive oxygen species) including superoxide ions (free radicals) and hydrogen peroxide. These molecules are a natural byproduct of the metabolism of oxygen and they play important roles in cell signaling and homeostasis. ROS, however, can also damage macromolecules important to cell function, including nuclear DNA, and ROS production may function in living organisms to reduce life expectancy. For example, typical mice and shrews (5–40 grams) may live three years in the wild. Compared to mouse and shrew mitochondria, the mitochondria of bats produce fewer ROS, although the heart, brain and kidney tissues of bats have the same levels of activity of powerful antioxidant enzymes as those of shrews and mice. The data from bats support the supposition that they have reduced production of ROS per unit of oxygen consumed during metabolism. This may make bats better at resisting oxidation and enhancing the stability of proteins compared to other mammals. This could protect bats from the destructive effects of ROS and explain, at least in part, their propensity for long life spans.

Genetic factors may also underlie the longevity of bats. Sequencing of the whole genomes of Black Flying Foxes (*Pteropus alecto*) and David's Myotis (*Myotis davidi*) revealed high concentrations of genes in the DNA damage checkpoint. Bats are quite different from other mammals in this regard and these differences may underlie protection of bats from ROS. The genetic differences between bats and other mammals also may explain the relative resistance of bats to some diseases. (See page 207.) The same explanation may apply to other long-lived mammals such as Naked Mole Rats (*Heterocephalus glaber*).

Reproduction

In most aspects bat reproduction is typical of placental mammals. But, the prevalence of small litters among bats is more like primates and other larger mammals than smaller mammals such as rodents. Males provide sperm, while females bear the brunt of the costs of reproduction—from becoming pregnant to producing milk and providing most of the parental care. Most female bats are monoestrus, reproducing just once a year. Some bats are dioestrus, giving birth twice a year, and a few are polyoestrus, in that they may reproduce more than twice annually, such as Indian Pygmy Pipistrelles (*Pipistrellus mimus*). When females come into heat (oestrus) immediately after giving birth, an initial delay in development of the fertilized egg may provide time for the lining of the uterus to be prepared for implantation.

Copulation takes place in the usual mammalian way and transfers sperm from male to female. (Figure 7.3) During ovulation the female releases an egg (or several eggs) anfertilization may follow immediately. The fertilized egg(s) begins to divide, and the developing embryo implants in the endometrium, the well-developed lining of the uterus. Implantation connects the developing embryo to its mother's circulatory system, providing access to food and oxygen as well as removal of metabolic wastes.

Implantation

Sites of implantation vary considerably among bats, making them quite different from most other mammalian orders in which sites of implantation are typically less varied. Implantation in bats involves considerable invasion of the uterine lining by the developing embryo (trophoblast). In

Seba's Short-tailed Bat and Greater Bulldog Bats, different maternal blood vessels supply blood to each half of the placenta. This adaptation permits the development of larger young.

Bats vary in the details of uterine structure. Many species have uteri with two horns (one associated with each ovary), while others have a simplex uterus (with a single midline chamber) like humans. During the course of evolution, the two horns have fused into a single (simplex) uterus, a change that has occurred more than once in different lineages of bats. In uteri with two horns, implantation may occur in the left horn or right horn, or alternate in successive ovulations. Although it seems reasonable to expect that eggs produced in the right ovary would implant in the right horn of the uterus, this is not always the case. In some Bent-winged Bats (Miniopteridae), ovulation usually occurs in the left ovary but implantation occurs in the right horn. Nobody knows why fertilized eggs make the long journey from one side of the uterus to the other before implanting—surely there must be a reason, or this system would not have evolved. This remains one of the mysteries of bat reproduction.

In most mammals, the embryos show little development during passage down the oviduct to the uterus (before implantation). Bats show surprising variation in this aspect of reproduction. Implantation occurs four to five days after fertilization in many bats, *e.g.*, species of Vesper Bats, Free-tailed Bats, Horseshoe Bats and Mouse-tailed Bats. But in Sheath-tailed Bats such as Greater Dog-like Bats (*Peropteryx kappleri*) and in New World Leaf-nosed Bats and Bulldog Bats implantation occurs twelve to thirteen days after fertilization. In these species secretory cells lining the oviduct produce glycogen (a glucose) and lipid droplets (fatty acids), providing energy required by the developing embryo.

If implantation does not occur successfully at the appropriate time for that species, the reproductive system resets and pregnancy must wait until another season. Menstruation, the elimination of the endometrium (lining of the uterus) on a regular cycle when pregnancy does not occur is typical of mammals with low reproductive potential. We know most about this topic in humans and other primates, but at least some New World Leaf-nosed Bats and Free-tailed Bats also menstruate.

Gestation

Gestation (duration of pregnancy) in bats ranges from about 40 to about 150 days. Not surprisingly, there are few details for most species of bats. Pups are usually quite large at birth (parturition) and are typically born during a season when food is plentiful. For bats of the north temperate zones parturition usually occurs in May or June, depending upon the latitude and prevailing weather conditions.

In bats, as in other placental mammals, most of the increase in embryo size occurs in the latter part of pregnancy. For example, in Seba's Short-tailed Bat, the gestation period usually lasts ninety days. (Figure 7.4) By forty days after conception the developing embryo weighs about 4.3 mg. and is about 3.4 mm. long. The most dramatic changes in size occur after seventy days post conception when the embryo weighs 727 mg. (17.1 mm long). By eighty days the embryo weighs 1527 mg and is 20.3 mm. long, increasing to 2327 mg (21.2 mm. long) at birth. This pattern of growth and development minimizes the period during which the pregnant female carries the burden of a large fetus.

In developing embryonic Seba's Short-tailed Bats the wings are obvious by sixty days post conception. In embryonic mice, apoptosis (programmed cell death) occurs in both forefeet and hindfeet, resulting in separate digits on both. In contrast, in bats apoptosis occurs only in the hind feet, leaving the forelimbs with skin between the digits that subsequently forms the flight membranes of the wings. Bone morphogenetic proteins trigger apoptosis in chicks and

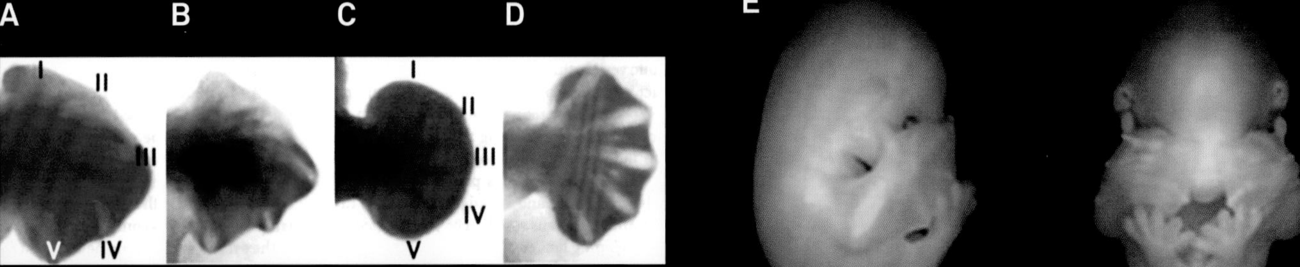
A
B
C
D
E
I
II
III
IV
V
I
II
III
IV
V

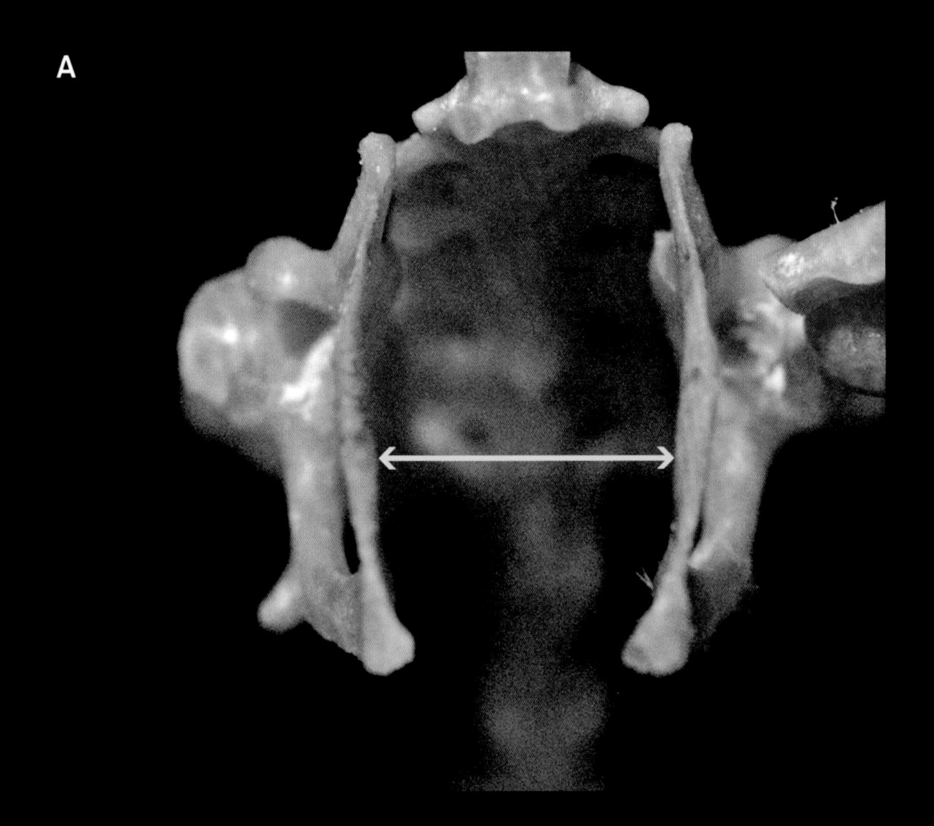
A

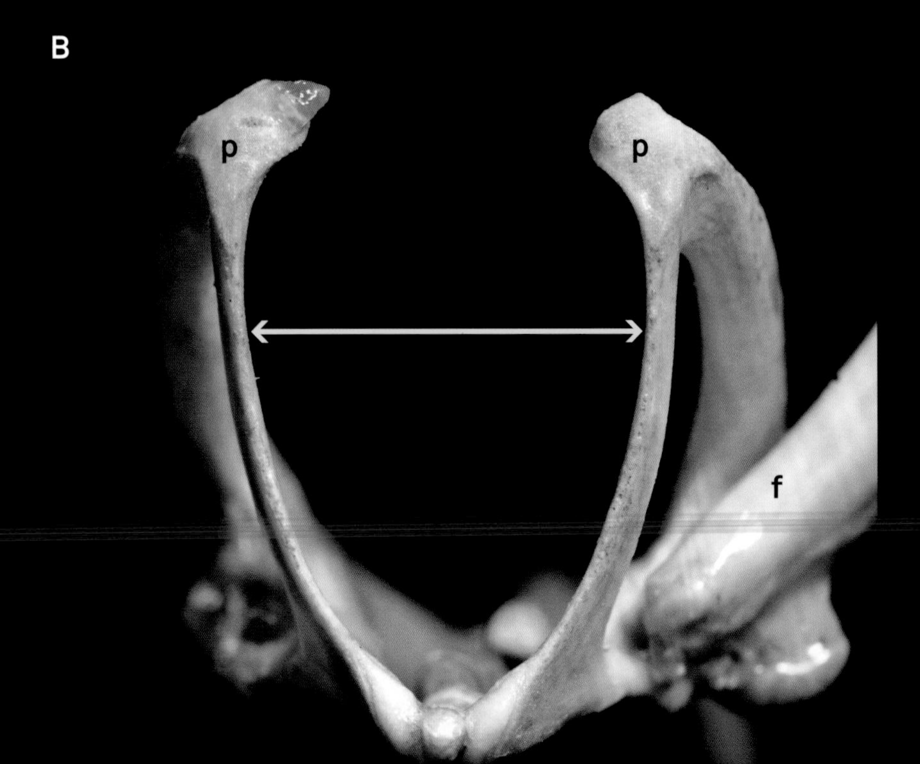
B
p
p
f

in mice (which have separate digits), but not in ducks, which have webbed feet. In embryonic bats, one of these proteins—bone morphogenetic protein 2—apparently stimulates the growth and proliferation of cartilages in the hand, accounting for the increased lengths of hand and finger bones that support the wings.

As the end of pregnancy approaches, granular bodies form in the glandular yolk sacs of some Free-tailed Bats. The grandular bodies are rich in glycogen and lipids. These may provide an energy supply for the mother, reducing the time she needs to spend flying towards the end of pregnancy.

Birth

The very large size of young bats at birth makes the process of parturition (giving birth) difficult. Young bats are born rear-first, a position that is called breech presentation in humans. Unlike humans, however, breech presentation is the norm for bats. One plausible explanation for breech births in bats is minimizing the possibility of the elongated forelimbs become tangled during birthing. The only other mammals known to typically have breech presentations are whales. In whales, breech presentation is thought to keep the young in contact with the placenta through more of the birth process, thus facilitating prolonged contact with the oxygenated blood flowing through the placenta. As soon as birth is complete, the calf must swim to the surface and breathe air.

To accommodate the large size of young bats at birth, the two halves of the female pelvic girdle are not joined or fused across the pubic bones (the front or ventral end of the pelvis) as they are in most other mammals. Instead, the right and left pubic bones are connected by ligaments allowing the diameter of the birth canal to greatly expand the first time a female bat gives birth. In Brazilian Free-tailed Bats, the space between the right and left pubic bones expands from 2 mm. in diameter to 35 mm. in diameter (Figure 7.5A). The stretching of the interpubic ligaments that allows this expansion is facilitated by hormone relaxin, which makes the ligaments more flexible and allows them to elongate sufficiently.

Bats normally roost in a head down position, but just before giving birth most female bats turn head up. This is thought to ensure that gravity assists the birth process. Bats may also help each other when they are in labor. A few years ago at the Lubee Foundation in Florida, a captive Rodriguez Flying Fox (*Pteropus rodricensis*) was having difficulty giving birth when another female bat came to her assistance. The "midwife" bat showed the laboring bat the correct birth posture, and the laboring bat then assumed it. The midwife licked the other bat's vulva. Eventually the new mother managed to successfully give birth to her pup. Under normal circumstances, Rodriguez Flying Foxes give birth in about forty minutes, but in this instance the process took closer to three hours.

Figure 7.5.
Giving birth to a very large young places a large demand on the female's reproductive tract. These pictures show the birth canal (yellow arrows) of two female bats, a Brazilian Free-tailed Bat (**A**) and a Hammer-headed Bat (**B**). During birth the canal of the Brazilian Freetailed Bat expands from 2 mm to 35 mm. **A** is the view from below, **B** a view from behind. Note that the two halves of the pelvic girdle do not join at the pubis (**p**). The head of the left femur (**f**) is shown in **B**.

Lactation

Female bats feed their young milk produced from a pair of mammary glands that are often located almost in their armpits. (Figure 7.6) In some species, accessory nipples that do not produce milk are located in the groin area. The functioning nipples serve as attachment points for the young, which wrap their hind legs around their mother's neck and hold on to the nipple with their mouth when their mother moves from one roost to another. While mother bats often move their young between roost sites, they rarely carry them about while foraging. Leaving the young in the roost most of the time probably serves many purposes including keeping them safe from predators, the mother safe (a bat burdened by carrying a large offspring may not be able to fly fast enough or be maneuverable enough to evade predators herself) and reducing the energetic demands on the mother (carry heavy weights during flight is energetically costly).

Bat milk varies considerably from species to species, partly reflecting differences in adult diet. The milk of insectivorous bats tends to be higher in fat and protein than that of frugivorous bats, which is more like the milk of other small mammals. Carbohydrate content of milk tends to be the same across bat species. Triglycerides (fatty acids) are important components of bat milk, and their composition usually reflects a mixture of those ingested by the mother and other fatty acids that are synthesized by the mother. Like other mammals, lactating female bats mobilize calcium from their long bones, which causes the bones to thin. In bats, this also occurs in hibernating males and females, which suggests that calcium is important in the development and maintenance of teeth and bones in bats as it is in other mammals.

Female bats show considerable variation in the amounts of milk that they produce. Eighty-four gram Greater Spear-nosed Bats (*Phyllostomus hastatus*) have one offspring at a time and produce 15.7 grams of milk per day. In contrast, much smaller Big Brown Bats (15–20 g.) typically have twins, and lactating females produce 20.2 grams of milk per day. Among bats for which there are data, Little Brown Myotis are the champion milk producers: on average, females produce an amount of milk per day that is equivalent to about 75 percent of their body weight. This is the equivalent of a lactating 45 kg. woman producing 33 kg. of milk—per day!

Although lactation is a diagnostic trait of mammals, it is typically a female function. Not surprisingly, the discovery of lactating male Dyak Fruit Bats (*Dyacopterus spadiceus*) attracted a great deal of media attention. Expression of 4-6 microliters (1 microliter = one millionth of a liter) of milk from the nipples of two male Malaysian Dyak Fruit Bats was unexpected. Compare this to the 50 microliters of milk extracted from the right nipple of one lactating female. The males showed some development of mammary tissue, but the report provided no significance behavioral observations about "lactating males" and no data about the hormonal state of the males. Lactating males have also been reported among Bismarck Masked Flying Foxes (*Pteropus capistrastus*). Gynecomastia is the development of mammary tissue in male humans, while galactorrhea refers to production of milk by male mammals. Both conditions are usually associated with hormonal changes at puberty. Interpretation of the significance of lactating male Dyak Fruit Bats and Bismarck Masked Flying Foxes remains a topic of discussion.

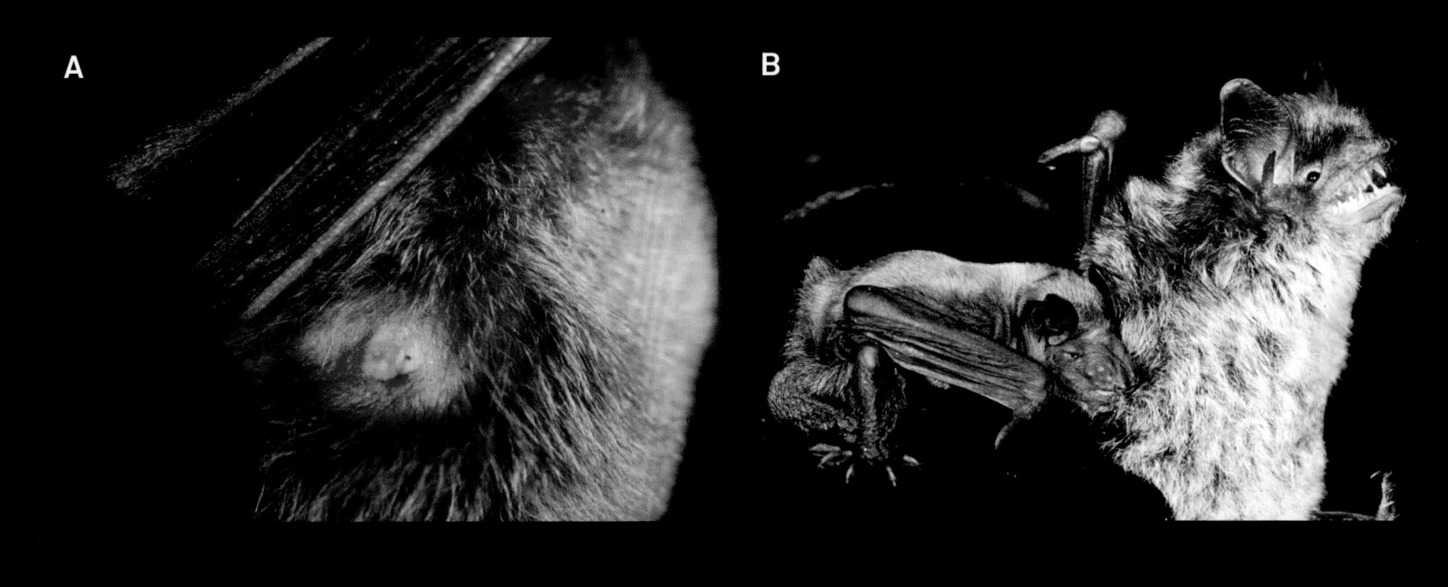

Figure 7.6.
A nipple of a lactating Yellow-shouldered Bat (**A**), a five-day-old Little Brown Myotis nursing from its mother (**B**), and a female Egyptian Rousette carrying its offspring in flight (**C**). Note the large size of the baby in (**C**) compared to its mother. Baby bats hang on to their mother's nipple with their teeth and grasp her fur with their hind feet in order to stay on board when she flies.

A

B

Growth After Birth

In many species of bats, pregnant and lactating females roost in warm to hot settings (30° to 40°C) where high temperatures promote rapid growth of the young and lactation. In the Queen Charlotte Islands off the west coast of British Columbia, female Little Brown Myotis and Keen's Myotis (*Myotis keenii*) roost in rock crevices heated by hot springs, or at least they did until an earthquake in the fall of 2012 disrupted the flow of hot water to the springs. Under ideal conditions, Little Brown Myotis typically reach adult linear dimensions (body and forearm length) about eighteen days after birth. This coincides with weaning and the first flights, as well as replacement of the milk (deciduous) teeth with permanent adult dentition.

The age at weaning in bats varies from about eighteen days (less than three weeks) to 285 days (over nine months). Common Vampire Bats are the only species in which weaning occurs beyond eighty days. Common Vampire bats are exceptional among bats in terms of both their diet of blood diet and the level of maternal care as reflected by the timing of weaning. Feeding young with milk for over nine months places a heavy burden on the mother, but this pattern would not have evolved without a purpose. Apparently weaning begins at three months in Common Vampire Bats, but the young do not become fully independent until much later. This may reflect the difficulty of obtaining blood meals on a nightly basis, with ongoing nursing ensuring growth and survival of young bats until they are capable of reliably finding their own food.

Sexual Dimorphism

Sexual dimorphism—pronounced anatomical differences between males and females seen in some species of bats—results from different developmental trajectories. Male Grey-headed Flying Foxes and Hammer-headed Fruit Bats (*Hypsignathus monstrosus*) are markedly larger than females. (Figure 7.8) In the case of Hammer-headed Fruit Bats, males compete with each other to attract females at communal mating and display roosts called leks, and sexual selection for larger size and ability to make louder calls may have been responsible for the observed dimorphism. (See Chapter 8.) Larger males that produced louder calls were likely more successful (produced more young) than smaller males. This advantage would have been the selective pressure responsible for the evolution of sexual dimorphism in this species.

The larger males take longer to complete growth and mature than females, so sons must be nursed longer by their mothers. In most species of bats, males are smaller than females. In extreme cases, such as Little White-shouldered Bats, males are so much smaller than females that they were once thought to be separate species. Sexual dimorphism in cases like this may serve to reduce competition between the sexes for resources like food, *e.g.*, males may prefer smaller fruit than do females, or may reflect different selective pressures on the sexes, *e.g.*, larger females may be able to carry larger young.

Figure 7.7.
Rock crevices and spaces (**A**) around geothermally heated hot springs in Haida Gwaii National Park in British Columbia are used as summer day roosts by Keen's Myotis (**B**) and Little Brown Myotis. The numbered green band indicated that the bat was at least thirteen years old when the photograph was taken.

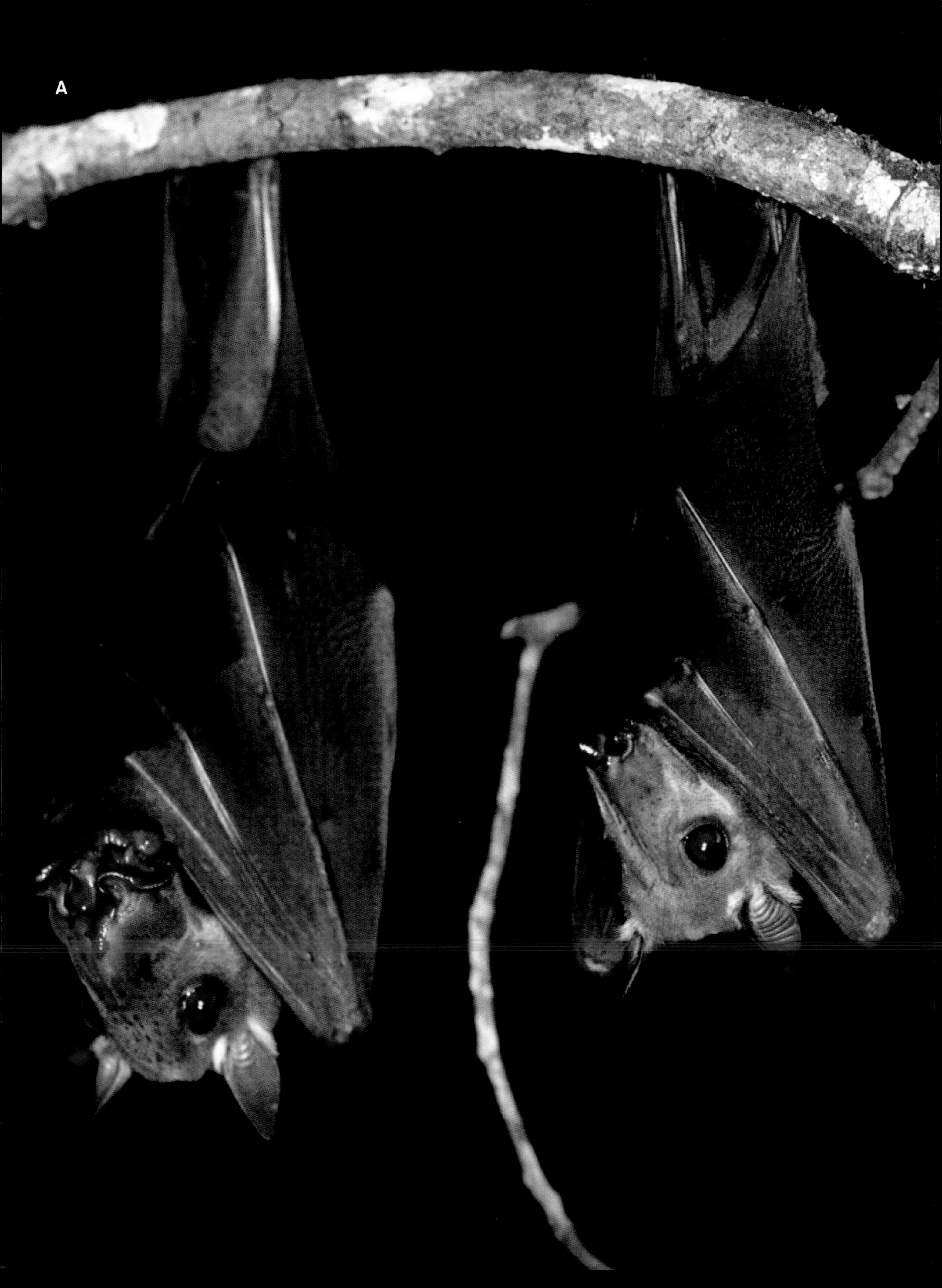
A

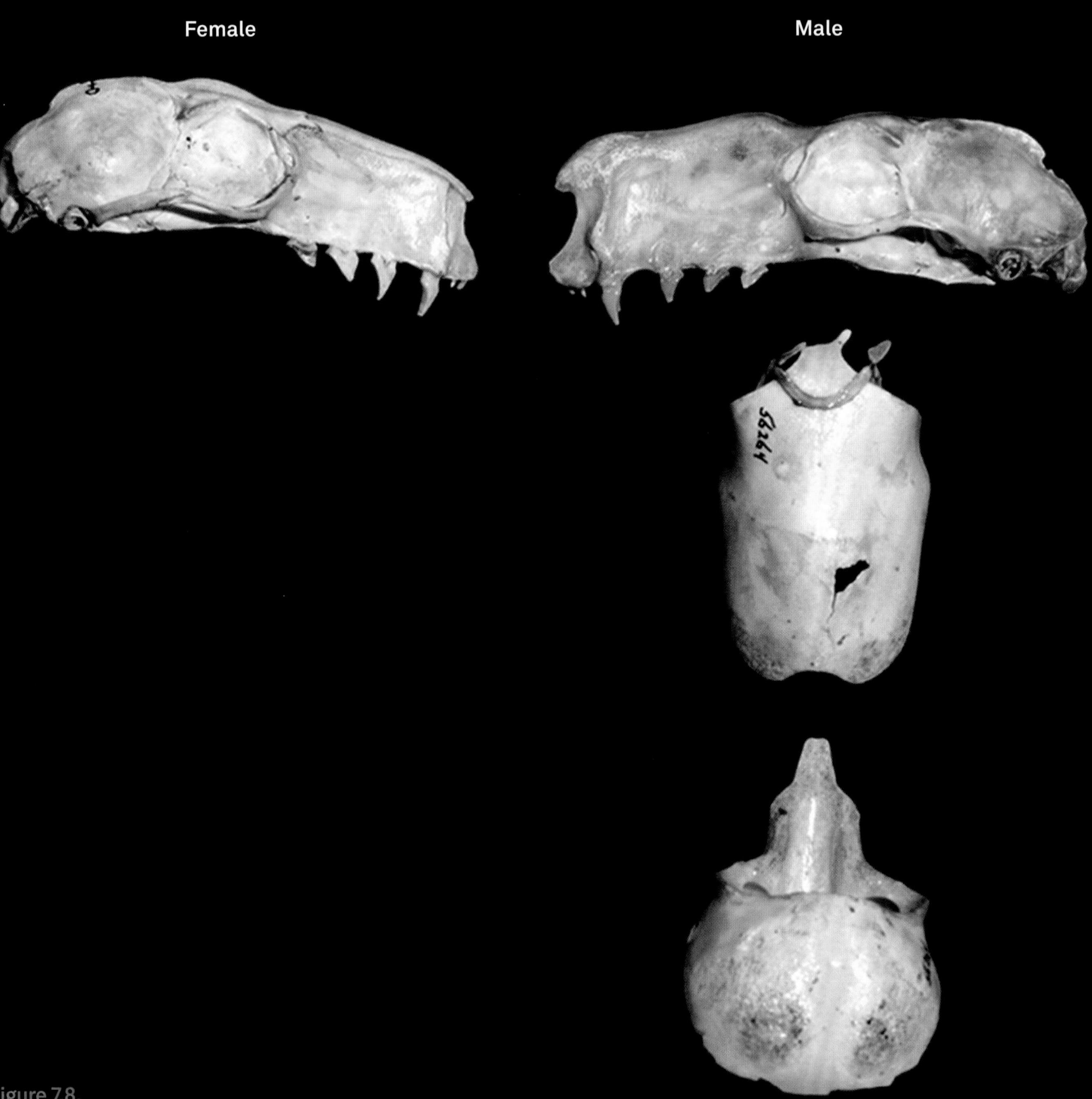

Figure 7.8.
Sexual dimorphism in Hammer-headed Bats (*Hypsignathus monstrosus*). In **A**, a male (left) and female (right) roost side-by-side. The differences include both size (males are larger) and facial appearance. The differences in the skulls of females and males (**B**) also is obvious, including both size and appearance. Furthermore, in males there are two large ossified chambers along the vocal tract which contribute to the males' distinctive call and which do not appear in females. Photograph (A) courtesy of Jakob Fahr.

Delays to Get the Timing Right

Male and female bats living in temperate regions of the world may be in prime condition at the end of the summer, making it a suitable time to mate. But with gestation periods of ~60 days, mating in late summer or early autumn would result in young born as winter begins. In this sense, the annual cycle of the parents is not best for successful production of young.

Vesper Bats and Horseshoe Bats are common in temperate zones. In these species, females use delayed fertilization to give birth in spring and early summer when food is abundant. Mating occurs in late summer and in early fall, when males and females are in prime body condition. But if pregnancy were to occur immediately, the pups would be born in winter when females would have little access to food. So the female stores the sperm inside her reproductive tract over the winter, and ovulation and fertilization do not occur until the following spring. For bats living about 40° N in North America, mating occurs in August, September and early October, while ovulation and fertilization occur the following April, with young born in June. The specific timing depends upon the latitude and local weather. Delayed fertilization appears to take advantage of both the best time for mating and the best time for giving birth.

In other bats, notably some species of tropical New World Leaf-nosed Bats, mating is followed immediately by fertilization. Some bat species exhibit post-fertilization delays in the reproductive process, either delaying implantation of the fertilized egg in the uterus or delaying development of the egg after it has been implanted. These mechanisms help to ensure that the timing of birth coincides with the season of optimal resources for lactating mothers and growing young bats. In the tropics, the timing of the rainy versus dry seasons is more important because food availability is often tied to the flowering and fruiting schedules of tropical trees and shrubs.

In Seba's Short-tailed Bats, females are able to suspend the development of the embryo for at least sixty days. (Figure 7.10) This strategy appears to be relatively common in some tropical species, where bats may show two periods of oestrus that coincide with an abundance of food resources. In captivity, other factors may lead female Seba's Short-tailed Bats to delay development, sometimes for over 100 days. The details remain unknown but cues that trigger delayed development of embryos could include the social setting as well as availability of food.

Figure 7.9.
This transmission electron micrograph shows sperm stored in the uterus of a female Little Brown Myotis. The three sperm head (cyan arrows) are obvious, as is the line (blue arrow **ul**) denoting the lining of the uterus. The sperm heads tend to be oriented towards the lining of the uterus.

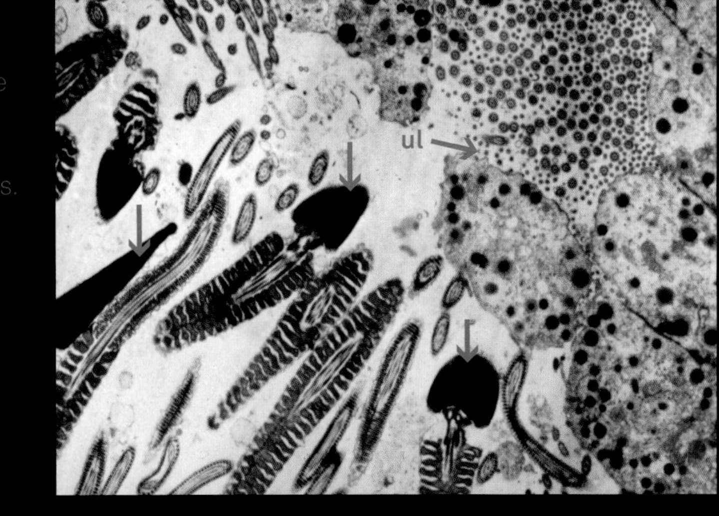

Figure 7.10.
A comparison of two embryos of Seba's Short-tailed Fruit Bats at day fifty post mating. The normal development is showed in **A**, delayed development in **B**. Scale bar in **A** is 200 µm.; in **B** 1 millimeter. Courtesy John J. Rasweiler IV.

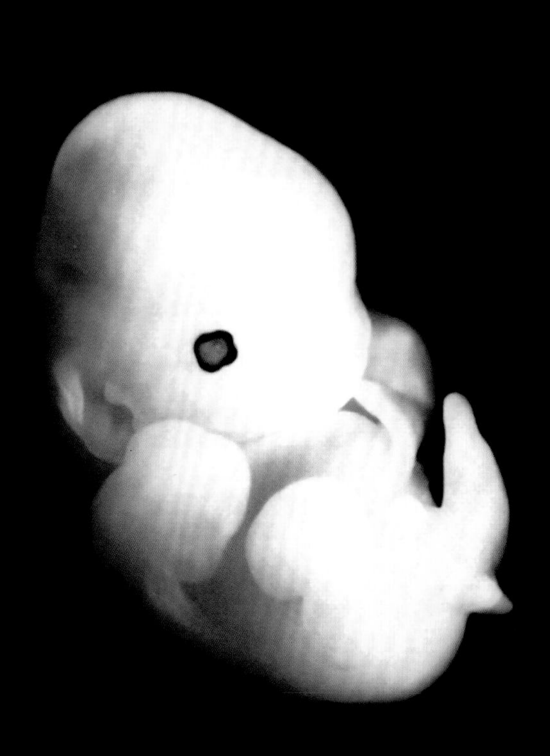

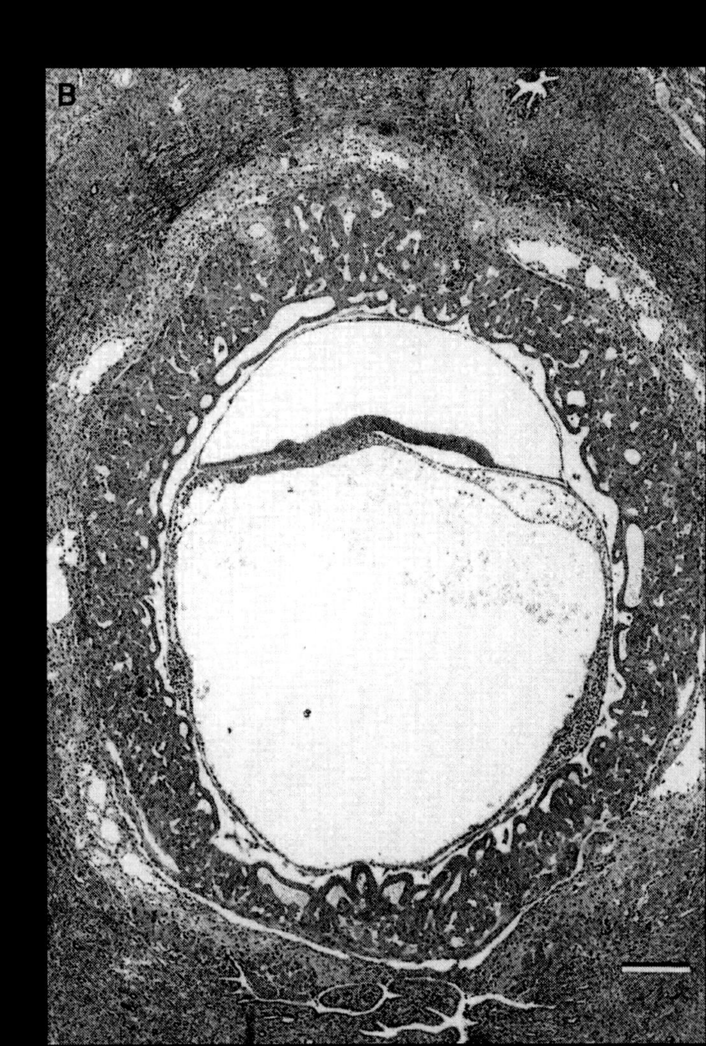

Figure 7.11.
Bats leaving cave roosts are vulnerable to predators. Inside St. Clair Cave in Jamaica there are sites used by feral house cats (**A**). These are marked by piles of moldering bat wings. Outside Great Windsor Cave in Jamaica, a Jamaican Boa (Epicrates subflavus) swallows a bat it has caught (**B**). Photograph (**A**) by Brock Fenton; (**B**) by Susan Koenig.

Bat Mortality

The available evidence suggests that more than half of the bats born in any year do not survive to their first birthday. For example, in the United Kingdom in some years, 60 percent of young Greater Horseshoe Bats do not survive a year. The relatively high levels of mortality experienced by bats before their first birthday probably reflects some combination of accidents and immaturity/inexperience at finding sufficient food and appropriate, safe roost sites. At times of the year when bats initially take to wing, it is all too common to find them tangled in the burrs of Lesser Burdock (*Arctium minus*) or impaled on barbed-wire fences. Close examination of the finger joints of these bats usually identifies them as yearlings. Why bats come to be entangled remains unknown. The propensity of some young bats to be accident prone may reflect inexperience or the effects of dehydration and starvation. Some young surely fail to put on enough weight to survive hibernation. (See Figure 7.12. and 7.13)

As with all other mammals, the cause of death for adult bats varies from individual to individual, and patterns vary from species to species. Some bats die from disease, others are killed by predators and some die in accidents (See Figure 7.12.). Perhaps relatively few live long enough to die of old age, although scientists know little about this phenomenon in bats. (See page 179.)

As small mammals, bats are potentially bite-sized for many predators, but their secretive roosting, ability to fly and nocturnal habits may put them out of reach of many would-be predators. Although the list of predators that occasionally take bats is long, this predation often takes place as bats, especially large numbers of them, emerge from roosts. At entrances to roosts (and in confined areas within roosts) bats may fall prey to house cats and snakes, *e.g.*, Jamaican Boas (*Epicrates subflavus*), as well as to birds of prey. (Figure 7.11) Some predators congregate at the entrances to roosts with large populations of bats in anticipation of the predictable emergence. Boas congregate around cave openings on some Caribbean islands, waiting for bats to emerge. The snakes hang from branches and rock outcrops, and strike at bats as they fly past. In some places in Mexico and the Caribbean, this spectacle has become an attraction for ecotourists.

Around bridges that serve as roosts for some bats in Kruger National Park, raptors such as Wahlberg's Eagles (*Aquila wahlbergi*), African Goshawks (*Accipiter tachiro*) and African Hobby (*Falco cuvierii*) roost patiently watching for the first bats, their signal to take flight and begin hunting. These birds only hunt bats when there is enough light. In Africa and at sites in the East Indies, ~600 g. Bat Hawks (*Macheirhamphus alcinus*) appear to specialize on bats. A 1 kg. Wahlberg's Eagle (*Aquila wahlbergi*) or 500 g. Peregrine Falcon (*Falco peregrinus,*) may take only a minute to perch and consume a bat after catching it. A Bat Hawk eats on the wing, swallowing bats whole and taking only about seven seconds from capture to swallowing! Analysis of cast pellets collected beneath a Bat Hawk nest in Zimbabwe revealed that these birds preyed on different species of bats in about the same proportion as the bat species were caught in mist nets, suggesting that the birds are not picky but simply capture bats as they encounter them. The first specimen of a Chapin's Free-tailed Bat (*Chaerephon chapini*) known to science was removed from the crop of a Bat Hawk shot in Uganda. Bat-hunting raptors have also evolved in the Neotropics, where the most common species is the aptly named Bat Falcon (*Falco rutigularis*), which routinely hunts bats and dragonflies at dusk.

mortality not caused by disease. In January 2014, perhaps as many as 100,000 Flying Foxes died as a result of an extreme heat wave in Queensland, Australia. Over the course of two days, the temperature hovered at a point over 43° C and bats literally fell from the trees as they succumbed to heat exhaustion. In Alberta (Canada) there is a record of over 1,000 Little Brown Myotis dying from poisoning caused by a toxic alkaloid produced by a blue-green alga (*Anabena flos-aquae*). Both the bats and several ducks were found dead in a lake, apparently after drinking from the water surface that was covered by a scum of algae. The long-term effects of such massive die-offs remain to be determined, but for animals that reproduce only once a year (like the flying foxes) such events may be catastrophic in terms of reducing population size and genetic diversity.

Other occurrences of mass mortality in bats result from human activities. For example, mass die-offs in Grey Bats and Brazilian Free-tailed Bats have been documented during fall migration. High levels of pesticides, such as DDT, DDE and dieldrin, apparently killed the bats. The pesticides had been stored in body fat of the bats, and released when the fat was metabolized to fuel migration. The bats were effectively poisoned by pesticide accumulation after the consumption of their insect prey. Evidence of Little Brown Myotis being killed by pesticides also includes exposure to poisons after the bats' roosts were sprayed as an effort to rid them from a building in which they roosted.

Reproduction in Big Brown Bats

Several intriguing aspects of reproductive biology and survival of bats emerge from research on Big Brown Bats, a species that occurs widely in North America and the West Indies. In some areas female Big Brown Bats typically have twins, but elsewhere single young are more common. In Kansas, in early May 1969, females averaged 1.2 embryos in the left horn of the uterus, 1.8 in the right horn. By 27 May that year, there were fewer embryos—0.9 embryos in the left horn, 1.0 in the right. By 11 June, the average was one embryo in each horn and the litter size was two. These findings indicate that not all fertilized eggs implanted and went to term. This could reflect the fact that implantation may only occur at a specific location in the uterus.

At this Kansas study site, the average time of lactation was forty days. Male and female young achieved adult forearm size by age two months, more or less coinciding with the end of lactation. At this time young weighed about 45 percent of adult body mass. Young bats made straight line flights by age three weeks, and were capable of negotiated turns at four weeks of age. This pattern of development may prove to be typical among bats.

Figure 7.12.
Bats and accidents. A Grey-headed Flying Fox dead on a barbwire fence in Queensland, Australia (**A**), and a Big Brown Bat dead on a burdock in southern Ontario (**B**).

Figure 7.13.
Many bats do not survive hibernation and, like this (**A**) Little Brown Myotis, end up entombed in ice. The cylindrical Hitchcock cage (**B**) for holding bats is ingenious because there is no door. One drops bats (Little Brown Myotis here) into the cylinder. They cannot climb out because of the slippery sheet metal, nor can they fly out because of the diameter of the cylinder (25 cm.). The calls of bats in a Hitchcock cage attract others and unattended cages can almost fill up, potentially resulting in the deaths of many bats. Photograph **B** courtesy Liam McGuire.

Figure 7.13.

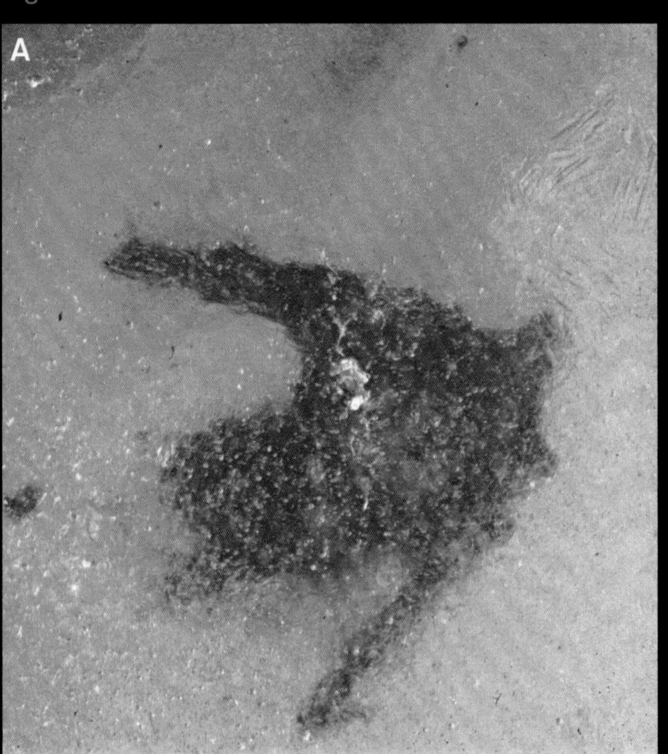

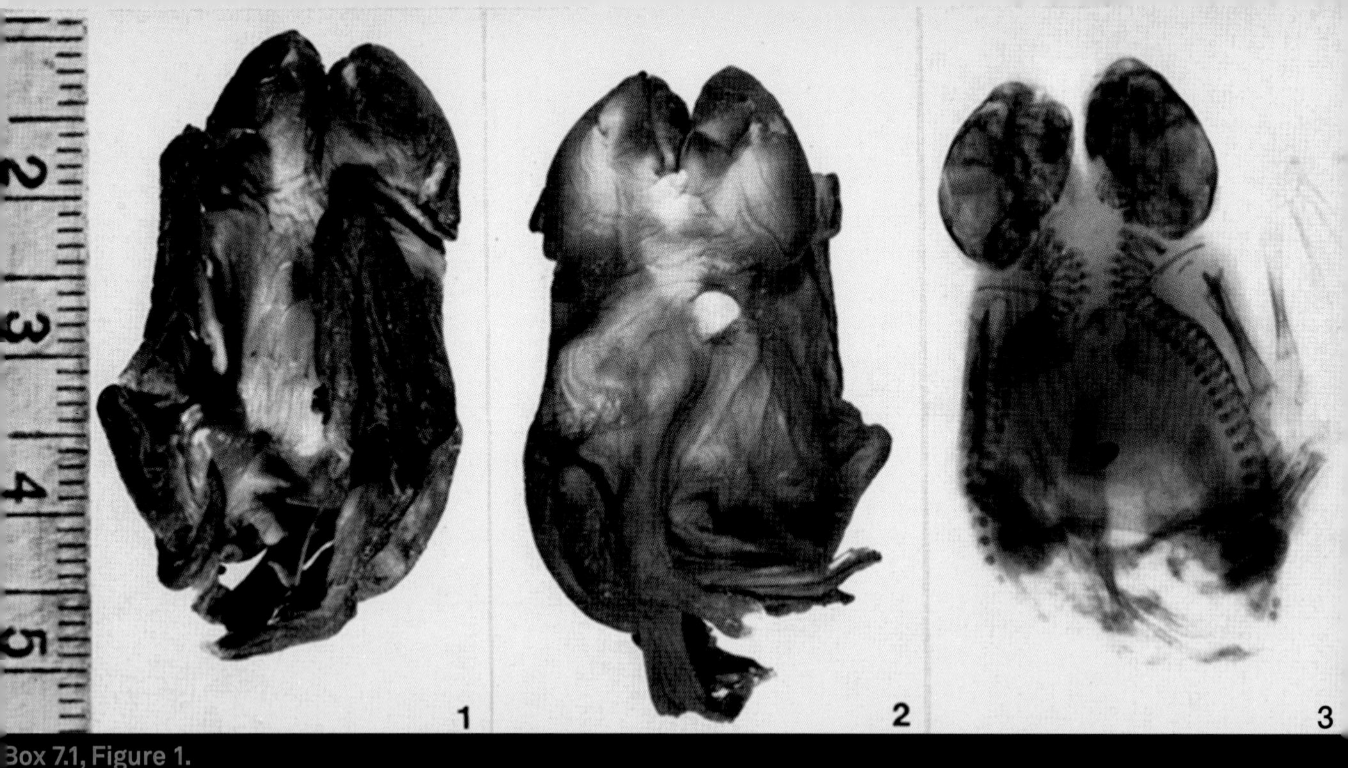

Box 7.1, Figure 1.

How can it be determined if twin bats share the same father or have different fathers? Analyses of genetic data revealed about half of Big Brown Bat litters involve two fathers. This confirmed that at least some females mate with more than one male, and also suggests post-copulatory mechanisms may exist for determining which sperm fertilizes the eggs. These findings are even more intriguing because female pups tend to be born earlier in the year than males, and females born earlier tend to survive longer and reproduce successful of selection on sperm, but this remains only a possibility.

An extensive study of Big Brown Bats in Fort Collins, Colorado revealed variation in patterns of survival with age, gender and season (Box 7.1, Figure 1). For example, about 67 percent of female young survived a year after weaning, while the annual survival of adult females was about 79 percent. Survival varied among roosts and between years. Survival of young was reduced in years of drought. The important parameters for survival of Big Brown Bats were similar to those of much larger mammals. Age at first reproduction was the exception because bats reproduce much earlier (at younger ages) than larger mammals. It is ironic that in spite of so much research, little is known about the mating systems of Big Brown Bats. The evidence that females mate with more than one male during any mating season suggests a mating system that does not involve extensive pair bonds between males and females, but that is about all that is known. (See page 195.)

Box 7.1, Figure 1.
Figure 1. Here are three views of conjoined twin Big Brown Bats were found outside a school in Newboro, Ontario. The first is a ventral view, the second a dorsal (from the back) view, and the third an x-ray view of the specimen. The mother probably roosted in the attic of the schoolhouse. Scale is millimeters.

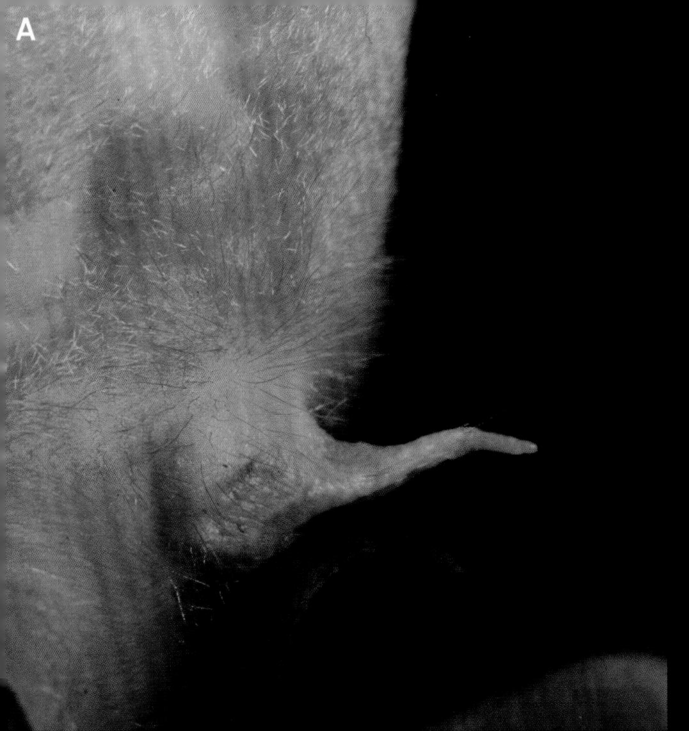

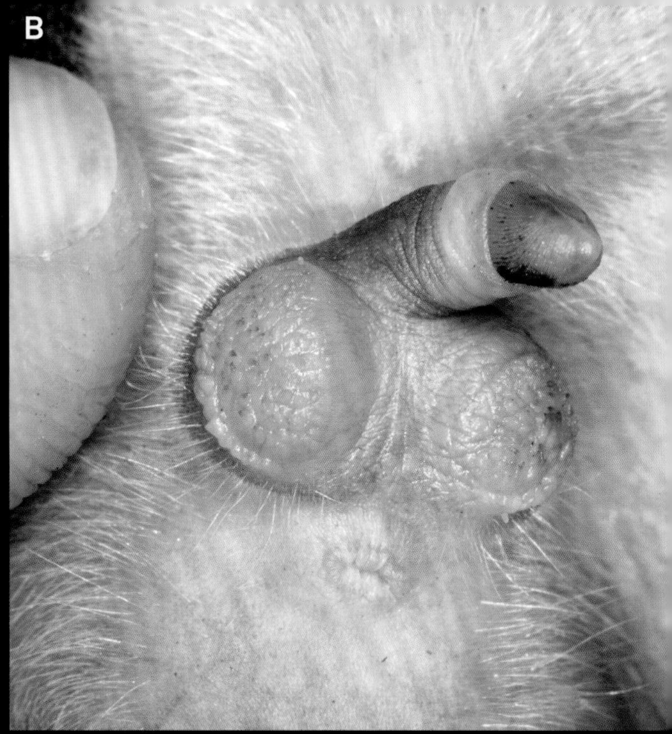

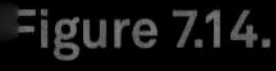

BOX. 7.2

Attracting Bats

People who try to catch bats, whether bat biologists or people taking bats to sell for food, know that the calls of one bat may attract others. (See page 188.) This can have fatal consequences for the bats. The description of an "unintended" bat trap provides a good example. On 21 June 1961 the late Middlebury College biologist Harold B. Hitchcock was collecting bats in the attic of an old hotel near Salisbury, Vermont. There he found a china slop jar containing many dead Little Brown Myotis. He proposed that the call of the bat that had initially entered the slop jar attracted others. The smooth sides that sloped towards the small opening of the jar prevented bats from escaping by climbing or flying out. Judging from the numbers of bats in the jar, it had been an all-too-efficient bat trap. Why did the first bat enter the jar in the first place? Nobody knows, but it could have been simple curiosity. Regardless, it ended badly for the bats. A similar combination of calling bats, slippery sides and narrow openings explains why bats can end up in wood stoves, having been attracted down the stove pipe or chimney by the calls of those that went before.

Greater Bulldog Bats Surprise Brock and Nancy

Although we expect that telling the males from the females is usually easy in mammals, this is not always the case. Both Nancy and Brock are intrigued by the external genitalia of Greater Bulldog Bats. (Figure 7.14) What is the significance of female genitalia that appear masculine? In Spotted Hyenas (Crocuta crocuta) a similar occurrence prevails. In those animals the appearance reflects the impact of circulating levels of testosterone in pregnant females. The topic has not been explored in bats.

Figure 7.14.
Two Greater Bulldog Bats illustrate the challenge of telling male from female. Which one is the "real" male? The one on the right. Photograph (**A**) by Brock Fenton; (**B**) by E. L. Clare

8

Behavior of Bats

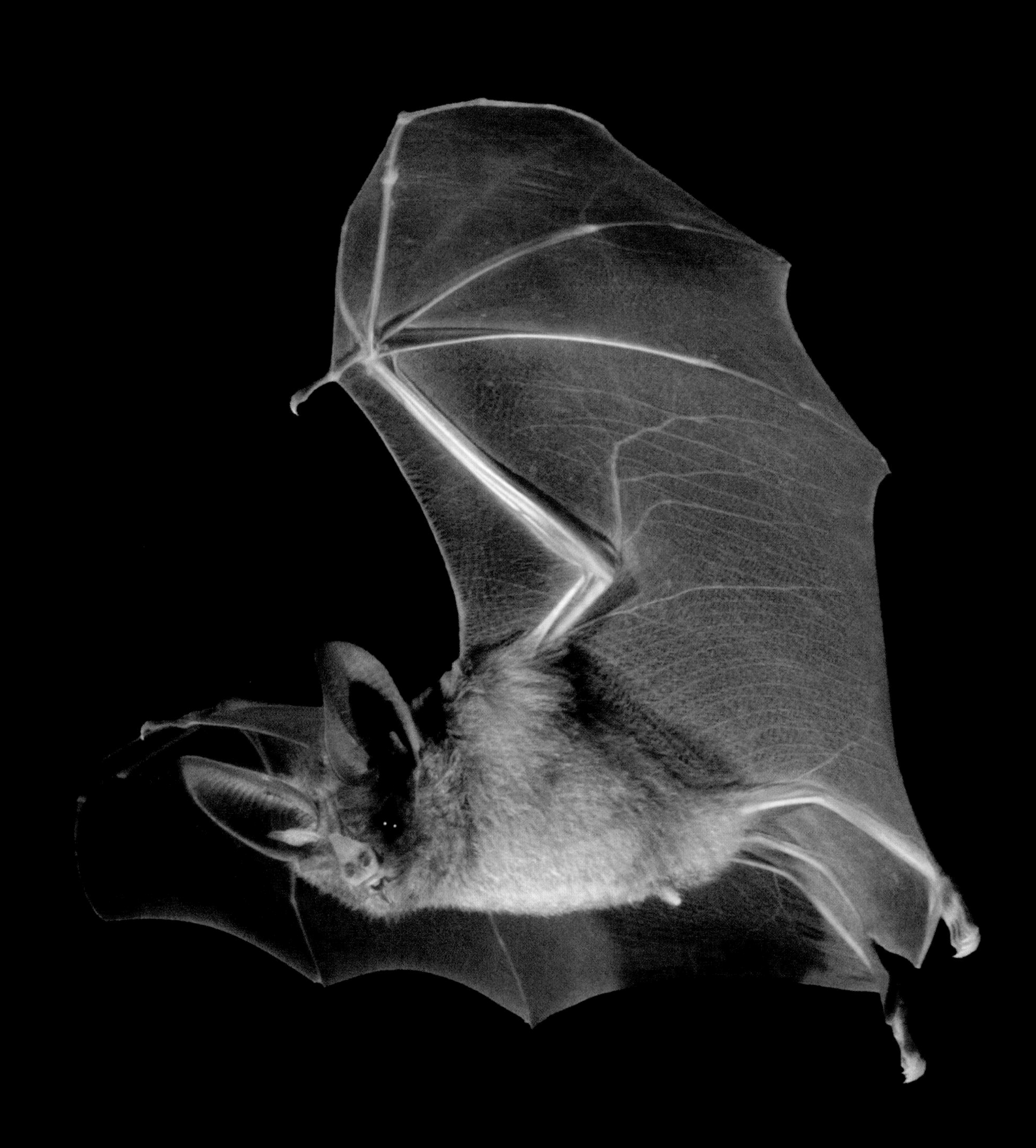

Figure 8.1.
Honduran White Bats in a tent made from a False Bird-of-Paradise leaf (**A**, **B** and **D**) as well as a group of seven Tent-making Bats in a tent (**C**). The White Bats' tent is viewed from below (A), from above (**B**), and the bats are shown both in digital (A) and thermal (**D**) image. Bat bites on the leaf are obvious in A and B. Photograph **C** by Sandra Peters and photograph **D** by Greg McIntosh.

What is a Colony of Bats?

Following Common Vampire Bats radio-tagged around La Selva Biological Station in Costa Rica provided an interesting view of what constitutes a "colony" in this species. Tracking tagged bats for several days revealed that each individual used several different hollow trees as roosts. Although after a month virtually all of bats had roosted with one another at one time or another, on any given night only some members of the colony were roost-mates. Following radio-tagged White-winged Free-tailed Bats (*Tadarida australis*) in Australia revealed a similar pattern of roost occupancy and associations between roost-mates.

The fact that all members of a colony of bats might not roost together most of the time illustrates the challenge of defining what is meant by a "colony" of bats. Is it the bats that roost together on any given night? Or those bats that roost together at least sometimes over a month? Are summer groups the same as winter groups? Do the groups found together during the breeding season correspond to the groups found at other times of the year? These are some of the tantalizing questions that remain unanswered for most species of bats. All associations with other bats of the same species are probably important to the bats, but exactly how important is not always known. (Figure 8.1)

Can a colony of bats be defined as a social unit? Biologists who study bats have uncovered some surprises. A statistical analysis of the flight order of almost 250,000 emergences by pit-tagged Big Brown Bats in Colorado suggested some underlying social structures. For example, some tagged individuals consistently emerged from the roost in the same order. Pit tags are embedded transponders that allow identification of individuals as they pass over a receiver such was installed at the roost entrances. But making sense of social structure among the ten million Brazilian Free-tailed Bats living in a single cave is currently beyond the abilities of science. It can be hard enough to identify social organization(s) among the many people emerging from a hotel or a high-rise apartment. Even if scientists could decode the structure of such large groups, it is unlikely that the definition of a "colony" would be the same across species, or even consistent within a species.

Brock Explores the Question: One Bat One Roost?

The advent of radio tags small enough to put on 8 g. bats opened my eyes to what had been secrets of bats. (Figure 8.2) Other researchers had already used radio tags to find bats roosting in unexpected places. (See Chapter 6.) Catching a bat while it was foraging, tagging it and then following it home also showed that some species (and then many) regularly switched roosts, something not previously suspected. Home today might be different than home tomorrow, let alone next month or next year. Furthermore, disturbance at a site could affect bats' behavior with respect to roost choices.

In Zimbabwe, along the Zambezi River in Mana Pools National Park, I regularly found Large Slit-faced Bats roosting in disused military bunkers, the roofs of buildings and in hollows in trees. In November 1979, a bat that I caught roosting in a bunker <50 m. from the shore of the Zambezi River immediately stopped using the bunker. This bat spent the next four days roosting in a hollow acacia tree about 2 km. from the river. Each night this bat foraged along the shore. On the fifth night it moved back to the bunker and stayed there for the next ten days. Roosting in the bunker meant a shorter commute to its preferred foraging area, but my capture of the bat apparently had made it shy about using the bunker roost for a few days after our initial encounter.

Bat biologists now know that a bat may use several roosts and, over a month, may roost with several different roost-mates. Whether we are trying to understand the roosting behavior of bats to better understand their social organization or to promote their conservation, we must appreciate the dynamics of roosting patterns.

Daily Activity

The behavior of bats as they emerge from their roosts and begin their nightly activities marks a milestone in bat daily activity. Observing bats emerge provides a view of the behavior of the population, not necessarily of individuals. Some species often forage well before dark when they are clearly visible to human observers. In temperate regions, especially early in spring and late in fall, Big Brown Bats and Eastern Red Bats commonly forage in full daylight. Nancy has seen bats in the Grand Canyon foraging in swarms along with swallows in the late afternoon. Emerging early gives the bats access to insects that are mainly diurnal as well as to the nocturnal array of insects. Flight activity of insects is predictably high at temperatures above 10°C, and more likely to occur in late afternoon early and late in the spring and fall. But foraging in daylight can make bats more vulnerable to predators.

To the observer standing outside a bat roost, there often appear to be two peaks of activity, the emergence at dusk and the return at dawn. But when female bats are nursing their young and when young begin to fly, activity around a bat roost can be almost continuous through the night. Nursing females may visit their young several times over a night. When young are practicing flying, they may repeatedly emerge from and then return to a roost.

Brock Studies Bats and their Responses to Predators

In 1993, at the Letaba River high water bridge in Kruger National Park I observed birds of prey attacking Little Free-tailed Bats (*Chaerephon pumilus*) and Angolan Free-tailed Bats (*Mops condylurus*) as they emerged from crevices in the bridge. (See page 177.) Hundreds of these bats roosted in crevices under the bridge and a simple experiment demonstrated what triggered the birds' hunting behavior. The birds waited in trees at either end of the bridge or on the bridge railing. If I released a bat, its flight stimulated the waiting birds to take to the air and begin hunting. The first bats to emerge were never caught by the birds, but the bats that followed them were sometimes caught. When a bird successfully captured a bat, the time it took to handle and eat it gave other bats a free exit pass. Each night at the bridge, a few bats were caught by the birds, but the bats continued to begin their emergence about twenty minutes before dark.

Predators were less common around buildings whose attics were roosts used by smaller numbers of bats. Even there, bats were aware of the dangers posed by predators. When I placed an artificial predator—a bucket suspended below the entrance to the roost—I usually caught the first one or two bats that emerged. But then the rest of the bats in the roost delayed their emergence for about 100 minutes. Others just stayed home for the night.

These observations suggest that some bats consciously adjusted their behavior according to their perception of risk. When hundreds emerged at the same time, the risk to any one bat was diluted by the effect of the numbers of bats in the air. At smaller roosts, any risk of predation seemed unacceptable to the bats. The advantage to early emergence, however, was access to more insects (both diurnal and nocturnal). This may have meant not having to fly as far to feed, which often may be the force driving them to emerge early.

Swarming Behavior

At sites in the north temperate zones (North America, Eurasia) on nights in August and September, there are high levels of bat traffic in underground sites such as caves and mines that will later serve as hibernation sites. (Figure 8.3) This "swarming" behavior appears to play at least three roles in the lives of these bats. First, the high proportion in these swarms of young born earlier in the year suggests that swarming introduces young bats to hibernation sites, which are located in different places than the nursery roosts where they were born and raised. (See Chapter 7.) This explanation implies that young bats follow adults to swarming sites. Second, as swarming progresses through August and September, the bats begin to mate, so swarming appears to be, in part, prenuptial. Third, banding studies have revealed that the

Figure 8.2.

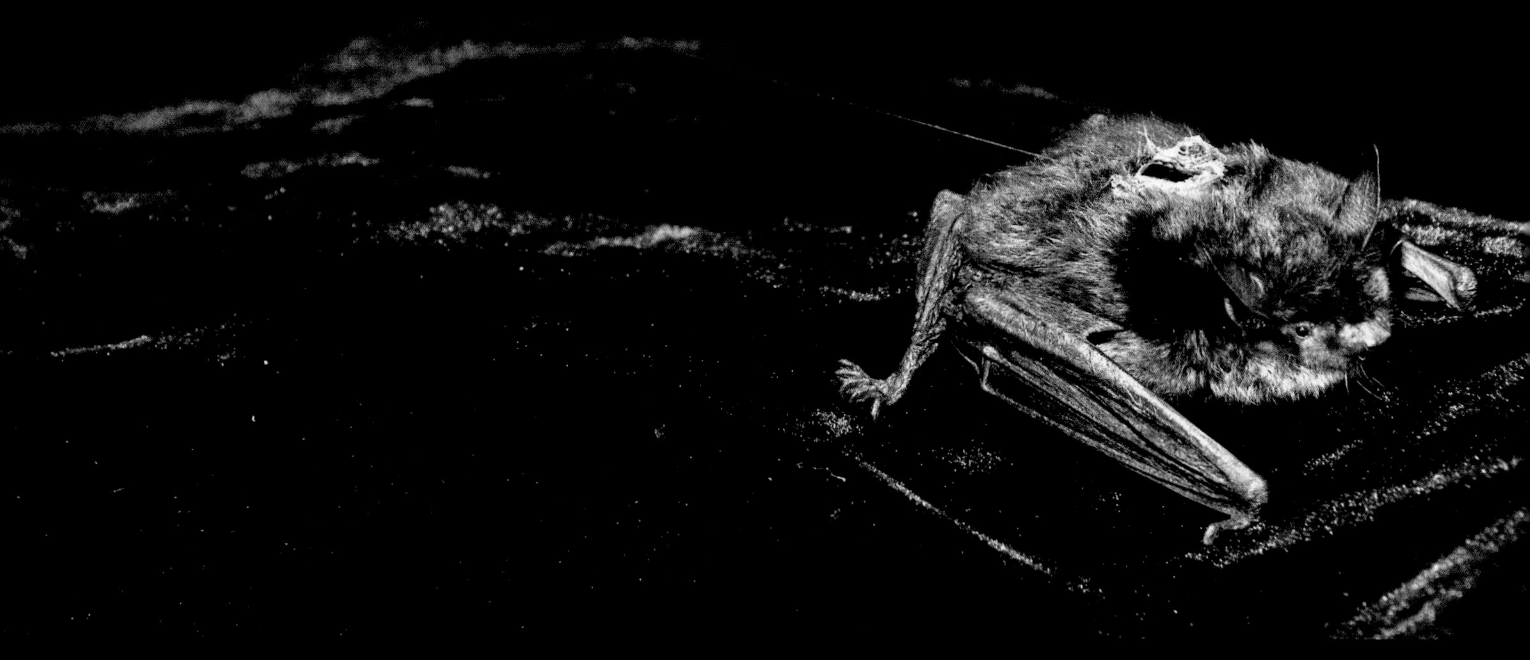

Figure 8.2.
A Little Brown Myotis with a radio-tag attached with surgical adhesive on its back. The bat weighted 8 grams, the transmitter 0.6 grams. At 0.6 g. the transmitter is heavier than the preferred target of less than 5 percent body mass. Note the bat also is banded on its left wing.

Figure 8.3.
Swarming bats in an abandoned mine near Renfrew, Ontario, on a night in August 2004. The bat at the bottom right is a Big Brown Bat, the others are Little Brown Myotis.

Figure 8.3.

Figure 8.4.

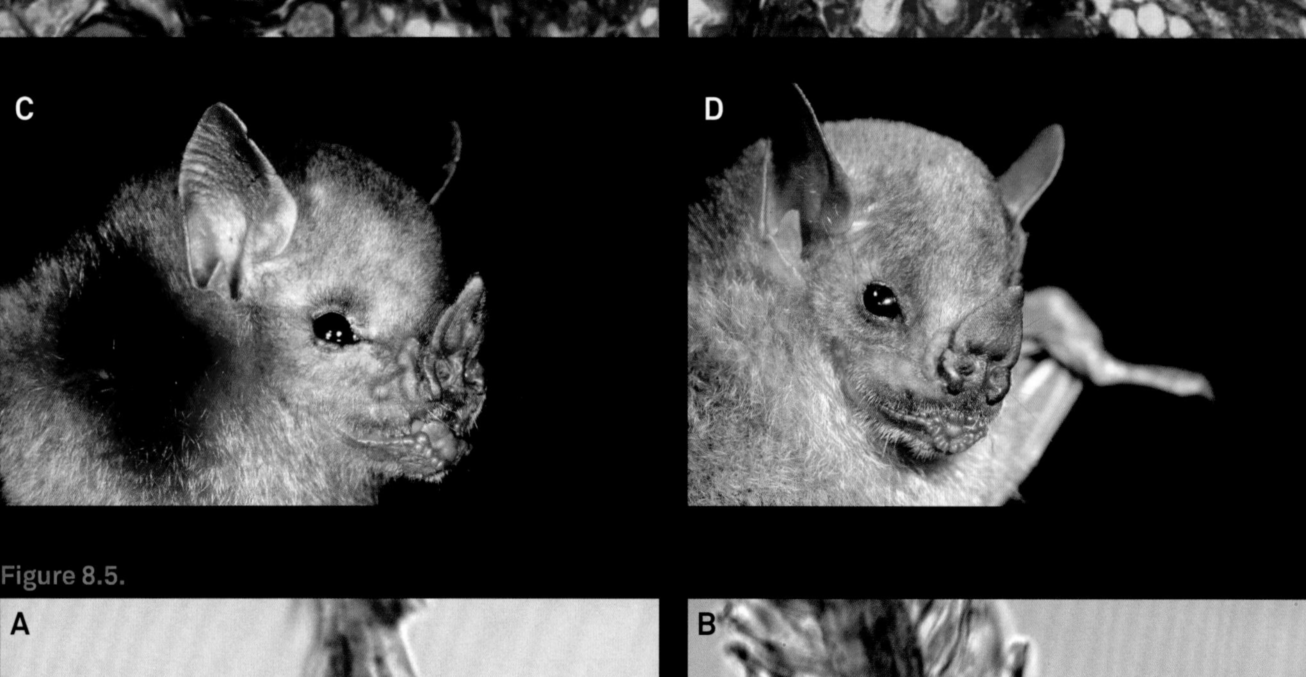

Figure 8.5.

Figure 8.4.
Histological sections of the neck and shoulder area of a male (**A** and **C**) and female (**B** and **D**) Little Yellow-shouldered bat. In the male, sebaceous glands (**se**) are associated with the hairs (**h**). Note that the sebaceous glands dominate the male's shoulder area where they coincide with conspicuous hairs (**C**) not seen on the female (**D**). Histological sections courtesy of Bill Scully.

Figure 8.5.
Osmotrichia, hairs specialized for dispersing scent, are larger than normal body hairs. In this comparison the body hair (**A**) of an adult male Little Collared Fruit Bat (*Myonycteris torquata*) is about 20 percent of the diameter of the hair from the neck ruff (**B**) which are used to disperse scent. The patterns of the scales on the hairs also are strikingly different.

bats swarming on any night are different from the ones that were there the night before or are there the night after. Swarming bats visiting a cave or mine may have come from tens to hundreds of kilometers away. The mixing of populations during swarming coincides with mating, and may ensure high levels of outbreeding among bats. Swarming appears a key factor in the spread of White Nose Syndrome, which since 2006 has killed literally millions of bats in eastern North America. (See page 200.)

As the season progresses, the number of swarming bats diminishes, coinciding with increases in the numbers of bats hibernating at underground sites. But there is some activity of bats through the winter. The best evidence of this comes from monitoring the echolocation calls of bats. These data show considerable bat activity even in the dead of winter in the Canadian prairies when temperatures at night are well below freezing. What are these bats doing out in the middle of winter? The mystery remains, but there are several possible explanations. Bats may be looking for mates, or moving in response to a change such as disturbance or a rock fall at the site they had been using for hibernation. In any event, their energy budgets for surviving winter may not be as limited as bat biologists had thought.

In August and September, Common Pipistrelles (*Pipistrellus pipistrellus*) visit some building roosts in Europe. This behavior, referred to as "Invasionen", (or invasion behavior) may prove to be a kind of swarming behavior since large numbers of bats may visit a site not used as a summer roost. This area of bat behavior surely deserves more attention. Bat biologists would like to know whether it involves mating, and if individuals visit more than one site.

Communication Signals of Bats

Like other mammals, bats communicate using some combination of sound, smell, touch and vision. The signals used by any species or individual depend upon the setting, *e.g.*, roosting, flying, foraging as well as the message, *e.g.*, greetings, alarm, mother/offspring recognition. For a human observer, recognizing and interpreting the communication signals of bats can be challenging. Much of what some bats say is beyond the range of human hearing, and what bats vocalize often occurs in poorly lit or dark roosts. (See page 184.) The small size of bats means that an observer must be very close to them to see the details of what they are doing, which is hard to arrange.

Olfactory Signals

Skin glands are common in bats and some are highly specialized. Glandular secretions are important in a variety of contexts, from mating displays to mother-young interactions. Hairs often are associated with glands, for example, shoulder glands of male Little Yellow-shouldered Bats. (Figure 8.4) Mature males in breeding condition have conspicuous shoulder glands that produce a spicy-smelling odor. The secretions stain the hairs of the shoulder region, giving these bats the epaulettes for which they are named. This is an example of sexually dimorphic glands (present in one gender but not in the other). Hairs specialized to retain and disperse glandular secretions are called "osmotrichia" and they are common in bats. For example, some Old World Fruit Bats have a ruff of such hairs that forms a conspicuous collar around the neck in males. (Figure 8.5; see also Figure 1.12A)

Bat-to-Bat Acoustic Signals

Just before dawn along the Verde River in Arizona piercing directive calls herald groups of Pallid Bats circling around the entrance to the roost they will use during the coming day. These calls make it easy for group members to find and then join their roost-mate. Spix's Disk-winged Bats show similar calling behavior as they choose an unfurled leaf as a roost for the coming day. While Pallid Bats choose among hollows and crevices in rock that may be used repeatedly, Spix's Disk-winged Bats are searching for roosts that are usually good for just one day. The calls they make allow every bat to find the group or individuals with which it wants to roost—not an easy task given the ephemeral nature of their roosts and the complex rainforest environments in which these bats live. Playback experiments have shown how the calls of Noctules attract other Noctules to roosts, typically in hollow trees. (Figure 8.6) These findings, coupled with those for Pallid Bats, make clear how bats may find roosts by listening to the vocalizations of other bats.

Brock Encounters a Conundrum

It turns out chiropterologists are not the only ones wondering how bats initially find and evaluate potential roosts. In Ottawa one morning early in October, a caller once asked me about bats in her unit on the fifth floor of a small apartment building. The bats, she said, had flown in through an open sliding door to the balcony and gone straight into a closet. I found six Big Brown Bats in the closet, all apparently in good condition. Bats had never been a problem in the building before, and none of the other residents had experience with them. I never did figure out what attracted the bats to that particular roost site.

Figure 8.6.
An inquisitive Noctule.

Figure 8.7
A European Free-tailed Bat.

Air Traffic Control

When two hunting Greater Bulldog Bats are on a collision course, one or both "honk", changing the design of their echolocation calls to alert one another to a possible collision. Jamming (the confusion of a bat's echolocation system by overlap of too many incoming signals) may be a problem when multiple bats are flying or foraging in the same area, yet somehow bats of many species frequently effectively navigate and hunt near other bats. Over the Negev Desert, when two or more European Free-tailed Bats (*Tadarida teniotis*) are flying together (within earshot of one another), the bats change the frequencies of their echolocation calls and minimize overlap among the calls. (Figure 8.7 and 8.8) Brazilian Free-tailed Bats show the same behavior, as do many other species of bats. While the Greater Bulldog Bats avoid one another, effectively showing a form of air traffic control, changes in the echolocation calls of the European Free-tailed Bats may instead reflect jamming avoidance. It seems likely that many bats may engage in both forms of behavior: listening for the calls of other bats and using this information to change their own behavior.

Other Bat-to-Bat Signals

Greater Sac-winged Bats (*Saccopteryx bilineata*) typically roost in small groups (five to fifteen) on the trunks of trees, particularly between the root buttresses of large rainforest trees. They also roost on the walls of buildings so they can be a familiar sight to those who live in the areas where they live. Groups of Greater Sac-winged Bats roosting together typically consist of an adult male and some adult females and their young. There may be several groups on the same tree trunk or building. Under these circumstances, it is sometimes possible to observe the display flights of males and their "salting behavior." Male Greater Sac-winged Bats,

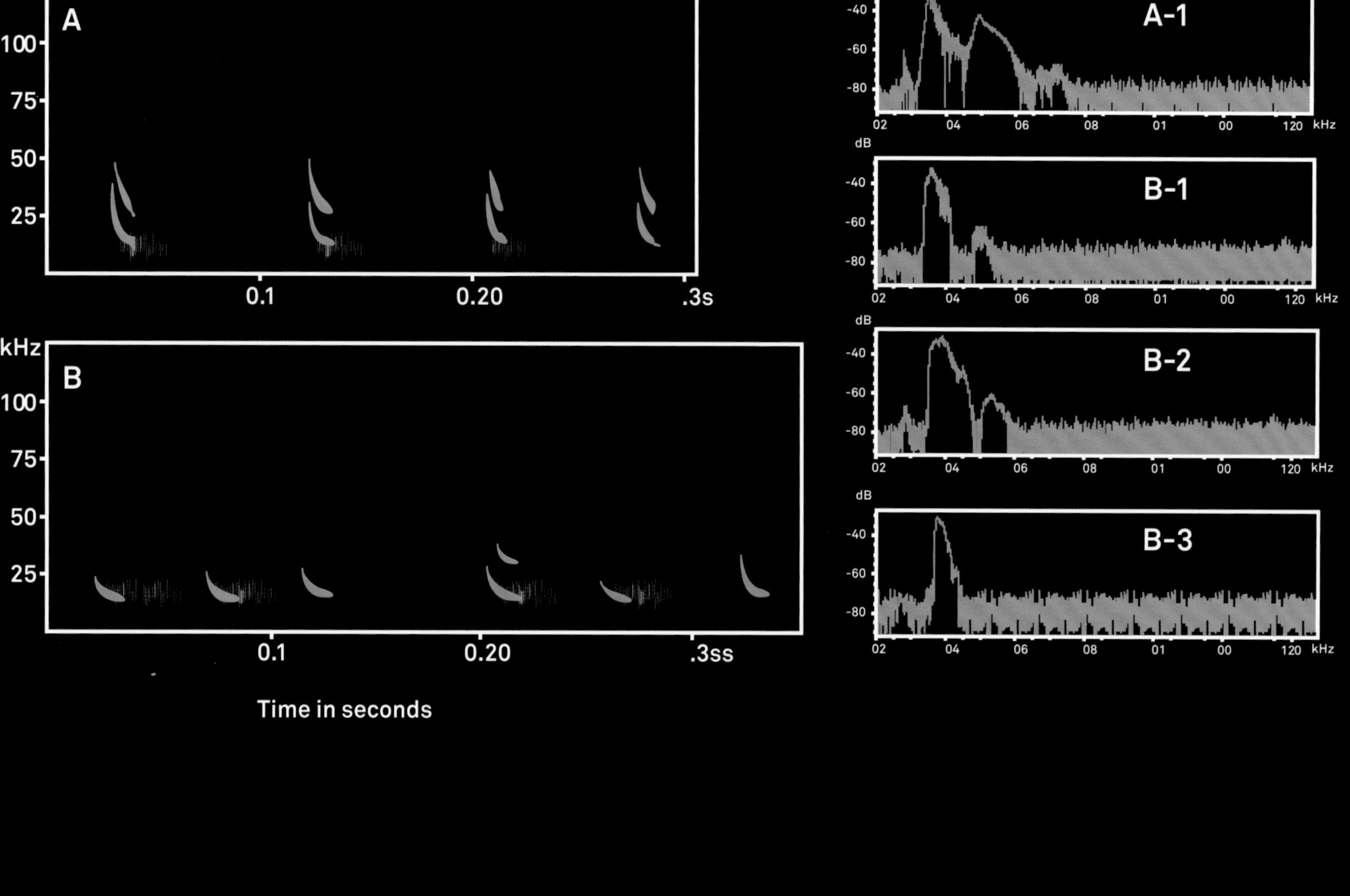

like some other Sheath-tailed Bats, have sacs in their wing membranes just in front of their upper arm or elbow. (Figure 8.9) Although these have sometimes been referred to as "glands", close examination reveals that they contain no glandular tissue. Rather, the sacs serve as fermentation chambers. The bats put body secretions such as saliva and urine into the sac, where the mixtures ferment to form a kind of perfume. Each male has his own distinctive scent, and he uses a behavior called salting (in reference to how humans use salt shakers) to deposit his perfume on roost surfaces and females. Salting involves extruding the sac and shaking the folded wing over the surface (or bat) to be marked. This is often done while the male is hovering.

Salting behavior and hovering flights are often accompanied by distinctive songs. Male Greater Sac-winged Bats advertise their presence and their virility using a combination of olfactory, acoustic and

Figure 8.8.
The echolocation calls produced by a single European Free-tailed Bat foraging alone (**A, A-1**) and in a threesome (**B**, **B-1**, **B-2** and **B-3**). Flying alone, the bat produced consistent calls, each the same as the others. When flying with others, each bat produced slightly different calls. The differences are more obvious in the power spectra (**A-1**, **B-1**, **B-2** and **B-3**). Time scales in **A** and **B** do not show exact intervals between calls.

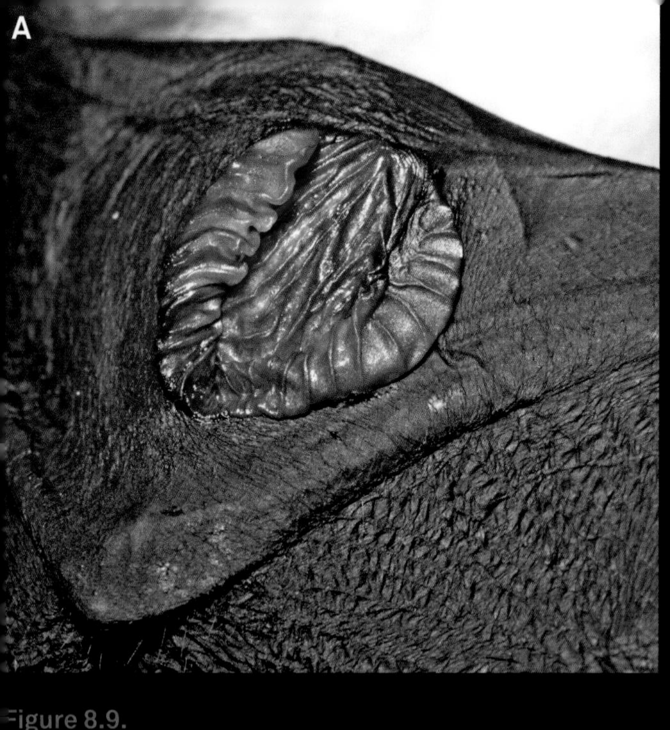

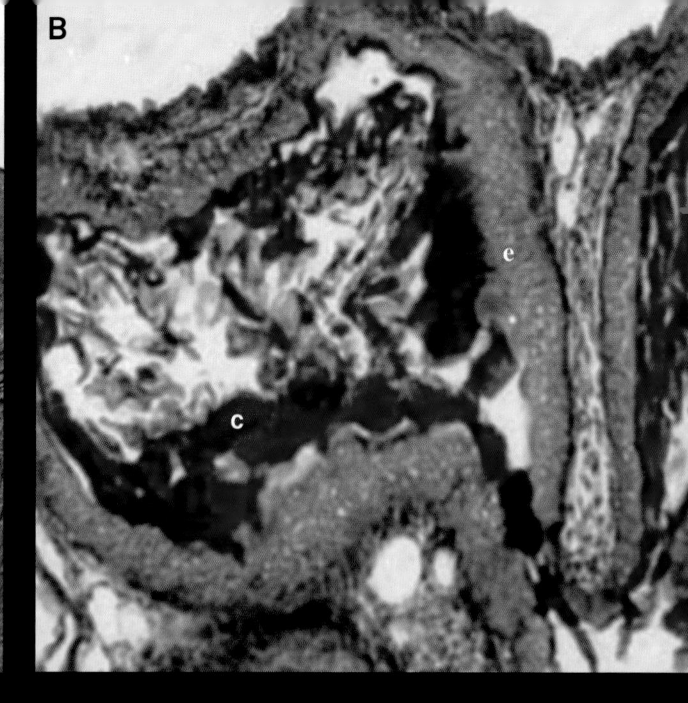

Figure 8.9.
The wing sac of an adult male Greater Sac-winged Bat (**A**) and a histological section of the sac (**B**). The lack of glandular tissue associated with the sac demonstrates that it is a holding sac rather than a gland. In the section, **c** = cornified layer and **e** is epidermis (skin). Histological section (**B**) courtesy of Bill Scully.

visual displays. Their displays function to attract females and repel other males. The display of a male to his harem also can affect the behavior of the young bats living within view or earshot, providing an introduction to the behaviors and vocalizations they will encounter later in life. A harem male is thought to sire most of the young born to the females roosting with him, so any nearby juveniles may be his offspring, but this assumption needs to be tested with genetic data.

Sometimes the displays of male bats are loud and conspicuous. Starting just after dark, male Hammer-headed Bats gather on tree branches along rivers. (See Figure 7.8.) Spaced about ten meters apart, the males flap their wings and produce a deep, resonating, monotonous call. The lines of displaying males attract females that fly up and down each line, apparently sampling the local talent. Eventually, one of the males, usually the one with the most central position, proves to be the common choice of females who visit and mate with him. This kind of group behavior is known as a "lek", during which males aggregate to engage in competitive displays, and females visit to take their pick. Females apparently identify the most fit or desirable males by the qualities of their displays, a mechanism thought to allow females to ensure that their offspring are likely to have a good genetic heritage. Although the males do not physically battle one another, they compete for mates through lek displays, and it is thought that only a few males (presumably those with good genes and good health) sire the majority of offspring in any given season. This, however, has yet to be tested with genetic data. Other African Old World Fruit Bats also have mating displays that involve calling and wing flapping, but usually the males are more spread out around the landscape.

Mother–Young Behavior

The large investment that mother bats make in pregnancy and giving birth, together with the cost of producing milk, suggests that mother bats should normally nurse only their own pups, or perhaps the young of their sisters. In the early 1960s some bat biologists proposed that mother Brazilian Free-tailed Bats behaved like a milk herd, providing milk to any young that approached them. This hypothesis seemed reasonable when observing cave roosts housing literally millions of young bats and millions of mothers. The challenge to a mother trying to find "her" pup when returning to the roost cave was presumed to be beyond the abilities of mother bats. Further research, however, revealed that mother Brazilian Free-tailed Bats recognize both the unique voices and the odors of their young. These cues, combined with her memory of where she left her young within the roost, made mothers very good at finding and recognizing their young. Within a cave used by several million bats, females remember where they have left their young, narrowing the challenge to recognizing their individual offspring among several hundreds of others. Studies of Brazilian Free-tailed Bats and Little Brown Myotis revealed that females vigorously and sometimes violently resist attempts by pups that are not theirs to "steal" their milk. Experiments in which mother bats had to choose between two pups invariably demonstrated that mother bats have a strong preference for their own young. Playback of recordings of the calls of pups and Q-tips coated with their odor made it possible for researchers to establish which cues were important. So far this research has not established the source of the odor perhaps it is produced by the young, perhaps from the milk it had drunk or from the mother's saliva transferred during grooming. Or maybe it is a combination of all three.

Considering the genetic consequences of having young survive and prosper, it is easy to understand why evolution may favor behaviors that facilitate mother-pup recognition and limit nursing behavior so mothers feed only their own pups. The close genetic relationship among sisters provides an explanation of why females might occasionally nurse their nieces and nephews.

Some female bats that roost in colonies, for example Egyptian Slit-faced Bats, actually move their young away from the group as they leave to forage. This simplifies the problem of recognizing her young (she only has to remember where she left it). Moving offspring to a different location also may make pups less vulnerable to predators if a group of young bats is more conspicuous than a single pup.

Female bats that roost alone do not face the same problems of recognizing their young as females that are part of a colony. An Eastern Red Bat or Hoary Bat, for example, need only remember the roost site where she had left her young because she does not have to sort among other pups to find her own. To the surprise of many bat biologists, young Hoary Bats call often when left alone by their mother. Their calls are quite audible to human observers and you would have expected them to be quiet until the mother returned.

Bat biologists studying Pallid Bats have reported what appear to be "baby sitters", adult females remaining with young left in a nursery roost when their mothers are out foraging. This behavior is particularly intriguing because it implies a level of social organization not necessarily expected from bats. Baby sitting also has been reported in Indian False Vampire Bats at colonies in India.

There are some reports of infanticide among bats. Male Indian False Vampire Bats have been reported to cannibalize young that were not theirs. Male White-throated Round-eared Bats left in roosts with young sometimes attacked and bit them. Sometimes the attacks were not intense, but other attacks resulted in the young falling out of the roosts. These attacks did not occur when adult females (mothers) were in the roost. Some caves harbor several discrete groups of Greater Spear-nosed Bats and there are reports of adults attacking and killing young from other social groups.

Territorial Bats

In East Africa, Heart-nosed Bats (*Cardioderma cor*) usually roost in social groups in hollows of Baobab (*Adansonia digitata*) and other trees. (See Figure 6.3.) The groups include adult males and females as well as young. When foraging, adults disperse from the roost hollow and move out to the surrounding savannah. Adult males produce distinctive vocalizations ("songs") as they move around their foraging areas. The reason for this is not clear; more details of the songs, and the movements of other individuals (adult males and females), must be gathered before scientists can fully understand this behavior. Males could be advertising and defending territories in which "their" females could forage, but this remains an untested hypothesis.

Greater Spear-nosed Bats are widely distributed in South and Central America. These bats often form social units within cave roosts or large hollow trees, typically one adult male with a group of females and their dependent young (fathered by the resident male). Away from the roost, groups produce distinctive vocalizations that appear to serve in maintaining contact with other group members. These calls are lower frequency than typical echolocation calls and are within the range of human hearing. To date there is no evidence of these group calls serving a territorial function.

The interactions of male Greater Sac-winged Bats involve vocalizations and scent marks that serve to maintain the integrity of their roosting territories. It is less clear if these bats also are territorial in their foraging areas. Lesser Sac-winged Bats (*Saccopteryx leptura*) have been said to be territorial on feeding grounds, but, again, more research will be necessary to determine whether this is the case.

Many animals are territorial when defending some resource, such as access to mates or food. At this time researchers have no clear evidence that population sizes of bats are limited by their access to food, so the lack of territoriality on feeding grounds may be expected. The question about whether bats compete for food (or roosts) is important when trying to understand their evolution and diversification. Does the appearance of several species in a genus (or family) reflect somewhat different approaches to feeding or roosting? The advent of DNA barcoding for identifying the species of prey taken by bats may shed some light on these questions. (See page 110.) For example, researchers could test the hypothesis that two anatomically similar sister species living in the same place must be feeding on different insect species in order to coexist.

Nancy Hears the Bats Coming

Although most of the night sounds in the tropics are made by insects, sometimes it is possible to actually hear bats "talking" to one another. This surprised me the first time I heard it, because I had always thought that most or all bat calls are ultrasonic. Not true! In French Guiana, I often heard flocks of Greater Spear-nosed bats from quite a long distance away through the forest understory. The bats chattered to each other as they flew, apparently keeping track of one another by voice. Once I set nets in a banana plantation, hoping to catch bats attracted to the banana fruit and flowers. I wasn't sure if any bats would come until suddenly I heard a flock of Greater Spear-nosed Bats approaching through the nearby forest. They got louder and louder (to my ears) as they approached, and then suddenly there they were. Before I reached the nets I knew what species I had captured because of their distinctive—and highly audible (to humans)—communication calls. Interestingly, they ceased calling as soon as they became entangled in the net. This behavior probably affords protection from predators that could otherwise find them simply by listening if these bats called when not on the wing.

Male Contributions.... Or Lack Thereof

The difference between the investment that male and female bats commit to reproduction makes bats "typical" mammals. The principle male investment consists only of sperm. Typically males do not contribute to parental care, and in fact do not usually interact with females or young at all after mating. Females must devote the energy to grow large embryos, give birth and feed their offspring with milk until they grow to nearly adult size. This disparity leads to the prediction that during the mating season, bats will form polygynous harem groups consisting of an adult male and several adult females. The genetic payoff (fitness) of being a successful harem male—passing genes into the next generation—means that there will be competition among males for the position of harem male. Only the strongest and healthiest males can win harems, hence females maximize their chances of having their offspring fathered by a male with superior genes. The lek system seen in Hammer-headed Bats is a variation on this theme. The genetic advantage to females is breeding with the "best" male(s).

When mating is followed directly by fertilization and development to term, Seba's Short-tailed Bats live in harem groups. (Figure 8.10) Here, males advertise and attract females to a suitable roost while deterring other males. In captivity, female Seba's Short-tailed Bats often move among harems. When females have a post-partum oestrus and delayed development, the social system may change. For a male to attract females that are already pregnant by another male appears to offer no genetic benefit. But this circumstance could allow females to consort with and become pregnant by one male, and then move to a more attractive roost location controlled by a different male. In other words, a postpartum oestrus and delayed development could allow females to separate the business of mate choice from the business of finding the best living conditions in which to raise offspring. Fortunately techniques of molecular genetics (DNA fingerprinting) offer biologists tools for investigating this strategy.

For many species of bats, the social composition of a large group of bats may be fluid over time. It seems likely that extended social groups will be the rule rather than the exception for bats. (See page 185.) In other animal groups, particularly primates, including humans, this kind of social organization is known as "fission-fusion" (referring the splitting and merging of subgroups over time). This type of social structure and has been observed in several species of bats, *e.g.*, Common Vampire Bats, Big Brown Bats and White-winged Free-tailed Bats. (See page 180.) Other bat species appear to be effectively monogamous, showing long-term pair bonds between individual males and individual females. Spectral Bats are one example, but there are very few details about their social behavior. Observations of bats roosting together, particularly when supported by genetic evidence, will be necessary to confirm the occurrence of monogamy.

Figure 8.10.
A Seba's Short-tailed Bat.

Bats Demonstrate Learning to Brock

In the 1980s for several months several students and I had some captive California Leaf-nosed Bats in the lab at York University where we were studying their behavior. (Figure 8.11 shows a closely related species) We usually fed the bats mealworms (beetle larvae) or crickets, and the bats quickly learned to glean their snacks from dishes that we put out for them. It was common at midday for the lab group to be sitting around having lunch while two or three bats flew around the room. The bats typically finished their food first and then began checking out our dishes (and cups) for more food. This setting made for some interesting lunches.

One of our captive California Leaf-nosed Bats also "learned" about doors. When it was finished eating, it would fly immediately behind anyone who moved towards the door and then followed the person out of the room. Not all of the people on the floor of the building appreciated having bats in the hallway, so we had to be alert to keep this bat from taking off down the hall. These anecdotes suggest that bats are capable of learning. Perhaps biologists should not be surprised by this, given that bats can live a long time and survive a range of circumstances and conditions.

In 1980 Carleton University biologist Connie Gaudet and I found that inducing bats to feed in captivity meant giving them food and then making chewing noises that seemed to encourage them to eat. Working with Big Brown Bats, Little Brown Myotis and Pallid Bats, we demonstrated that these bats learned from one another. These experiments involved Connie training a bat to fly across a room and land on a target where there was a piece of a meal worm. It did not take Little Brown Myotis, Big Brown Bats or Pallid Bats long to learn this task. What was more interesting was that we found that when we put a trained bat in the room with a naive one, the

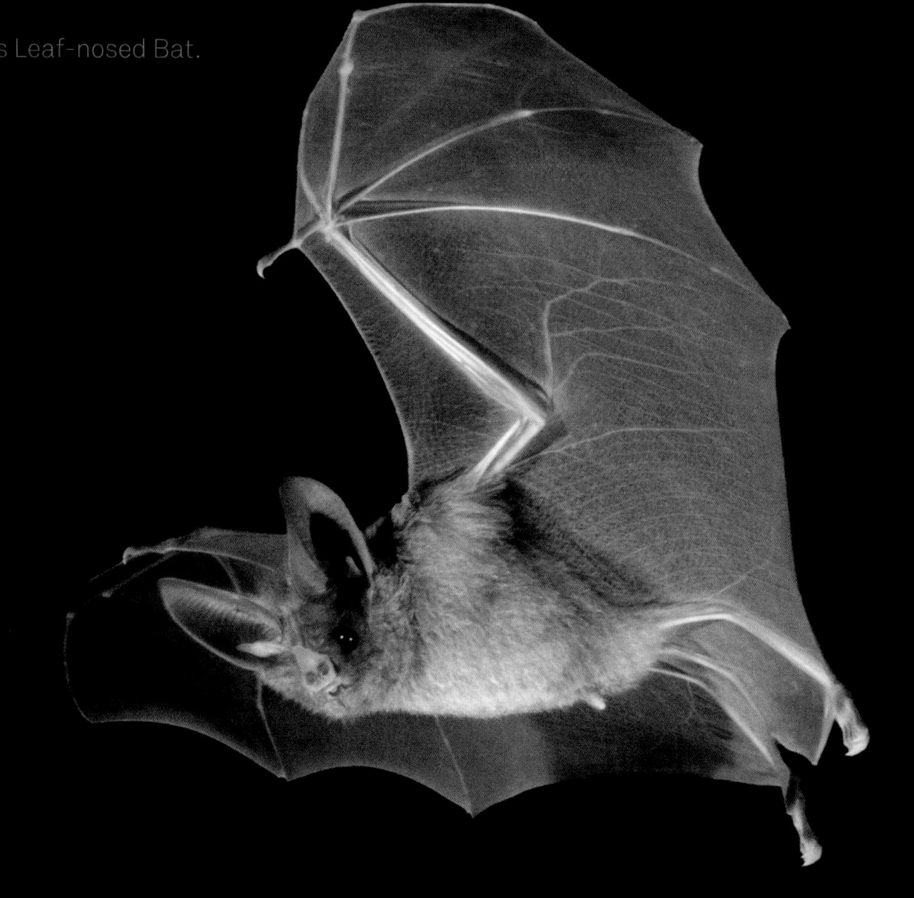

Figure 8.11.
A Waterhouse's Leaf-nosed Bat.

naive one quickly learned the task. Furthermore, all three species would learn from one another. But, left alone, Big Brown Bats, Little Brown Myotis and Pallid Bats never learned the task.

Working in Zimbabwe in 1982, Connie and I used chewing sounds to prompt captive Large Slit-faced Bats to take unfamiliar food. We had six of these bats captive, and one female was clearly dominant over the others. At feeding time, we gave all of the bats their food (fish or frogs) at the same time because when the dominant one was not chewing, she would take food from others that were chewing. It seems likely that chewing sounds provide evidence that bats use to determine that something edible is nearby, leading them to search for it (or attempt to steal it from another bat!)

Since then biologists Rachel Page, Elisabeth Kalko and their colleagues at the Smithsonian Tropical Research Institute in Panama reported how Fringe-lipped Bats located, caught and ate singing male Tungara Frogs (*Physalaemus pustulosus*). The same bats avoided poisonous frogs and toads. Further research demonstrated Fringe-lipped Bats could be trained to respond to almost any acoustic signal associated with food, including recordings of Bob Marley songs.

At some sites in the United States and Europe, recovery of banded individuals revealed long distance migrations between roosts used in summer and winter. (See Figure 7.1.) In the Southwestern United States and adjacent Mexico, two species of nectar-feeding bats—Lesser Long-nosed Bats (See Figure 5.8.) and Mexican Long-tongued Bats—regularly migrate north in spring and south in autumn. The fall migrations are more conspicuous because literally hundreds of both species visit hummingbird feeders in Arizona and New Mexico, presumably to refuel on their trips south. Bat biologists believe that some species of bats migrate at least hundreds of kilometers between summer and winter areas. For the most part, however, recovery of tagged bats usually involves a few hundred kilometers, probably not indicating the full range of the bats' movements.

What cues do bats use in navigation? In the 1960s at sites in Trinidad, pioneering researcher Donald Griffin and two students outfitted Large Spear-nosed Bats with radio transmitters, some with goggles and others with blindfolds. The researchers then transported the bats different distances from their home cave and released them. The bats with goggles could see their surroundings, but the blindfolded individuals could not. Bats that could see returned to their home caves more quickly and directly than blindfolded bats. But even the blind-folded ones eventually made it home, demonstrating that these bats could use a combination of visual and auditory (including echolocation) cues. The experiment did not take olfactory (odor) cues into consideration.

Radio transmitters have been instrumental in opening biologists' eyes to some aspects of the lives of bats. (See Figure 8.2.) Most recently, transmitters including tiny GPS tracking devices have greatly extended our knowledge and appreciation of bats and the distances they can cover. Bat biologists also have made more use of transponders, enty Egyptian Rousettes—each weighing on average 130 grams—with 11 gram tracking devices and individually tagged with transponders. (Figure 8.12) This experiment allowed researchers to determine how bats navigated long distances in the wild.

They transported the bats 94 km. south from their home roosts and released them at a deep natural crater in the Negev desert (Mitzpe Roman). Tracking the bats revealed that these bats used visual landmarks to return to their home roosts and foraging areas. Bats released on the lip of the crater homed directly, while those released within the crater took time to determine the direction of their home. This research clearly demonstrates that Egyptian Rousettes have a large-scale navigational map of visual landmark features.

able there. The departure times of bats from these sites are equally synchronized and conspicuous. How do the bats know when and where to go? That is one of the mysteries not yet solved. Another mystery is where they go next.

In Southeast Asia, an international team of scientists led by Craig Smith of the Queensland Biosecurity Sciences Laboratory equipped Malayan Flying Foxes (*Pteropus vampyrus*) with satellite tags. These bats, which typically weigh about 700 grams, readily carried 20 gram battery-operated or 12 gram solar-powered transmitters. The tagged bats were tracked for hundreds of kilometers between roosts in Malaysia, Indonesia and Thailand, sometimes crossing large bodies of water such as the Straits of Malacca. This work extends our knowledge of the details of the movements of bat across international borders and seaways, and also highlights concerns for the conservation of these animals. With Flying Foxes crossing international borders, conservation efforts must extend to more than one country to be effective.

Other studies have depended upon innovative techniques to document migrations by bats. Erin Fraser and a group of colleagues from the University of Western Ontario and Royal Ontario Museum analyzed stable hydrogen isotopes to demonstrate latitudinal migrations by Tricolored Bats. The fur of mammals contains the levels of hydrogen isotopes prevailing in the environment (food, water, *etc.*) where the animal was living when the fur was grown. By comparing the levels of hydrogen isotopes found in the fur of Hoary Bats with the background levels of these isotopes where the bat was captured confirmed the supposition that these bats moved north in spring, and south in the summer. The same may be true of other species such as Tricolored Bats, although the pattern there is less clear. (See Figure 8.13.)

Body size is a principal hurdle facing bat biologists who investigate bat migration. Satellite tags have revealed details about the astonishing migrations of some birds. The same technology has been used to advantage with other large animals, from Great White Sharks to Blue Whales, African Elephants to Caribou. But these tags are not small enough yet to be used with most bats (recalling that most bat species weigh <60 grams). As a result there is a paucity of information about the movement of bats between summer and winter roosts.

Scientists are familiar with data from banded birds and even marked Monarch Butterflies (*Danaus plexippus*) flying from sites in North America to wintering locations in the tropics. The same data are not available for North American bats (Hoary Bats, Eastern Red Bats, Silverhaired Bats) thought to migrate long distances between summer and winter ranges. Some recoveries of banded individuals have demonstrated that bats of these species disappear from their summer range in winter and return the following summer. But there are virtually no records of any banded individuals traveling from sites in eastern North America to locations further south. We do not know where these bats go when they disappear. Isotopic data for Tricolor Bats and Hoary Bats suggest that these bats made long distance migrations.

Meanwhile recoveries of banded individuals have repeatedly demonstrated how Little Brown Myotis, Grey Bats and Cave Myotis (for example) regularly move between known summer and winter roosts. There are many comparable records of banded bats from Europe, including a 2013 record of a 7.6 gram Nathus's Pipistrelle (*Pipistrellus nathusi*) that was banded in Somerset, England, and recovered 870 kilometers distant in Pietersbierum, The Netherlands. Clearly even tiny bats can travel very large distances!

Figure 8.13.
A Tricolor Bat roosting in foliage.

9

Bats and Diseases

Figure 9.2.
Three Little Brown Myotis, one (middle) showing characteristic signs of infection by the fungus *Pseudogymnoascus destructans* (formerly *Geomyces destructans*) that causes White Nose Syndrome (WNS) in bats that hibernate underground in the United States and Canada. Photograph by Lesley Hale.

victim is a bat, a dog, a raccoon or a human. It remains to be seen if bats suffer from other diseases not currently known or recognized, but the more that researchers survey bats for unknown viruses, the more they find. Yet few bats ever appear sick when captured. One thought-provoking proposal is that genetic changes associated with the evolution of flight have changed bats' capacity for repairing damage to DNA. Associated with this may have been changes in the innate immune systems of bats, affording them protection against some diseases. These changes may also partially explain why bats live remarkable longer lives than most other mammals of comparable body size. (See page 154.) Currently this is an untested theory, but one of great interest to researchers investigating the nature of disease and the aging process.

For the most part, little is known about the diseases of bats, except those to which humans are susceptible. (See below.) White Nose Syndrome is a dramatic exception. (Figure 9.2)

White Nose Syndrome

White Nose Syndrome (WNS) affects bats hibernating underground and its impact has put some species of North American bats at risk of extinction. Beginning in January 2007, unusual winter flight activity and die-offs of bats in a few caves near Albany, New York, attracted the attention of bat biologists and conservationists. Thousands of dead bats were found, and many survivors had white noses—indications of fungal growth on their faces. Cave surveys indicated that only bats in a 15 km^2 area were affected, and examination of photographs taken in preceding years showed that white-nosed bats first appeared in local caves in March 2006. Biologists hoped initially that this was a singular event that would not spread beyond the affected area, but this was not the case. Subsequent surveys found WNS spreading on a yearly basis, with more and more bats affected every year. The fungus has subsequently spread north, south, east and west and has killed literally millions of bats in the eastern parts of the United States and Canada. (Figure 9.3) In 2013, the threat that WNS posed to bat populations was directly responsible for three bat species–Little Brown Myotis, Northern Long-eared Myotis and Tricolored Bats–being listed as Endangered in Canada. It was heartbreaking for Nancy to see bags of dead bats brought into museums for archiving—whole populations essentially wiped out in a single winter.

Bat and fungus experts quickly mobilized to study both the disease and its effects. By 2012 it was clear that a European strain of a soil fungus (*Pseudogymnoascus destructans*) was the causative agent of WNS. This fungus was presumably introduced inadvertently to North America, perhaps by bat biologists, cavers or tourists who visited caves in Europe and unknowingly transported fungal spores to New York State on their clothing or equipment. Many North American bats have succumbed to the European strain of WNS, which apparently does not kill European species of bats. It is not known why European bat species are less affected by the WNS fungus, but it may be because they coevolved with it over tens of thousands of years there.

The WNS fungus apparently kills bats by interrupting their rhythm of hibernation. (See page 154.) Rather than arousing from their winter sleep (torpor) perhaps once a month, infected bats arouse much more often and exhaust their stores of body fat by January or February, instead of March or April as is normal. Energy-depleted bats emerge from their hibernacula in the dead of winter, apparently looking for insect prey when none are available. This hypothesis is corroborated by direct observations of bats being seen flying during the day in January and February in areas where WNS has been reported. The starving bats don't last long, soon dying due to cold exposure and lack of food. Many individuals never even make it out of their caves, and those that do usually do not travel far, judging from the numbers of dead bats around the entrances to hibernation sites in northern New York State.

The fungus that causes WNS is a cold-loving fungus (psychrophilic), adapted for cooler environments such as those prevailing in the underground habitats (caves and abandoned mines) where bats hibernate. (See page 154.) The fungus only grows on bats when they are in hibernation and have depressed body temperatures due to torpor. If a bat survives the winter, its higher body temperatures while active during the spring and summer causes the fungus to die back—but some fungal hyphae remain in the bat's body. In the later summer and fall when bats are mating during swarming behavior in warmer weather, bats visit many potential hibernation sites and bat-to-bat contact, perhaps associated with mating, can spread the fungus between infected and uninfected individuals. (See page 186.) The rapid spread of WNS in the United States and Canada demonstrates the extent to which bats move around during the swarming season.

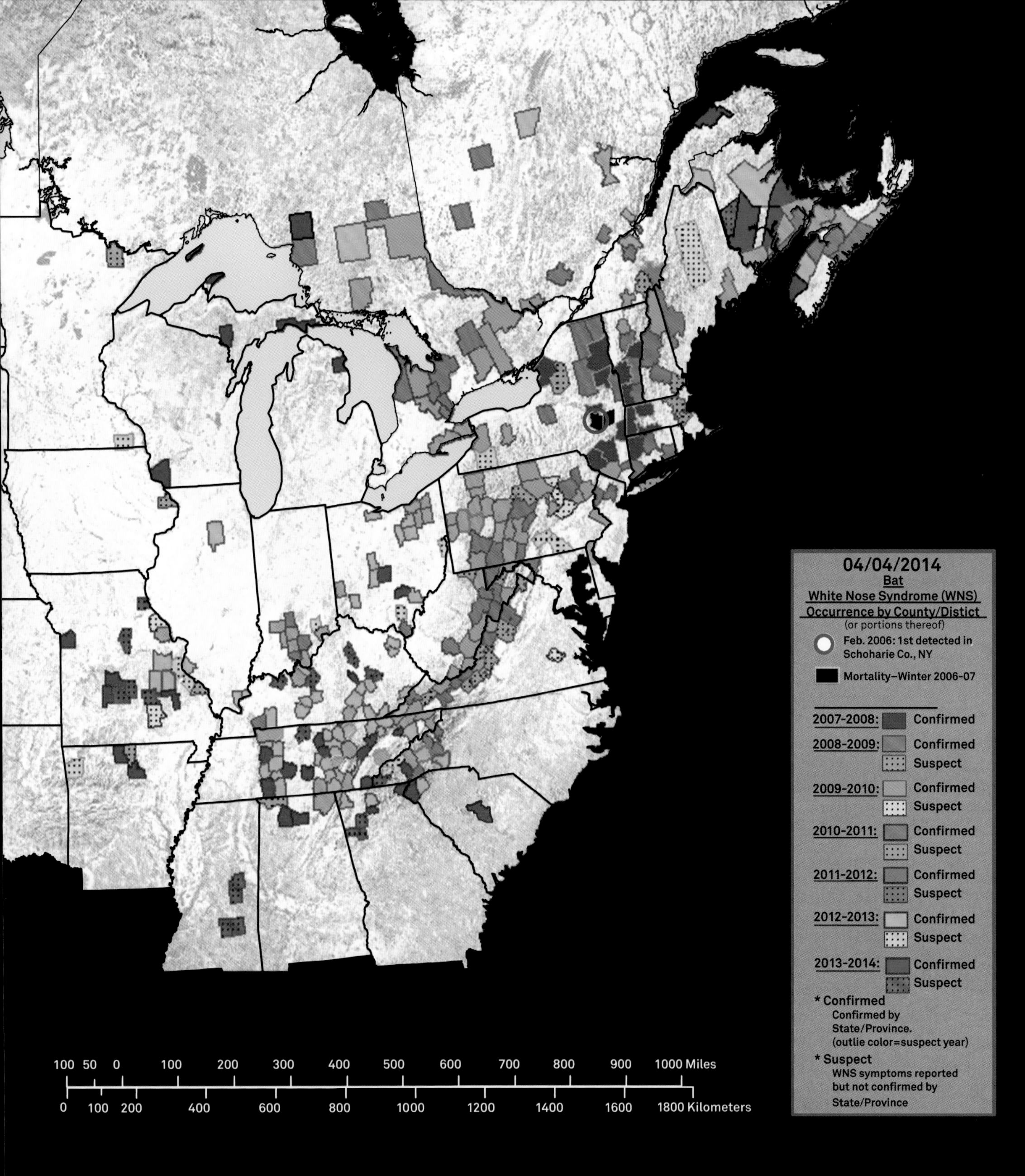

Figure 9.3.
The distribution of WNS in eastern North America (Canada and the United States) as of 4 April 2014. Note the progressive spread of the disease in all directions from its epicenter in New York State. Courtesy of Lindsey Heffernan, PA Game Commission.

Transfer of Disease-causing Organisms from Bats to People

Contagious diseases that can be transmitted from animals to humans are called zoonotic diseases. While there is no indication that the fungus that causes WNS in bats poses any threat to people (who do not hibernate or go into torpor), this is not true for some other zoonotic diseases associated with bats. Along with rabies, histoplasmosis is the disease most often associated with bats. These are quite different afflictions, but both can and do affect humans. Neither is restricted to nor characteristic of bats, but both profoundly affect people's perceptions of bats.

Histoplasmosis is a fungal disease of the lungs that can cause flu-like symptoms in humans. In some cases, infection from the fungus can be debilitating or even fatal. The spores of histoplasmosis are usually found in the feces of birds and bats. People become infected when they are exposed to fungal spores, usually while visiting caves. Between 1957 and 1966, one estimate is that there were five million cases of histoplasmosis in people in the United States. Contact with droppings containing the spores of the fungus (*Histoplasma capsulatum*) can occur in settings where people are working, for example cleaning up an area with feces, or during exploration or inspection of places used by birds (usually chickens or pigeons) or bats. (Figure 9.4) Spread of fungal spores does not necessarily involve direct animal-to-people contact.

Rabies is a neurological disease caused by a viral infection. (Figure 9.5) Typically, rabies virus is spread by bites because the virus concentrates in the saliva of an infected mammal. Once an individual shows clinical symptoms of rabies, the disease is almost always fatal.

While rabies is a relatively uncommon disease in the developed world, in third world countries the World Health Organization reports anywhere from 55,000 to 80,000 human deaths annually from rabies, 95 percent of them in Asia and Africa. World-wide, dog bites account for most cases of rabies in humans, but people can be exposed to the virus by bites from any mammal including raccoons, foxes, skunks and bats. Rabies is spread most often through the saliva of an infected animal making direct contact with an open wound. Although rare, handling dead animals infected with the virus can also result in transmission, but only if the person has an open wound in their skin or their mucous membranes that come into direct contact with saliva or blood containing live rabies virus.

Risks associated with exposure of people to rabies and histoplasmosis, and exposure of bats to WNS, have influenced the behavior of many people who study bats. Exposure to histoplasmosis can be avoided by wearing a mask that filters out spores of the fungus (anything larger than 10 microns). Biologists can obtain pre-exposure vaccinations to protect them from rabies. Bat biologists working in areas where WNS is a threat have adopted sterilization procedures for clothing and equipment to minimize the risk of transferring *Pseudogymnoascus destructans* from one site to another. (Figure 9.6) This includes treating shirts, pants, gloves, boots, mist nets, holding bags and calipers for measuring bats—essentially anything and everything that could spread the fungus.

Figure 9.4.
A church (**A**) in Willsboro, New York, showing evidence of a large colony of Little Brown Myotis, including formations hanging from the attic known as "pissicles" (**B**), a several year accumulation of guano (**C**). Spores of *Histoplasma capsulatum* may occur in the guano.

Figure 9.5.
A much magnified view of the rabies virus. From Shutterstock.

Figure 9.6.
The arrival of WNS has changed the way bat biologists and others operate. To minimize the chances of contaminating one site from another, a film crew from the Canadian Broadcasting Corporation (CBC) wear Tyvek suits (**A**). People handling bats wear latex gloves (**B**) that are changed between individual bats to minimize cross contamination.

Figure 9.6.

A

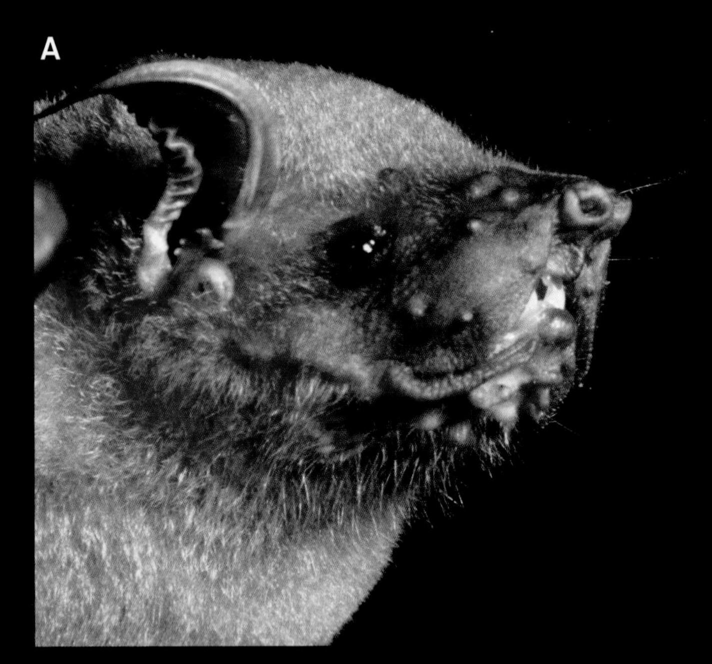

B

Figure 9.7.
A bat's facial expression can change depending upon how much tooth its showing. This Greater Bulldog Bat can look calm (**A**) but appears much more threatening with its mouth open and teeth showing (**B**). This bat weighed about 60 g. and was at least a "one glove" bat.

Brock Is Bitten

Bats have teeth and, like most mammals, will bite when threatened or frightened. (Figure 9.7) Risk of infection from rabies or another virus is perhaps the main reason why I try to avoid being bitten by bats, but there is also the issue of the bites themselves. Do bat bites hurt? The answer depends upon the size and behavior of the bat that does the biting, and where it bites! Recall that the majority of the >1300 living bat species are fairly small mammals, weighing less than 100 g. Even the very largest species, Indian Flying Foxes, weigh just 1500 g. As would be expected, the bite from a large bat is usually more painful than one from a small bat. In addition to body size, some bats (individuals and species) bite with much more enthusiasm than others. Almost every 5 g. Schlieffen's Bat (*Nycticeinops schlieffeni*) I have handled has been very annoyed about being captured and bites effectively and with vigor. Each time I remove one from a mist net I am glad it does not weigh 10 g., let alone a 50 g.! But even when a 5 g. bat bites you along the edge of your finger, the bite can be painful and draw blood.

Some bats change their disposition according to the situation. I have found that White-bellied Yellow Bats become frantic when tangled in a mist net and bite forcefully. These 15 g. bats nearly always manage to draw blood. But after you disentangle them from the net, most just sit in your hand and "smile."

Based on one instance, I have found that the defensive bite of a vampire bat does not hurt. (Figure 9.8) These 30 g. bats are fast and agile and their teeth are razor-sharp. They bite quickly and the wound bleeds profusely, but there is little or no pain.

Nancy Is Bitten

Once when I was handling a Fringe-lipped Bat, its upper canine tooth punched through my thumbnail. Needless to say, this was very painful! The bat continued to chomp away, puncturing the end of my thumb repeatedly with its lower teeth. Both the bat and I struggled mightily to disentangle from one another, which seemed to take forever. I felt badly about the bite (and the stress to the Fringe-lipped Bat I was handling) because the entire episode was my fault—I had not been paying sufficient attention while removing the bat from a cloth holding bag. It was a lesson that I have not forgotten, and I now pay much closer attention when handling bats, no matter what their size.

BOX 9.1

Wearing Gloves

Some bat biologists always wear gloves when handling bats, others never do. Nancy and Brock always carry at least a single leather glove when catching bats, and the decision about whether or not to wear it is depends on the species of bat we are after. When it's a "glove bat" (*i.e.*, one that we really do not want to be bitten by due to its size or attitude), we grab the head with the gloved hand and use our bare hand to remove it from a mist net. Small bats (<10 g.) are fragile and easily damaged if you are not careful when handling them, so we rarely wear gloves when untangling them. Larger bats including omnivorous and carnivorous species in the New World, and Flying Foxes in the Old World, are a different matter. These species are capable biting through some leather gloves so it pays to be especially cautious when removing them from mist nets. Biologists handling bats suspected of harboring infectious viruses—such as Common Vampire Bats or Flying Foxes—routinely wear latex gloves under their leather gloves for safety. (Box 7.1, Figure 1)

When 30 g. Large-eared Free-tailed Bats are sufficiently annoyed their bites can be quite painful. The slash from the razor-sharp upper canine tooth of a vampire bat does not involve any pinching of the flesh and, while it results in much bleeding, it is relatively painless. (Figure 9.9) The canines of Large-eared Free-tailed Bats deliver a pointed pinch that does not result in much bleeding but plenty of pain.

Bat Bites and Rabies

The best way to avoid being exposed to rabies virus is not be bitten by an infected animal. This is why public health officials always stress that lay people should not handle wildlife, from stray dogs to bats to lions. Unfortunately, the issue of preventing rabies infections is somewhat more complicated than this for at least two reasons. First, in the 1950s it appeared that breathing the air in a cave housing tens of thousands of Brazilian Free-tailed Bats could expose people to rabies virus. Although subsequent studies have demonstrated that airborne transmission of the virus most likely did not occur at that time, the possibility remains if only in some people's minds. Second, there have been at least three incidents in which donated organs contaminated with rabies virus have been transplanted to human recipients, rabies and all. In each case there was a failure to recognize that the organ donor had died of rabies, and regrettably each incident resulted in the deaths of the organ recipients. Otherwise transmission of rabies from one human to another has never been documented.

Another misapprehension concerning bats and rabies is the widely held belief that bats are asymptomatic carriers of the disease. In other words, bats have been said to harbor infectious rabies virus but otherwise experience no symptoms of the disease. Evidence supporting the idea that bats are asymptomatic "carriers of rabies", however, has been overturned by more recent studies. The bottom line is that the rabies virus is just as effective at killing bats as it is killing other mammals.

There are two ways animals infected by rabies "express" the disease, and both can occur from infection from the same strain of virus. Some bats exhibit paralytic ("dumb") rabies, meaning that virus kills the host without the host developing any aggressive tendencies. Animals suffering from this form of rabies have neuronal dysfunction, including swelling of the brain (encephalitis) and spinal cord (myelitis), paralysis, cessation of breathing (apnea) and death. The other form is known as encephalitic ("furious") rabies. Bat with furious rabies typically show symptoms of hyperactivity, aggressiveness, irritability and erratic flight. Bats and other mammals with these symptoms may make unprovoked attacks on other animals, including humans, though in the end the infected individuals also experience the

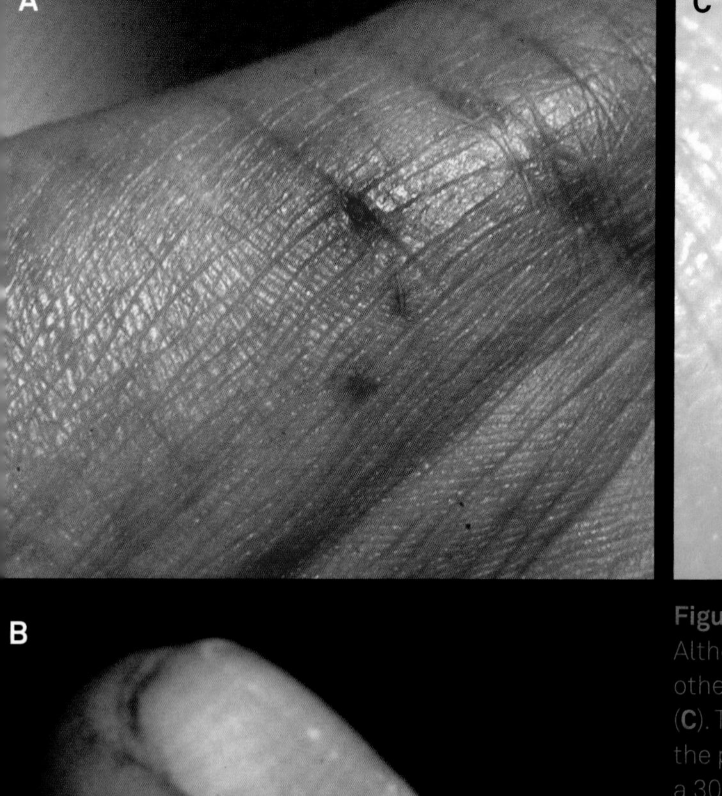

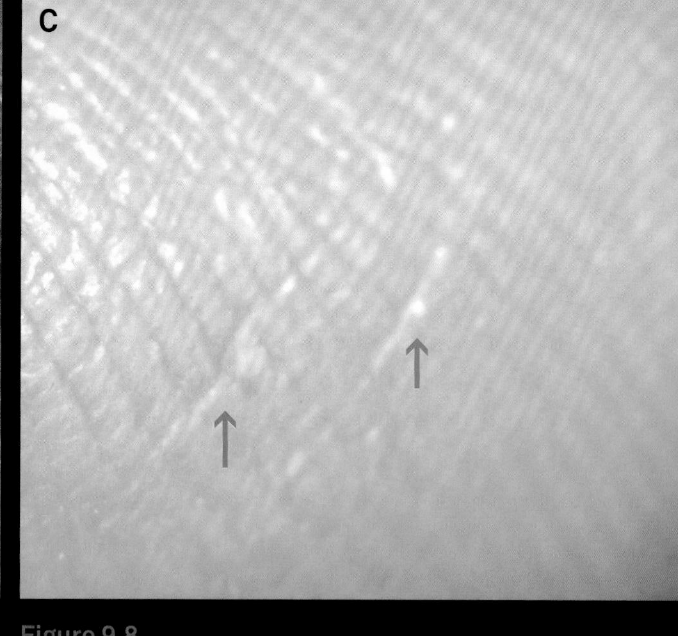

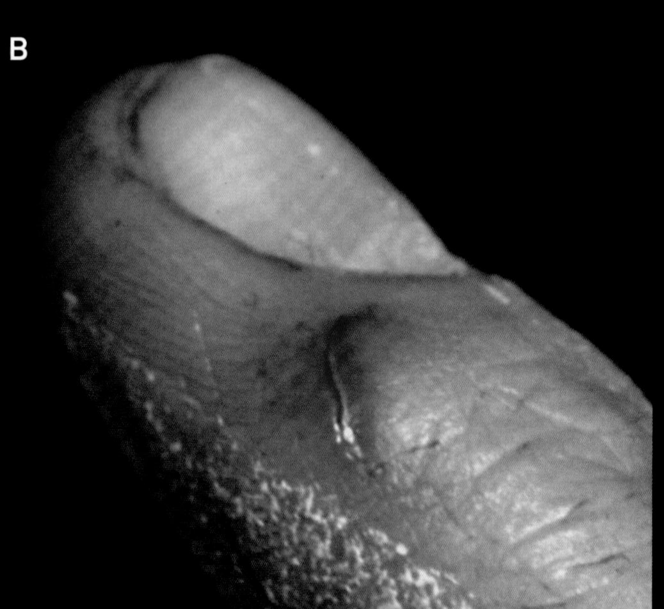

Figure 9.8.
Although bat bites are typically puncture wounds (**A**), others are slashes (**B**) and still others small scratches (**C**). The canine tooth of a 30 g. Jamaican Fruit Bat caused the puncture wound seen in (**A**). The defensive wound of a 30 g. vampire bat is shown in (**B**), and the scratches on Brock's fingers from the bites of an Elegant Myotis in (**C**). Bite (**A**) was painful, bite (**B**) did not hurt at all and bite (**C**) was barely noticeable.

same neuronal dysfunction leading to death. Both manifestations of rabies have been reported in bats, although paralytic rabies may be more typical of bats that live in colonies. In some cases animals with rabies may show both furious and dumb stages, but not always.

Most people are not familiar with the normal behavior of bats and thus do not recognize the aberrant behavioral patterns indicative of rabies. Lay persons rarely see healthy bats up close because they roost in secluded spots during the day and fly at night; however, it is not unusual for healthy bats to occasionally become trapped inside buildings, especially if their roost is located in an attic, or to be found roosting during the day behind shutters or in wood piles. Bats that use daily torpor to reduce energy expenditures may be very sluggish when woken from their daily slumber, but will fly away once they are warm and fully aroused. (See Chapter 6.) A bat trapped inside a building and disturbed may be frightened and will usually fly continuously while looking for an escape route. This is normal and is not evidence of illness. In contrast, a bat flying during the day for no apparent reason, or found flopping or crawling on the ground, is more likely to be sick and deserves caution.

Brock and a Rabid Lab Bat

I have had several occasions to observe the changes in bat behavior associated with the onset of rabies. I had trained a Big Brown Bat to perform in behavioral experiments in my laboratory. Unbeknownst to me and my colleagues in the lab, this bat had previously been infected by the rabies virus. As the disease took hold, the bat changed from being cooperative, passive and even "friendly" to being irritable and aggressive. It increasingly refused to eat and began to attack anything and everything that came near. Luckily, this bat did not bite any of us. I have also witnessed the apparent ostracism of infected bats by others in a colony.

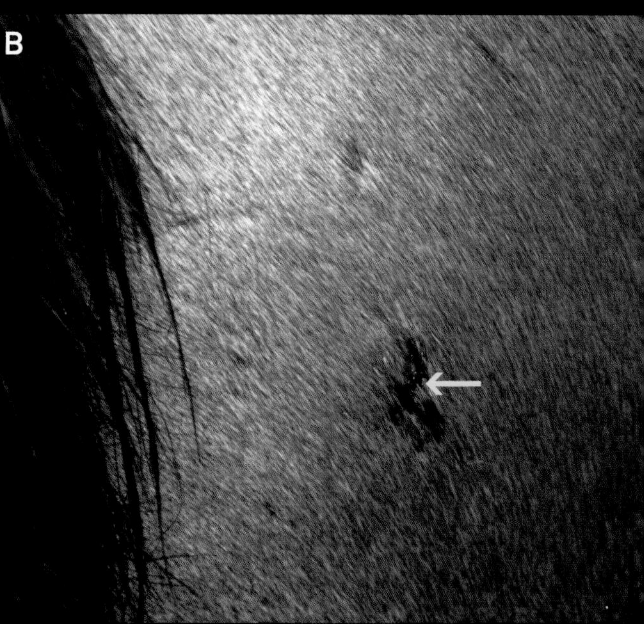

Figure 9.9.
Dried blood running from the bite sites of Common Vampire Bats on the shoulder of a cow (**A**) and scabs from a similar bite on the rear haunch of a horse (**B**, yellow arrow). These wounds differ in shape (a divot of skin removed) from defensive bites.

Public Health

From a public health standpoint, both forms of rabies in bats are important. Furious rabies may seem more dangerous because animals showing these symptoms may attack. Because the behavior of bats expressing furious rabies is so abnormal, lay people may recognize that there is something wrong and, if they are bitten, report the incident. On the other hand, bats expressing paralytic rabies may pose an even greater risk because a concerned person seeing an apparently placid or helpless animal may try to rescue or help it, increasing the probability of bat-to-human contact. The first human death from bat rabies reported in the United States resulted from such a situation in Texas in the 1950s. A woman tried to help a grounded Silver-haired Bat that was flopping around in distress during the day. She was bitten but was not worried about the bite, which was a fatal mistake.

The importance of knowing if you have been bitten or not raises another issue about bat bites. The small size of many bats means that their bite may not leave obvious marks on the skin. (Figures 9.8C and 9.10B) Immediately after a bite there may be a small drop or two of blood where a canine tooth has punctured the skin, but a short time later an obvious wound may no longer be visible. The Centre for Disease Control and Prevention (CDC) in Atlanta, Georgia recognizes this possibility in its guidelines discussing exposure to rabies in bats. Five scenarios are addressed. First, a person who has been touched, scratched or bitten by a bat, or has had contact with bat saliva in an open wound or onto mucous membranes. This constitutes a direct viral exposure. Second, a person who has contacted a bat but without evidence of a known exposure (*i.e.*, a bite or scratch). Third, a person who has been in the same room or dwelling as a bat, but with no known contact with the animal. Fourth, a small child or a sleeping, intoxicated, or mentally unfit person with no evidence of the presence or direct contact (*i.e.*, a bat bite or scratch) but who is unable to rule out the possibility of an exposure. Fifth, a person sitting or walking outside when a bat flew near. A conservative interpretation of the CDC guidelines would recommend people in the first four situations be given post-exposure vaccinations (prophylaxis) because the nearly consistently fatal effect of rabies often justifies the most cautious approach. But the costs of employing such caution are high; the vaccine for post exposure prophylaxis may cost about (US)$1000 per person.

There are many strains of rabies virus identifiable by the presence of monoclonal antibodies. In the New World, there are eight strains of bat rabies. Other strains have been reported from Europe, Africa and Australia. Several strains may be specific to a particular bat species, such as Big Brown Bats, Brazilian Free-tailed Bats, bats in the genus *Lasiurus* and vampire bats. One strain is known from Silver-haired Bats and Tricolored Bats. This is notable because the Silver-haired Bat/Tricolored Bat strain is more prevalent in cases of human deaths from rabies than others.

"What is the incidence of rabies in bats?" This is an important question to which there is not a clear answer. Brandon Klug from the University of Calgary and colleagues tested brainstem impressions from 217 bats killed at wind facilities (wind farms) in Alberta in 2007 and 2008 for the presence of rabies virus. Of these, antigen analyses revealed no rabies virus in the ninety-six Hoary Bats carcasses. In contrast, rabies antibodies were found in twelve of 121 (10 percent) of Silver-haired Bats tested. Other estimates of the incidence of rabies in bats in the United States vary considerably. Between 1984 and 2000 in the United Kingdom and Europe, 534 of 9754 bats (5 percent) tested positive for rabies exposure. The bats that most often tested positive were Serotine Bats (*Eptesicus serotinus*) (513), but an additional ten other species also tested positive for rabies exposure. Both sets of data demonstrate that the incidence of rabies in bats can vary among species. In reviewing statistics about the incidence of rabies it is important to recognize

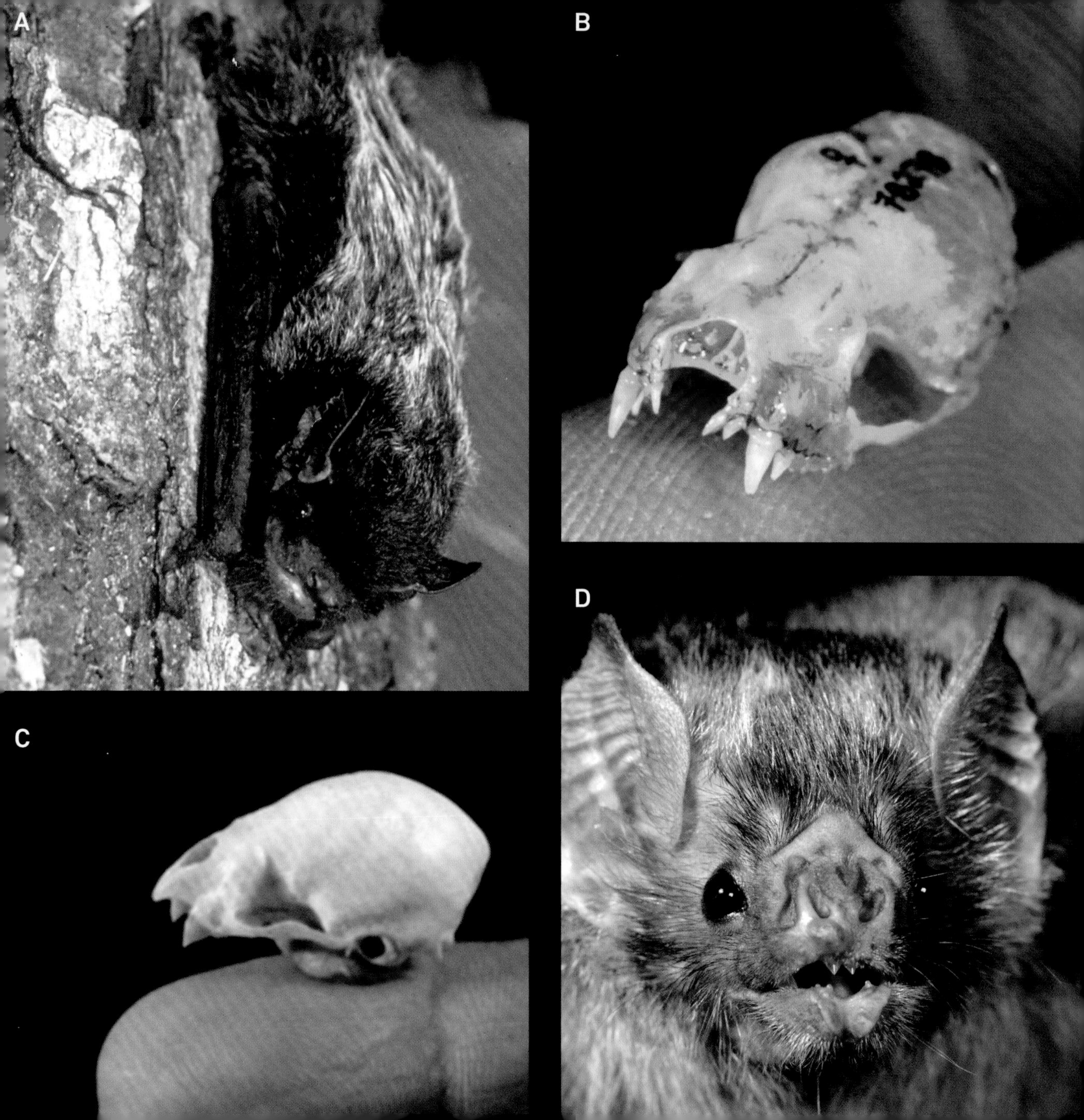
A
B
C
D

that reports based on animals submitted for testing at public health labs are undoubtedly overestimates of the incidence of the virus in natural populations. This is because it is usually only animals collected under suspicious circumstances, *e.g.*, in homes or found dead near dwellings, that are submitted for testing—not an unbiased sample.

Brock Hears a Sad Story

In September 2000, North American Symposium on Bat Research held its annual meeting in Miami, Florida. Colleagues from the CDC gave presentations about bats and rabies that aggravated some of the bat biologists. The morning after the presentations, I was talking to a colleague, Michel Delorme from the Biodôme in Montreal, who looked terrible that day. He explained he had just heard that the young son of one of his co-workers was in the hospital, apparently suffering with clinical symptoms of rabies virus infection.

The boy and his family had spent a holiday in August at a cottage north of Ottawa. Apparently, they had found two Little Brown Myotis in their cottage. One appeared to have difficulty flying, but after releasing both outside the family did not give the matter much thought. Two days later, the boy reported an inflamed irritation on his upper arm but no one connected his symptoms to contact with the bats. About a month later the boy woke up feverish and with a pain in his upper arm. Within two weeks he had died of rabies, the Silver-haired Bat/Tricolored Bat strain (which also sometimes occurs in other species).

Another colleague, Alan Jackson, a University of Manitoba neurologist who studies rabies, and I visited this cottage in December 2000. There was no bat colony inside the cottage and we found no evidence of bats in the attic; however, there were two bat droppings on one of the walls. The veranda was served by two screen doors, and two doors opened into the cottage from the veranda. It was not clear how the bats had entered the cottage since the screens were intact on all the doors and windows. For a bat find its way into the cottage, at least two doors would have had to have been left open. Perhaps the bats came down the chimney stovepipe, or were carried in on firewood. Alan and I spent several hours talking about how anyone could keep this from happening again. We found no solution.

Vampires and Rabies

Common Vampire Bats merit special mention in the context of rabies. Their blood-feeding habits make them ideal vectors for the disease. Predation on livestock by Common Vampire Bats can be responsible for spreading the virus to the livestock. The same appears to be true for spreading rabies virus from Common Vampire Bats to people. Yet, in many areas where Common Vampire Bats occur, there are relatively low levels of rabies in livestock, and transmission of rabies to people is an uncommon occurrence. Disease specialists now recognize two major epidemiological forms of rabies: urban rabies and sylvatic rabies. Urban rabies occurs in more urban areas and the main viral reservoir is dogs. Sylvatic rabies occurs in more remote areas, such as the Amazon rainforest, and several species of wild carnivores and bats apparently maintain independent strains of the virus.

A 2013 study conducted by researchers from the University of Georgia, the CDC and Peru's Instituto Nacional de Salud Centro Nacional de Salud Pública showed that rabies virus circulates in populations of Common Vampire Bats in Peru. The incidence of rabies in vampires ranged from 3 percent to 28 percent but was not predictable from colony size. Young and subadult Common Vampire Bats were more often affected than adults. Most importantly, operations to kill these bats did not affect the incidence of rabies, perhaps because culling was directed at adults rather than younger bats. A 2013 study led by Centers for Disease Control and Prevention researcher Rene Condori-Condori showed that there are at least four distinct lineages of rabies circulating in populations in Common Vampire Bats in the region, and that these move independently of one another through bat populations. Even if one strain of virus was wiped out in an area, another strain might move in to replace it. This understanding of the dynamics of the virus makes it clear that controlling rabies will never be a simple matter.

Other Lyssaviruses

Up until about twenty years ago, people thought that rabies in bats was a New World (Western Hemisphere) phenomenon, so bat biologists working elsewhere were under the false impression that rabies or similar viruses were not something about which they needed to be concerned. This has changed with the discovery of other strains of *lyssaviruses* that cause rabies-like symptoms and mortality. Lyssavirus is the genus of viruses that includes rabies as well as a variety of other related viruses. The genus is named after Lyssa, the Greek goddess of madness, and placed in the family Rhabdoviridae. Bats appear to be involved in maintaining the presence and spread of these viruses. Lyssaviruses have made at least two major switches from bats to terrestrial mammals. In addition to rabies, at least eleven lyssaviruses occur in Africa, Eurasia and Australia (the Eastern Hemisphere) and can cause rabies-like symptoms. Included are Duvenhage virus, European bat lyssavirus (two strains), Australian bat lyssavirus, Irkut virus, Ikoma virus, and Lagos bat virus and Mokola virus. All of these lyssaviruses are thought to be maintained in populations of bats except the Mokola virus, which may be maintained in populations of rodents and shrew-like eutherian mammals. The mode of transmission for lyssaviruses appears to be bites from infected animals. Lyssaviruses fall into two main groups, one that is neutralized by anti-rabies antibodies and another that is not.

Other Viruses

Hendra, Nipah and Cedar viruses in the family Paramyxoviridae cause respiratory illness and encephalitis, and both have been associated with bats. Hendra virus was first isolated in 1994 in a suburb of Brisbane, Australia. This virus caused human infections through direct exposure from infected horses, but Flying Foxes (genus *Pteropus*) appear to be the natural viral reservoir. Nipah virus is related to Hendra virus and was first isolated in 1999 from an outbreak that occurred in men that worked with pigs in the village of Nipah on the Malaysian peninsula. Again, Flying Foxes appear to be the natural reservoir. Nipah also spread to cats and dogs that were exposed to infected pigs. To date, there is no evidence that either Nipah or Hendra virus can be spread from person to person. Flying Foxes and other Old World Fruit Bats sometimes roost in fruiting trees, and they tend to chew and suck the juices from fruits and spit out the pith they cannot digest. (See page 122.) Fruit bats roosting in these trees in horse pastures (Australia) or over pig pens (Malaysia) probably account for the spread of Hendra and Nipah from bats to horses and pigs. Cedar virus has been reported from flying fox bats in Australia but its implications for human health remain unclear.

First discovered in 1978, the Ebola virus, with its widespread effects in the body including acute viral illness, fever and massive organ failure, has fatality rates of 50 to 100 percent. Prevalent in rain forest habitats, Ebola virus, and the related Marburg virus, cause a severe form of haemorrhagic fever disrupting immune and blood coagulative defenses that can cause victims to "bleed out." Ebola has been reported in a number of animals, notably various primates and rodents, and at least one species of shrew. Outbreaks of Ebola virus have involved humans as well as Great Apes (gorillas and chimpanzees). Ebola outbreaks have often been tied to reliance of local peoples on "bush meat"—the hunting of native animals (in-

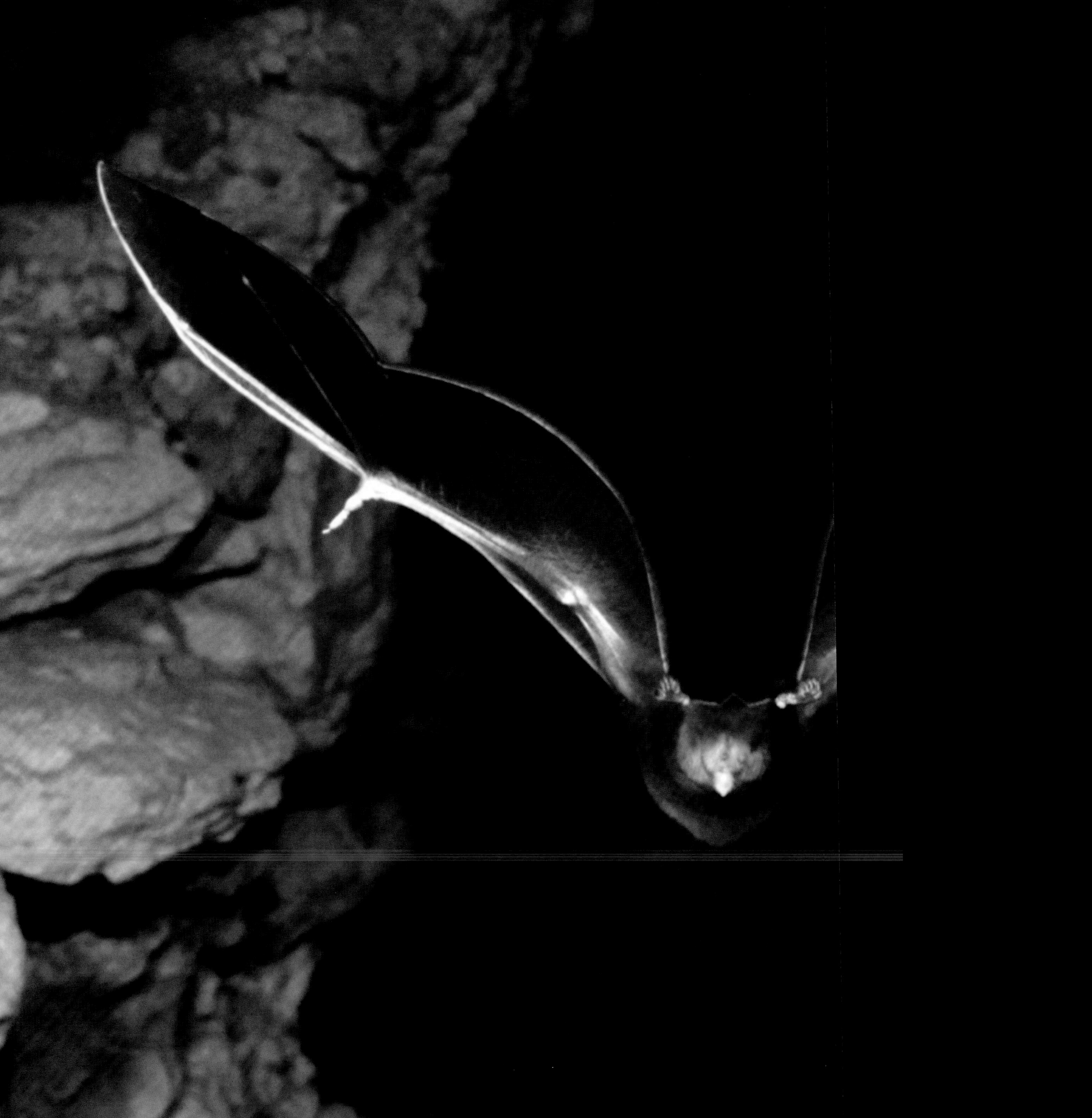

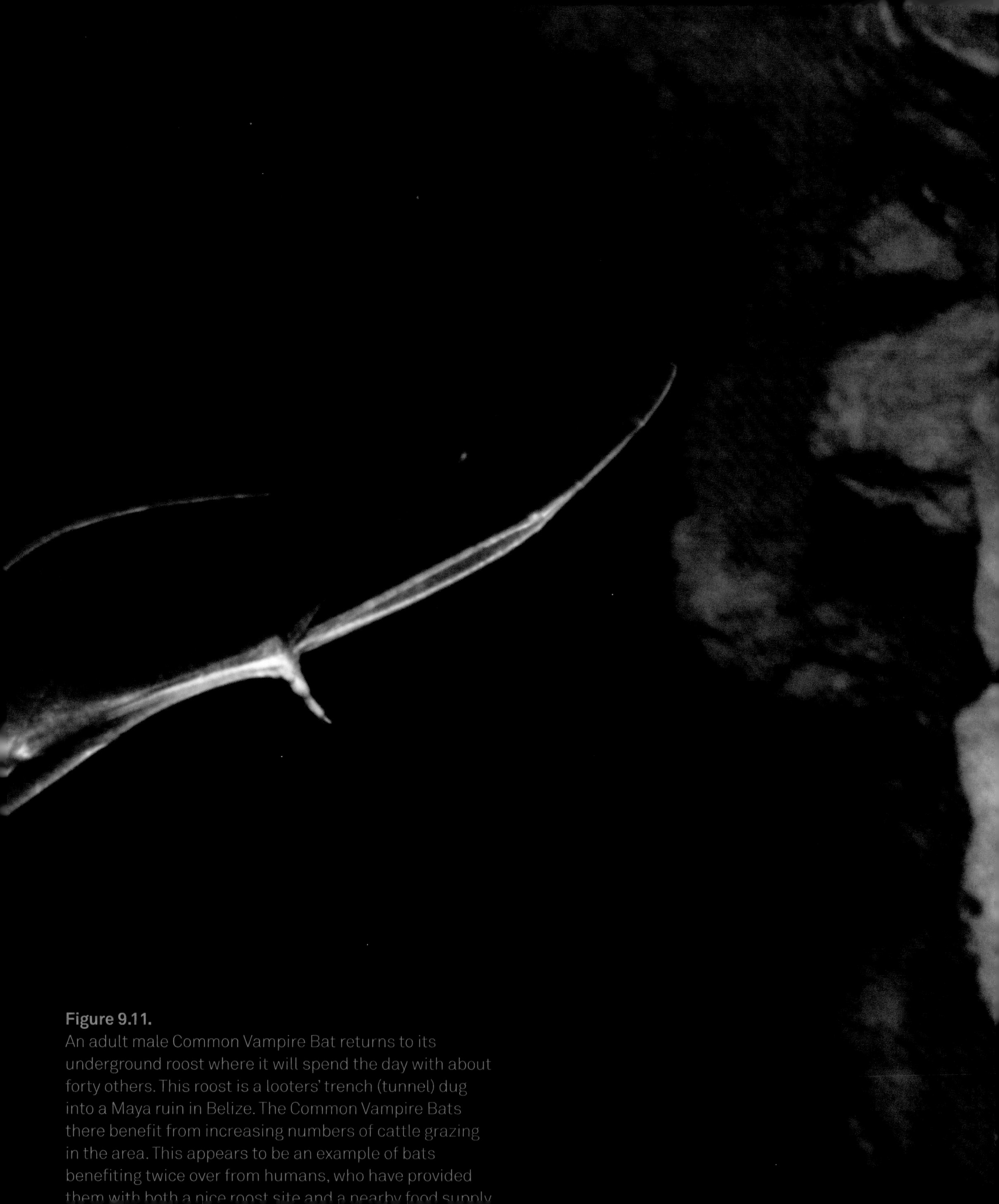

Figure 9.11.
An adult male Common Vampire Bat returns to its underground roost where it will spend the day with about forty others. This roost is a looters' trench (tunnel) dug into a Maya ruin in Belize. The Common Vampire Bats there benefit from increasing numbers of cattle grazing in the area. This appears to be an example of bats benefiting twice over from humans, who have provided them with both a nice roost site and a nearby food supply.

cluding bats and primates for food. Evidence of asymptomatic infection of some Old World Fruit Bats in Africa (Hammerheaded Fruit Bat, Beuttikofer's Epauletted Bat and Little Collared Fruit Bat) suggests that these species may be reservoirs for the Ebola virus. Two outbreaks of Ebola in the United States in 1989 and 1990 were associated with monkeys imported from the Philippines.

Coronaviruses cause respiratory diseases such as the common cold and viral gastroenteritis. They are especially important in public health because they can be spread from person to person. Although many strains of coronaviruses are known from animals, only five appear to affect humans. The most infamous coronavirus is the strain that caused SARS (Severe Acute Respiratory Syndrome). The symptoms of SARS include coughing, shortness of breath or difficulty breathing and may also include muscle aches, sore throat and diarrhea. In 2012 a new coronavirus (Middle East Respiratory coronavirus or MERS-CoV) was detected and subsequently reported from seventeen adult humans in five countries. Molecular analyses revealed that both the coronavirus causing SARS (SARS-CoV) and this new strain were most similar to coronaviruses found in bats. In the case of SARS, the similarity was closest with coronaviruses found in some Horseshoe Bats (*Rhinolophus*). The mechanism by which Horseshoe Bats might pass coronavirus to humans remains unclear. In 2002–2003, the SARS epidemic involved 8096 cases and 774 deaths. In 2005, ninety bat biologists who had worked in the field with bats (including Nancy and Brock) volunteered to be part of a Centers for Disease Control and Prevention survey for SARS-CoV. Blood serum of eighty-nine of the biologists was negative for SARS-CoV. Twenty of the volunteers had worked with Horseshoe Bats and at least one biologist, Brock, had visited caves in southern China—some with Horseshoe Bats, others with Rousette bats. Both Nancy and Brock were negative.

There have been considerable advances in using molecular methods to detect novel pathogens. These methods, when applied to bats, have revealed a substantial number of previously unknown viruses. From a biological perspective, it is fascinating to think about the ecology of the viruses. Why bats harbor such a diversity of viruses may reflect a combination behavior, immunology and physiology. The relationships between bat hosts and viral pathogens are ancient, preceding human activity such as urbanization and climate change. Concerns about public health and diseases caused by viruses need to be kept in context and perspective. For example, Influenza A viruses cause seasonal influenza in humans, and different strains, *e.g.*, H1N1, often spread quickly around the world. In 2012 a distinct lineage of influenza A virus was isolated from Little Yellow-shouldered Bats captured in Guatemala. It is not known if this strait has the potential to infect people, or if it is limited to bats. The potential importance to public health of this viral strain and others remains to be determined.

Bats and Diseases of Humans

All of us live in an ever-changing world. One important element of change is the ever-increasing human population that directly translates into increased rates at which natural habitats are being altered by human endeavors, including agriculture and resource extraction. Dwindling habitat refugia means that more people are coming into direct contact with more animals, offering increased potential routes for diseases to move from natural wildlife hosts to human. This phenomenon—emerging infectious diseases that make the leap into humans—is exacerbated by international trade in wildlife, including bats. Coupled with this are modern transportation networks that make it easy for someone to fly half way around the world within a few hours, and carrying a disease with them. These patterns appear responsible for the arrival of WNS in caves in the United States and Canada, and also explain the outbreaks of SARS in Canada.

What can be done to minimize the possible transfer of diseases from bats to people? More information and better public education are key to minimizing the impact of diseases transferred from bats to people. Being able to detect and identify uncommon or rare diseases is essential. Discovering new strains of viruses provides researchers and physicians with better chances of responding to potentially calamitous events. The worldwide health information system appears to have the capacity for doing this, as evidenced by the discovery and reporting of the 2012 strain of coronavirus. Research on rabies and Common Vampire Bats makes it clear that simply killing bats is not likely to be an effective solution. A quick test for the presence of rabies virus could allow doctors to avoid the tragedies associated with spreading the virus via transplants in humans.

Risk Assessment

In the final analysis, bat biologists, like anyone one else who drives or rides in automobiles, have a higher risk of being killed or injured in a traffic mishap than from anything they will encounter from working with bats. But bat biologists can use available knowledge to minimize their risks associated with handling bats, working in the field with bats or venturing underground to study them. Informed decisions are the key to prospering in the study of bats. In a sense, bat biologists could be the canaries in the coal mine when it comes to possible impacts of diseases people might contract from bats. Testing bat biologists for SARS-CoV has inspired a future model for investigative disease research. Arguably, people working regularly with bats are more apt to be exposed to diseases that can move from bats to people. But the topic of human health invariably attracts attention from the media and today's news may be tomorrow's false alarm. Middle East Respiratory Syndrome (MERS), first described in 2012, was initially associated with camels. Then, based on a sample size of one Egyptian Tomb Bat (*Taphozous perforatus*), it was linked to bats. By April 2014, the connection to camels was reaffirmed. Depending upon where you joined the story, MERS is either yet another human disease that may have originated in bats, or a disease that humans get from camels. In reality researchers are not yet sure what the transmission path is for MERS, nor what animal(s) may serve as the natural reservoir for this disease. This is an example of how quickly the picture of bats and human health can change.

10

Bats and People

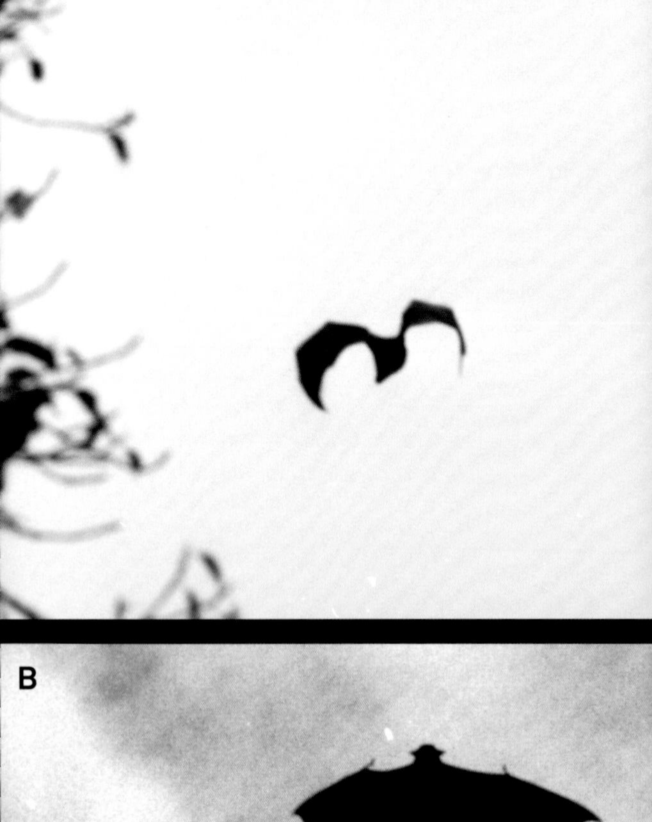

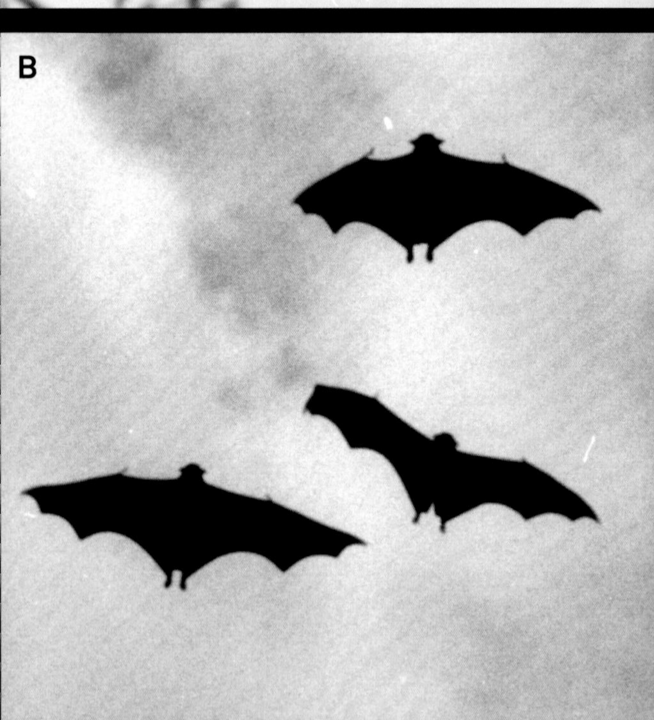

Bat Representations

Wings are the features (Figure 10.1) that usually allow people to recognize human representations of bats. (Figure 10.2A and B) Even when bat wings are stylized, the association to bats is usually apparent. (Figure 10.2C and D) This is less so when it comes to recognizing bat faces. To a bat biologist, the face in 10.2C is a Leaf-nosed Bat, perhaps with facial stripes. But to almost anyone else, the faces in B and D do not suggest bats, but instead seem to depict perhaps a bee or a mouse.

The miniature in 10.2A is a clay representation of from Hispaniola, the bowl in C is Maya in origin, recovered from a child's tomb at Altun 'Ha, a site in Belize. Both B and D are Chinese representations of bats. While the wings are obvious in 10.3A, a jade figure from China, only a bat biologist is apt to recognize the face in 10.3B as bat. The color and pattern of the wings in 10.3C (from a Chinese snuff bottle) resemble the wings of an actual bat. Painted Bats (*Kerivoula* spp.) occur from Afghanistan east through India, Pakistan and Bangladesh and China, including Taiwan, Japan and Korea. (See Figure 12.10.)

Human depictions of bats may be stylized (10.3A and C) or accurate in detail (10.3B), and sometimes are portrayed in color (10.3C). The Taironan representation (10.2B) shows details of ears and nostrils suggesting a Ghost-Faced Bat.

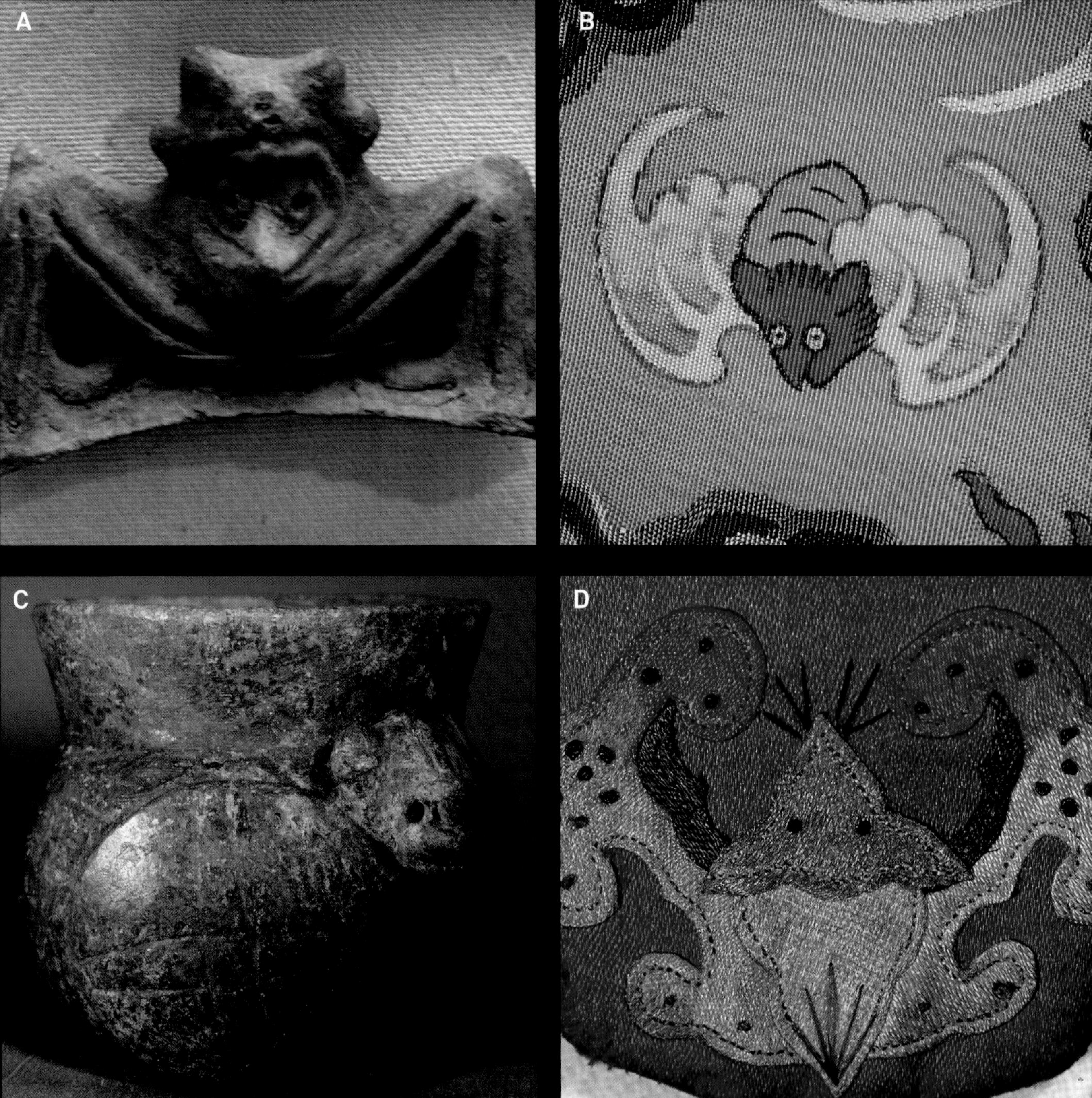

Figure 10.2.
Wings are variously rendered in these four depictions of bats. In (**A**) from Hispaniola and (**B**) from China, the wings are obvious. Wings are less obvious in (**C**) from the ancient Maya or (**D**) from China. More details about each image are presented in the text. The faces of (**C**) and (**D**) are distinctly unbat-like, while the one in (C) is clearly a

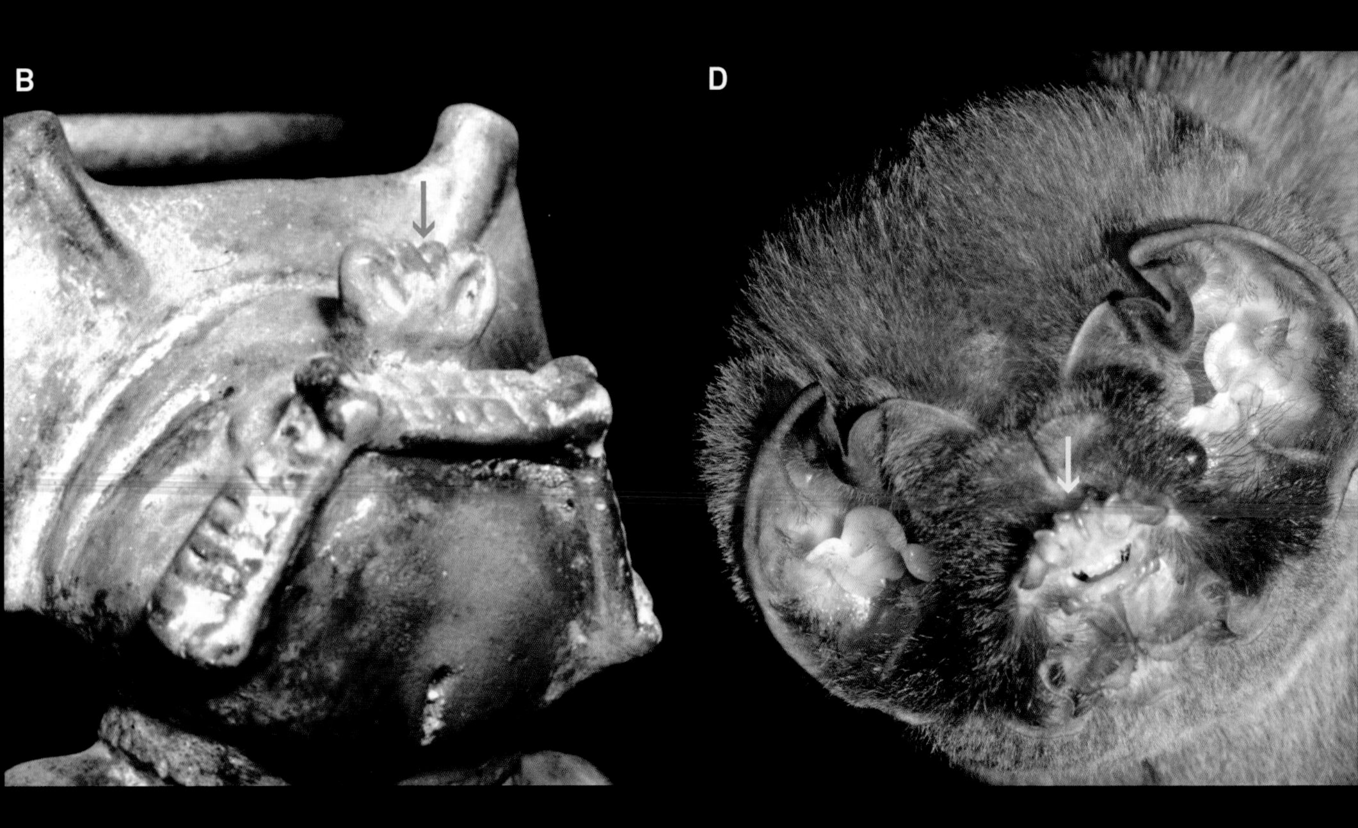
B
D

C

Figure 10.3.
A Chinese bat in jade (**A**) presents a stylized view of a bat, while a Taironan representation (**B**) is in more detail. (See Figure 10.8C for a view of the whole bat.) The colorful and stylized bat from a Chinese snuff bottle (**C**) provides a different view. The patterning on the wings is reminiscent of a Painted Bat (*Kerivoula picta*). The nostrils (arrows) are prominent features in both the Taironan depiction and in a living Ghost-Faced Bat (**D**).

Bats as Symbols

As noted above, the details depicted in10.3B strongly suggest a Ghost-face Bat, an insectivorous species that occurs today in what is now northern Colombia and Venezuela, the area where the Taironans lived. Anthropologists, however, do not know how Taironan people perceived the bat they sculpted on the statue shown in 10.3B. Much more is known about the significance of bats in some Chinese cultures.

In Chinese folklore, five bats, the wu fu (Figure 10.4), represent the five blessings: old age, wealth, health, love of virtue and a natural death. Bats are typically shown in red (Figure 10.5B and C), the traditional Chinese color of joy, and auguring money, blossoms or fruit (Figure 10.5B). Some bats carry red swastikas, a jarring image in the wake of the Nazi period in Germany. (Figure 10.5C) Yet, swastikas are ancient positive symbols known from many parts of the world, from the Americas through Asia. This is one reason the swastika was adopted by the Nazis.

In Chinese art, bats are often clearly identifiable by their wings. In other cases the symbolism is less clear, for example the bats around the inside of a rice bowl. (Figure 10.4) Diners in the United States are often surprised when they learn that the delicate designs on their rice bowls are bats! Although bats would certainly not be associated with food and dining in the United States, as good luck symbols they are considered perfectly appropriate decorations for serving ware in China.

Figure 10.4.
The five bats (wu fu) on these Chinese bowls represent the five blessings. Note the difference in the degree of stylization of the bats.

Figure 10.5.

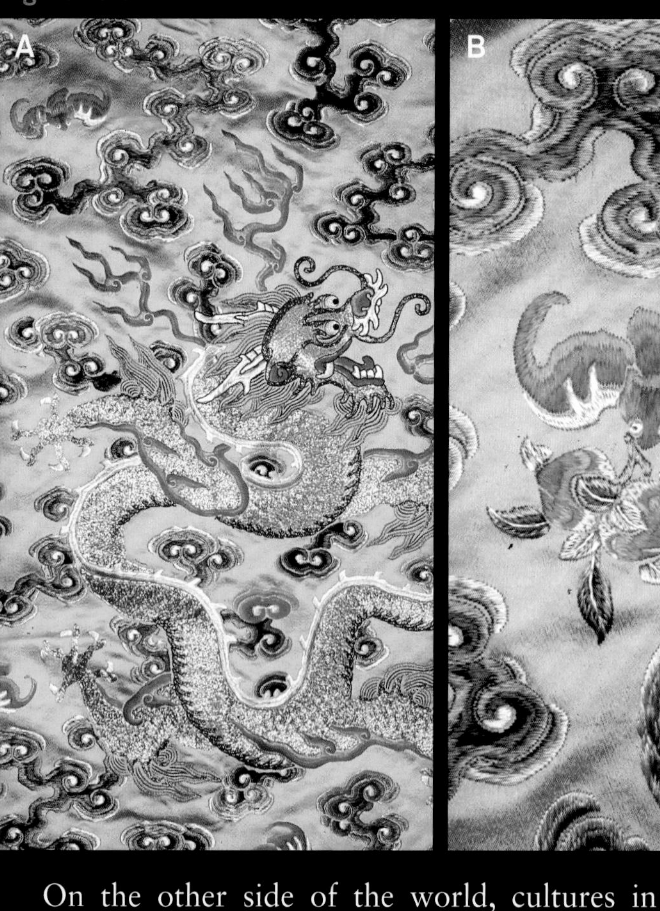

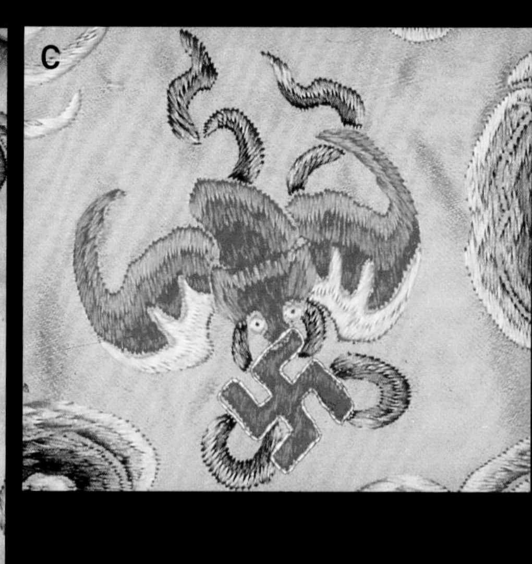

On the other side of the world, cultures in Central and South America and the West Indies feature a variety of representations of bats. (Figure 10.4) There is a striking contrast in the details between Chinese and Neotropical views of bats. In some cases, the Neotropical representations suggest at least the family of bats (Figure 10.6), while the Chinese versions are less specific (See Figure 10.2D).

The Taíno people of Hispaniola and the Caribbean originated from the Arawaks of the Orinoco Delta in what is now Venezuela. They had established communities in the Caribbean about 1500 years before Columbus encountered them for the first time. The faces of opías (souls of the dead), bats and owls associated with the spirits of the non-living (Figure 10.7), are commonly represented in Taíno art, sometimes in enough detail to allow scientists to identify them (Figure 10.7A). As in Maya representations, bats are shown in detail in spite of their small size.

Figure 10.5.
A twelve-symbol Chinese robe dating from the Ch'ing period (1636–1912) is dominated by a dragon (**A**). On the robe, bats are shown in red, some carry fruit (**B**), and others carry swastikas (**C**). In some Chinese dialects, the word for swastika is the same as the word for 10,000, and bat with a swastika can represent 10,000 blessings.

Figure 10.6.
Faces of Maya representations of a bat (**A**), and one from the Dominican Republic on the island of Hispaniola (**B**). The nostrils in (**A**), and the face in (**B**) suggest different species. The bat in (**A**) appears to have been a Moustached Bat. From its nostrils and ears, the bat in (**B**) may have been a Free-tailed Bat.

Figure 10.6.

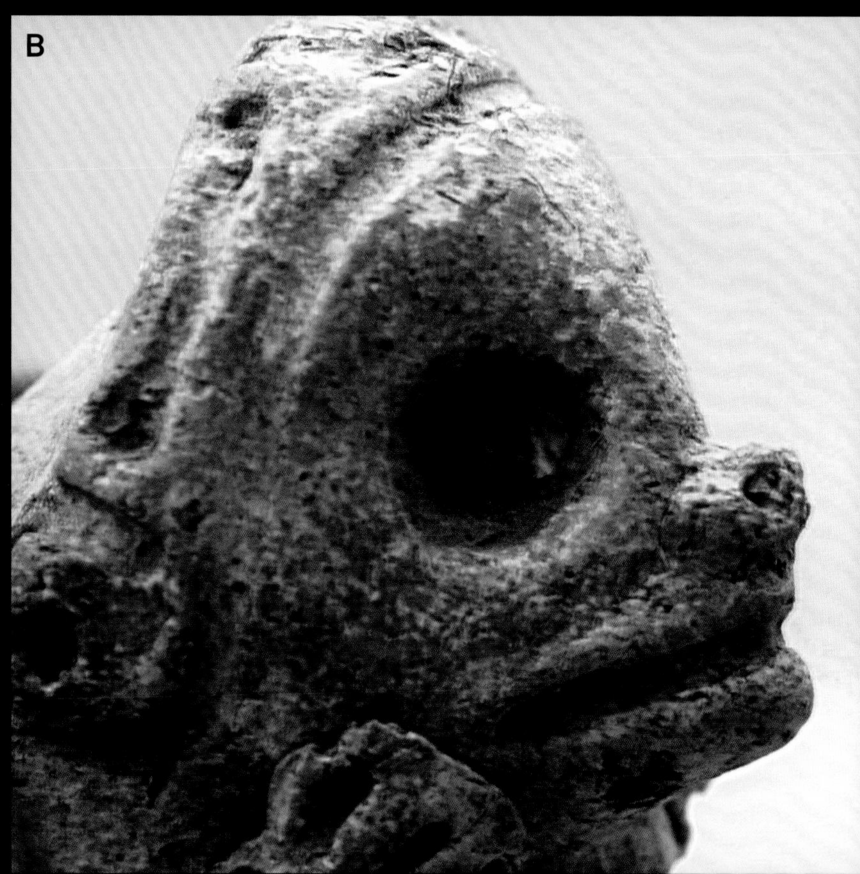

Figure 10.7.
The Maya glyph Sotz', a stylized bat with a conspicuous noseleaf. Compare with the faces and noseleaves of living bats shown in Figures 1.9 and 1.10. Sotz' is the logogram for the fourth month of the Maya calendar.

To the Maya, caves represented portals between their world and the other world where gods lived. Bats were a favorite motif, perhaps because they lived in caves and thus were connections between the two worlds. Some Maya representations of bats are presented in considerable detail. (Figures 10.2C and 10.6A) On both of these vessels, the bat faces and wings are obvious: a Moustached Bat (A), and a striped-aced leaf-nosed bat (B). The vessel in (B) was found in a child's tomb that dates from around 640 A.D. Other representations, such as the glyph (elemental symbol) for the month Sotz (= Zotz or Suutz) are more stylized. (Figure 10.8) The Maya used over 1000 glyphs from about the year 3 B.C. to the end of the Classic Period around 900 A.D. The noseleaf on Sotz (as in 10.7) makes it clear that it was not a vampire (See Figure 5.1F.).

To some people, the bat is a symbol of night and evil. This may explain why devils are typically depicted with bat wings. In Germany, John the Evangelist, purported author of the New Testament Book of Revelation, is often symbolized by a lectern with an eagle that supports a Bible, the symbol for light. On this lectern, a bat symboliz-

Elsewhere in the world, bats represent fertility. (Figure 10.8) The penis of many species of bats is prominent and conspicuous, perhaps overcoming the challenges of mounting the female from behind. (See Figure 7.3.) Many female bats have interfemoral membranes enclosing their legs and tails that complicate the juxtaposition of male and female genitalia during mating. The association with fertility of female bats and humans is less clear, but may reflect the human-like placement of mammary glands, or the fact that most bats give birth to a single, relatively large offspring just like people do. (Figure 10.8A) Alternatively, the connection may be blood well-known as part of the human female menstrual cycle. For the Taironan people, vampire bats symbolize the fertility of women.

A

C

B

Figure 10.8.
Representations of bats are often associated with fertility. A bat sculpture from New Guinea (**A**) is clearly female, while a fiber and shell bat representation from New Guinea (**B**) and a Taironan carving from South America (**C**) are clearly male.

are more like those of American Navy Seals.

U.S. Torpedo Squadron 27 based in the South Pacific during World War II had a Flying Fox on their Badge. A bat's ability to fly and navigate at night was the rational for using this animal as a symbol, combined with the alleged wile and cunning of the fox. Perhaps as important, local Flying Foxes could be killed and eaten by downed pilots on many Pacific Islands (hunting these bats is a local tradition on many islands). The 11e Escadrille from Belgium is a fighter squadron with a bat prominent on its badge both the older and the newer version. American 105th Ac&W Squadron is a radar unit whose badge is dominated by a bat.

In World War II, Project X-ray, a U.S. Marine Corps plan to release bats carrying incendiary bombs over Japanese cities, remains one of the more bizarre stories about bats and the military. The bats were to be placed in cages and dropped from bombers. The cages were equipped with pressure-sensors that would open the their doors

Bats as Heraldic Symbols

Bats appear on the coats-of-arms of many families, such as some Chauvet and Le Corre from France; Batzon from Belgium; Bateson from Ireland; Baxter, Martyn, Steynings, Bascom, Wakefield and Heyworth from the United Kingdom. Other examples include Wengatz and Krudener from Germany, Trippel from Switzerland and Yamamoto from Japan.

There is sometimes more than one story about bats used as heraldic symbols. Said to be the most famous bat in the world, the Bacardi bat is prominent on the label of many Bacardi products. (Figure 10.10) According to one story, the bat on the label represents Fruit Bats that lived in the attic of the Bacardi Rum distillery in Cuba. Another version claims that the bat comes from the coat of arms of Valencia, Spain, the hometown of the ancestors of the Bacardi family who owned the distillery. According to this story, on 28 September

Figure 10.9
Military badges that involve bats are shown here. A patch from the 360th Squadron of the Royal Air Force and Royal Navy in the United Kingdom (**A**) does not show a bat but rather a moth with radar-jamming signals. Badges from two Israeli units (669th from the airforce (**B**) and Shayetet 13 from the defense force **C**) show bats on their badges.

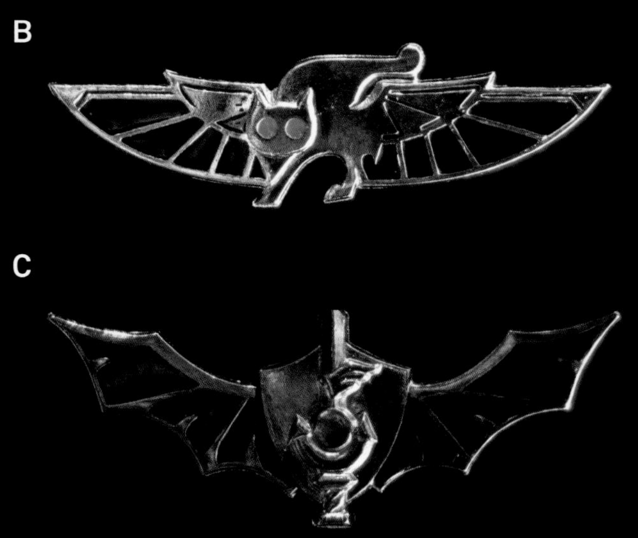

Figure 10.10.
The Bacardi bat.

1238, King James I of Aragon received the surrender of the Moors that had held Valencia. Before the battle, a bat had flown into King James' tent. The bat was thought to be a good omen, the Spanish won the day, and the bat was added to the coat of arms of Valencia and thence on the Bacardi label.

Information about bat biology may shed light on which version is correct. The Bacardi bat, although stylized, clearly has no noseleaf. This means that the bat is not based on the Fruit Bats in Cuba, all of which have noseleafs. The Bacardi bat appears to be a Plain-nosed Bat, which occurs in Spain (and in Cuba). The bat image is thus consistent with the second story but not the first (if the distillery bats were Fruit Bats).

Figure 10.11.
A Boy Scout Bat Patrol patch. Courtesy of Nancy Simmons.

Bats and Scouts

Just as bat symbols have been used as heraldic and military symbols, they have also served as symbols for smaller groups. Boy Scouts around the world use is known as the "Patrol System" grouping boys together in small groups of five to seven boys that work together to plan trips, learn skills and go on outdoor adventures. Each patrol chooses its own name and emblem, often an animal. Many Boy Scout Troops have a "Bat Patrol", a "Vampire Patrol", a "Twilight Patrol" or a "Nightwing Patrol", all of which are symbolized by bat patches that they boys wear on the sleeves of their uniform. (Figure 10.11) The appeal of spooky flying creatures for boys cannot be understated!

Myths about Bats

The diversity of bats and of humans makes it no surprise that there are many myths about bats. Here is a sampling.

Bats are said to become tangled in people's hair. The French have two quite different explanations for why this might be true. The first is that bats are embarrassed by the nakedness of their wings and try to steal hair with which to cover their wings. The second, and more sinister, is that some strange disease cause bats to lose the hair on their wings. It is imagined bats become entangled in a person's hair in an effort to spread this disease. Neither is true.

Bats are said to be blind, hence the expression "as blind as a bat." All bats have eyes and some species see very well, particularly in dim light. But a small bat tends to have small eyes, and one held in bright light will squint to reduce the light falling on its retinae. This may make bats appear to be blind. People who have heard about the echolocation of bats often believe that bats do not need to see because they have "sonar" (*i.e.*, echolocation). In reality most bats see quite well. (Chapter 1)

Bats are imagined to be "dirty" or unclean. This may be because bats usually hang upside down, raising interesting issues about personal hygiene. But anyone who has spent any time watching bats will know that they avoid soiling themselves when eliminating feces or urine by turning head up at the appropriate time. Others will have observed that bats spend a good deal of time each day grooming themselves rather like cats do. The combination of toenails, lower incisor teeth, which are often multi-lobed and somewhat comb-like, and tongues makes it relatively easy for a bat to groom itself. Brock and Nancy can both attest to the fact that bats keep themselves just as clean as cats do.... and are much cleaner than other familiar animals such as sheep and horses (and some dogs they have known).

Vampires

Vampires are the most infamous of living bats. As adults, the three living species of vampire bats eat only blood. People tend to be unenthusiastic about animals that eat blood, whether they are leeches, mosquitoes, bedbugs or vampire bats. Vampire bats occur only in the New World (Central and South America), and some are known as fossils from Cuba and the United States (California, New Mexico, Texas, West Virginia and Florida).

In traditional folklore, vampires were not bats, but humans that came back from the dead to feed on the blood of living people. The myths about vampires were widespread among human societies and are not restricted to any particular part of the world or group of people. In these mythologies, there was no connection between bats and vampires. European explorers in the New World used "vampire" to name the blood-feeding bats they found there. In short, Central and South American vampire bats were named after the folkloric vampire rather than vice versa.

The more recent association of bats with the vampires of human folklore is due in part to Bram Stoker's book *Dracula* published in 1897 that was loosely based on the life of Vlad Tepes (1431- Prince of Wallachia, which was part of present-day Romania (south and adjacent to Transylvania). He was known as Vlad the Impaler because it was said he took sadistic pleasure in impaling people on sharpened wooden stakes and watching while they died. Even so, there is no evidence that anyone thought Vlad was a vampire. While Stoker working on *Dracula*, vampire bats were in the news and he was inspired to write them into his book. It was, however, Bela Lugosi's portrayal of *Dracula* in Tod Browning's 1931 horror movie that appears to have cemented the connections in people's minds between *Dracula* and bats and vampires. (Figure 10.12)

Meanwhile, researchers know relatively little about how people living with vampire bats viewed these creatures. The Taironan people are an exception. As noted above, to them vampires symbolized the fertility of women. Saying that a woman had been "bitten by the bat" indicated that she had started her menstrual cycles and was able to bear children.

Draculin or desmoteplase is a glycoprotein in the saliva of vampire bats that promotes blood flow and efficient feeding. An anticoagulant, it inhibits blood clotting and constriction of peripheral veins and reduces the tendency of red blood cells to adhere to one another. Described in 1999, desmoteplase shows promise as a treatment for strokes because of its ability to break up blood clots. A Danish pharmaceutical company, H. Lundbeck, owns the worldwide rights to desmoteplase and has been conducting clinical testing, filing with health authorities for licenses for wider use is expected in 2014.

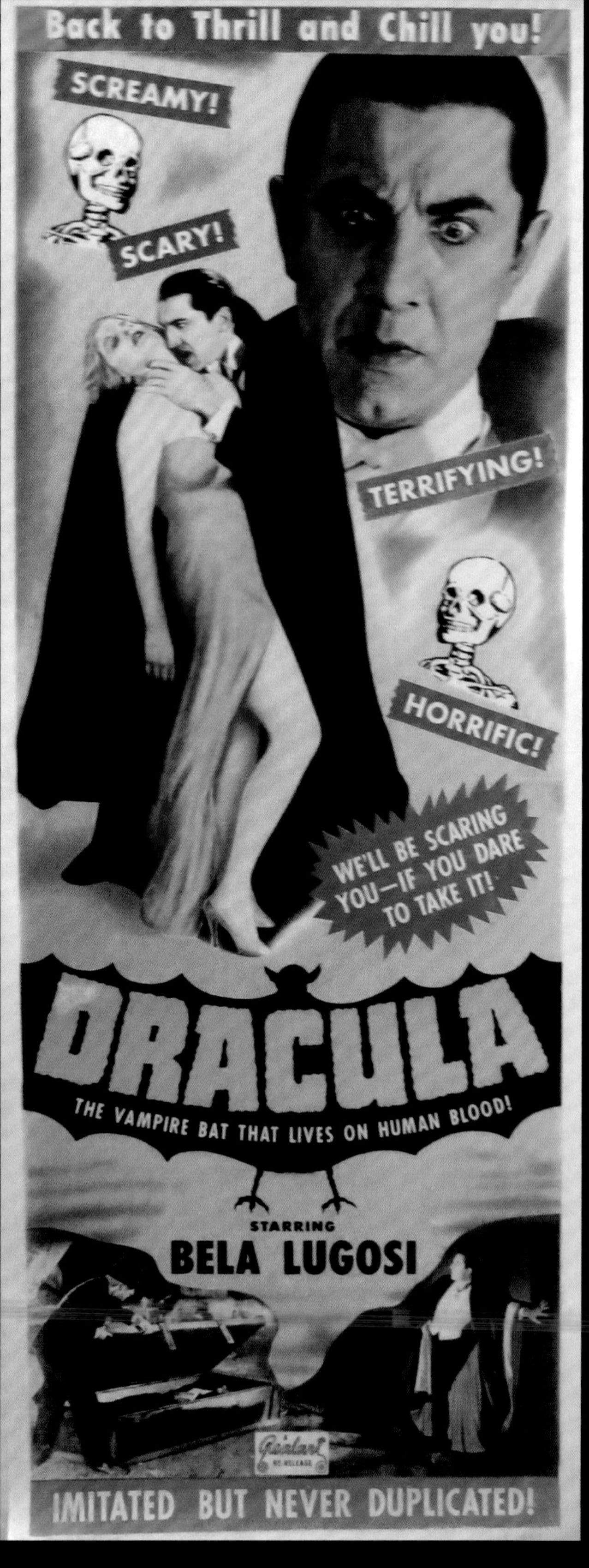

Figure 10.12.

Figure 10.12.
Tod Browning's 1931 horror film classic Dracula was based on the 1924 stage play *Dracula* by Hamilton Dean and John Balderston, which in turn was based loosely on Bram Stoker's 1897 Gothic novel *Dracula*. After some studio hesitation, the Hungarian actor Bela Lugosi, who had starred in the theatrical version, was chosen for the film's title role. Urban legend has it that Lugosi's eerie speech pattern was due to the fact that he did not speak English and had to deliver his lines phonetically.

Figure 10.13. (opposite)
This Polish 20 złoty coin was issued in 2010. It shows a Lesser Horseshoe Bat, a species whose conservation status places it on the Red List in Poland.

Figure 10.14. (See page 244)
This sampling of stamps showing pictures of bats includes four from the United States, one from Canada and one from Vanuatu (an island state in the South Pacific where bats are the only native land mammal). Merlin Tuttle's photographs of four bats appear on the American stamps, while the Canadian stamp was one of four recognizing animals thought to spend the summer in Canada and the winter in Mexico.

Figure 10.13.

RZECZPOSPOLITA POLSKA
2010
ZŁ 20 ZŁ
PODKOWIEC MAŁY - Rhinolophus hipposideros

A

B

C

Figure 10.14.

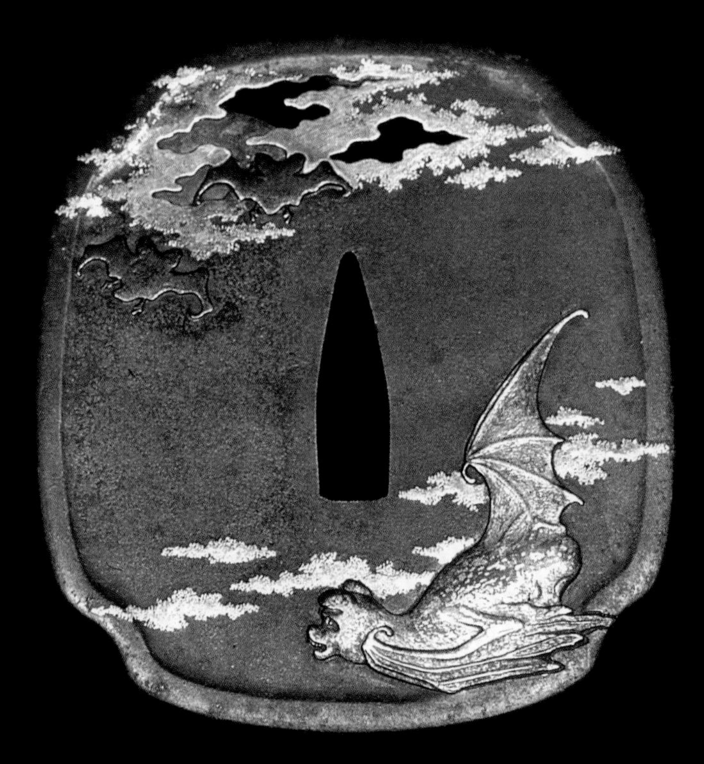

Figure 10.16.
Three bats (two different kinds) decorate this sword guard from Japan.

four migratory animals, the others being two birds and the Monarch Butterfly. The third stamp features the World Wildlife Fund panda logo, and identifies an endangered species of bat, a Flying Fox (genus *Pteropus*). (Figure 10.14C)

In some cases, it is difficult to identify the bat shown jewelry or other handicrafts. (Figure 10.15) The bat on an earring purchased at an airport shop in San Jose, Costa Rica, appears to be Central American, but lack of identifying characteristics doesn't allow a family or species identification.

A Japanese sword guard, placed between the haft (handle) and the blade to protect the users' hand, depicts three bats. (Figure 10.16) The larger, lower bat, rendered in gold color, is obviously a Flying Fox. There are good details of the head and wings, leaving no doubt about what bat it represents. The other two smaller bats are rendered as black silhouettes. They are clearly bats, but stylized. One presumes that bats on a sword guard are positive images, if only because the importance of the guard.

11

Conservation of Bats

People's Impact on Bats

The small body size of most bat species means that they have been relatively little harvested by humans for meat, fur or ivory. But people of some islands in the South Pacific use Flying Foxes as festive food, consuming them at weddings and funerals. Flying Foxes that roost in groups (camps) were usually the targets. When people used only nooses on the ends of long poles to catch bats (the traditional method), the rate of harvest was less than the bats' rates of reproduction. Therefore consumption did not put the bats at risk. The arrival of firearms and later mist nets made it easier to harvest more individuals, changing the balance between rates of harvest and rates of reproduction. In the Philippines, hunters use distress calls of captured bats to lure others into the nets. In this way they may catch fifteen to fifty bats a night, usually Large Flying Foxes (*Pteropus vampyrus*) and Golden-capped Fruit Bats (*Acerodon jubatus*). Even though these are internationally protected species, the "bush meat" trade in Flying Foxes continues, with vendors paying hunters ~US$1.25 per bat in 2005. In the markets, the same bats can sell for <US$3.00 each. Live bats sell for more than dead ones.

In the South Pacific, trade in Flying Foxes as food has resulted in drastic reductions in bat populations. But perhaps even more important to the survival of most bats is pressure from the ever-expanding human population. The impact of humans on bats is varied and largely negative. This is true whether one considers habitat lost due to agriculture and urban sprawl, or the application of pesticides that reach many bats through their insect prey. Features of the life histories of bats such as low rates of reproduction and high first-year mortality can combine to make them particularly vulnerable to many human-wrought changes. (See Chapter 7.) Any species with a small population can be vulnerable, especially when the habitat in which they live is under pressure.

The two species of bats known from Christmas Island (located off of the northwest coast of Australia) are examples of how quickly small populations can decline and even disappear. The Christmas Island Flying Fox (*Pteropus melanotus*) dwindled from camps of six to ten thousand individuals in the 1980s to only a few surviving bats by 2012. A major cyclone hit Christmas Island in 1988, and it apparently decimated the populations of these bats, which roost in relatively exposed places in trees. Sometimes natural phenomena such as cyclones can combine with human pressures to threaten bat populations.

In the mid 1980s, Christmas Island Pipistrelles (*Pipistrellus murrayi*) small insectivorous bats were widespread and common. But by 2002 the population was much reduced, and the last one was observed flying in the wild was seen in 2009. Before the crash in the population, monitoring of the echolocation calls of Christmas Island Pipistrelles indicated that they foraged mainly in along the boundaries between forest vegetation and clearings or fields. They roosted mainly in trees in primary forest, rather than in caves and buildings. Clearing of primary forest may have resulted in a decline in the numbers of these pipistrelles. Other possible contributing factors include the impact of introduced species such as Black Rats (*Rattus rattus*), Wolf Snakes (*Lycodon capucinus*), Yellow Crazy Ants (*Anoplolepis gracilipes*), Red-headed Centipides (*Scolpendra morsitans*) and house cats (*Felis catus*). Regardless, scientists agree that this species is now most likely extinct. Although threats to this species were recognized ten years before it vanished, conservation efforts came too late to save it.

The Seychelles Islands in the Indian Ocean north of Madagascar are home to five species of bats. One, the Seychelles Sheath-tailed Bat (*Coleura seychellensis*), is critically endangered. This bat roosts in small caves in boulder fields, areas overrun by some invasive plants. In the 1990s, only twenty-five of these bats were observed. This population increased to forty by 2009 in the wake of removal of alien plants in the areas where the bats roost. Population levels this low, however, may be terminal for a species due to loss of genetic diversity as well as vulnerability to stochastic events. It is not known in 2014 if these bats are still extant.

The preceding examples illustrate that mortality from predation, habitat destruction or extreme meteorological disturbance, as well as encounters with alien species, can threaten the survival of bats. The largest impact of an alien invasive species on bats documented to date is White Nose Syndrome. (WNS; see Chapter 9.) The European strain of the fungus *Pseudogymnoascus destructans*, probably inadvertently carried to North American caves by people, has wiped out literally hundreds of thousands of bats of at least five species that hibernate underground in caves and mines. Although it is a toss-up whether you consider this an alien invasion or an emerging infectious disease, the disastrous effects on bat populations is obvious.

Bats and Cities

At least some species of bats show remarkable opportunism and resilience. To appreciate the ability of bats to exploit urban environments, visit the Congress Avenue Bridge in Austin, Texas any evening from April to early October. In April, thousands of Brazilian Free-tailed Bats leave the bridge every night. (Figure 11.3) By August, it is over a million. The same scene is played out at many other sites in the world where thousands of bats congregate in roosts. In Abidjan, Ivory Coast, Kampala or Uganda, it is Straw-colored Fruit Bats. In Sydney and Cairns in Australia, it is Grey-headed Flying Foxes. At caves in Malaysia and Borneo, it is Wrinkle-lipped Free-tailed Bats (*Chaerephon plicata*). Although the aggregations of bats are not always tourist attractions, they are impressive and demonstrate that some bats are not deterred by human activities.

Brock Tracks Down Some Elusive Bats

Despite many visits to Africa from 1972 to 1999, I had only caught two Large-eared Free-tailed Bats. So, when I heard about these Free-tailed Bats roosting in houses in and around Durban, South Africa, it sounded too good to be true. (See Figures 11.4 and 11.15) These spectacular bats are widespread in Africa but are rarely caught, and nobody is sure of their population numbers. As a result it was a real treat when I and some colleagues found colonies of thirty to sixty Large-eared Free-tailed Bats in attics in and around Durban. In the South African province of Kwazulu-Natal, these bats have the same conservation status– "Endangered"–as Black Rhinos (*Diceros bicornis*). Yet some individual attics housed more of these bats than there are Black Rhinos in the whole province!

When I visited the area in 2000 and 2001, there was little native vegetation around Durban and most available land was used for growing sugarcane or was built over by humans. Even in this setting, the bats appeared to prosper. Large-eared Free-tailed Bats produce echolocation calls that are entirely audible to people but, which, because of their low frequency, should make the bats relatively inaudible to moths with bat-detecting ears. (See page 123.) Researchers look forward to hearing more about what insects they eat.

Figure 11.2.
Two views of recently cleared land in Belize. In 2010 this area was forested, but in 2011 when Brock took these pictures, the view was very different (**A**). the image below shows the junction between forest and cleared land as well as the scorching of fire (**B**). Remaining forest is visible in the distant background on the left in B.

A
B

Figure 11.3.

Figure 11.3.
Every night in the summer, tens of thousands of Brazilian Free-tailed Bats emerge from under the Congress Avenue Bridge in Austin, Texas. This view clearly shows the crowds of people that gather to watch the emergence. Photograph by Paula Tuttle.

Figure 11.4.
A group of Large-eared Free-tailed Bats in the attic of a farmhouse near Durban, South Africa (**A**). The distribution of these bats is shown in the map on the right (**B**). Arrows indicate known colonies of these bats; the other symbols show sites where they have been caught.

Figure 11.4.

Figure 11.5.
Just outside of Durban, some colonies of Large-eared Free-tailed Bats occur in houses in relatively new subdivisions.

Urban Bats

A surprising number of bat species can successfully live and thrive in urban settings. A survey of bats in parts of Mexico City revealed that Free-tailed Bats (at least two different species) are relatively abundant and widespread there. These bats tend to fly high where they are less likely to be hit by vehicles or taken by hunting cats. They also roost in roofs and crevices high buildings, again putting them out of harm's way. In Sweden, Particolored Bats (*Vespertilio murinus*) are expanding their range northwards as they exploit the roosting opportunities associated with apartment buildings and high rises. (See Figure 11.6.) The mating calls and displays of males make the bats obvious to the informed observer in cities though most people probably don't look up high enough to notice them.

Research has repeatedly demonstrated the pervasiveness of some species of bats in urban landscapes. The range of examples is impressive: from

Figure 11.6.
A Particolored Bat (*Vespertilio murinus*) photographed by Jens Rydell in Sweden.

Common Vampire Bats in Sao Paulo (Brazil), Wahlberg's Epauletted Fruit bats in Harare (Zimbabwe), Durban, and Petersburg (South Africa), Big Brown Bats in cities and towns in the United States and Canada and Common Pipistrelles in cities in Europe. Nancy has watched bats flying over Central Park in New York City, and has rescued Big Brown Bats found flying in exhibition halls at the American Museum of Natural History. Research in Calgary (Alberta, Canada) revealed the importance of opportunities for both roosting and foraging for the bats living there. As in Mexico City, Toronto or New York City, some bats are rarely encountered, but others may be very common. In some cases the roosting conditions in human structures may be more beneficial to bats than what they find in traditional roosts. Brazilian Free-tailed Bats roosting in bridges have been shown to be in better condition than those roosting in caves in some cases.

Bats and Agriculture

In Europe, bat biologists have documented correlations between the demise of some species of bats and the use of pesticides in agricultural operations. In many parts of the world, agrochemicals including pesticides are routinely involved in the intensification of agricultural operations. Intensive agriculture is intended to increase productivity (and profit) and is thought by many to be necessary to feed expanding human populations.

In southern England and Wales, bat activity has been shown to be higher at organic farms than at conventional ones. This was found to be true whether the measure was bat activity assessed by monitoring echolocation calls or foraging activity, over land or water. This included a range of species of vespertilionids and rhinolophids. Even the most common species, Common Pipistrelle and Soprano Pipistrelle (*Pipistrellus soprano*), showed this pattern of abundance and activity (Figure 11.7). Often organic farms differ from other farms in their reduced use of conventional agrochemicals, which may account for why bats seem to do better in this setting.

Bats and Climate Change

The fossil record repeatedly demonstrates that bat faunas have changed over time. On the Hawaiian Islands, the Hawaiian Hoary Bat or peapea (*Lasiurus cinereus semotus*) is currently the only endemic land mammal. There is evidence, however, that another, as yet un-named species of Vesper Bat also occurred there in the recent past, but is now extinct. Vampire bats are now restricted to areas in South and Central America, but are known as fossils from Cuba and Florida as well as California and West Virginia. Ghost-faced Bats (*Mormoops megalophylla*) previously occurred in Jamaica and Cuba but now are restricted to Central America. (Figure 11.8) Antillean Fruit-eating Bats, today known only from Puerto Rico, the Virgin Islands and Barbados, were formerly found in Jamaica. The reasons for these changes in the distribution of bats remain unclear. In some cases, such as in the Caribbean, changing sea levels due to climate change, together with stochastic event such as hurricanes, may be responsible. In other cases the reasons remain more mysterious.

Some species of bats once thought to be extinct are later shown to still exist alive in the wild, although this is very rare. Bulmer's Fruit Bat (*Aproteles bulmeri*) from Papua New Guinea is one of the best known examples. Originally described from 12,000 year-old fossil remains found by archaeologists in food middens, living specimens were "rediscovered" 1975. This may be an example of a bat that almost succumbed to human hunting pressure. The rugged terrain of Papua New Guinea helped to protect these bats but at the same time has made it difficult to obtain details about them.

Bats appear to be excellent indicators of changes in climate and habitat. Different bat species have different dietary (Chapter 5) and roosting requirements (Chapter 6), which means that they may respond differently to changes in their environment at both local and regional levels. The varying mobility of bats is another factor. Some bat species are specialized for efficient long-distance flight, while others are better suited to shorter flights inside dense forest. (See Chapter 3.) If habitats change in one region, the original species may be able to more easily exploit suitable habitats elsewhere in their range, while others may not. The fact that bats live a long time and reproduce slowly, *e.g.*, have only one or two young a year, is important because it may limit the ability of populations to "bounce back" after catastrophic events. In the United Kingdom, Grey Long-eared Bats (*Plecotus austriacus*) appear to be in decline along the edges of their range, while populations in the Balkans appear to be healthy and expanding. Knowing where bats forage, what they eat and where they roost allows biologists to use mathematical models to predict where they will occur. Data from genetic analysis provide an historical picture of bat populations. Taken together

Figure 11.7.
A Soprano Pipistrelle emerging from its roost in Sweden. Photograph by Jens Rydell.

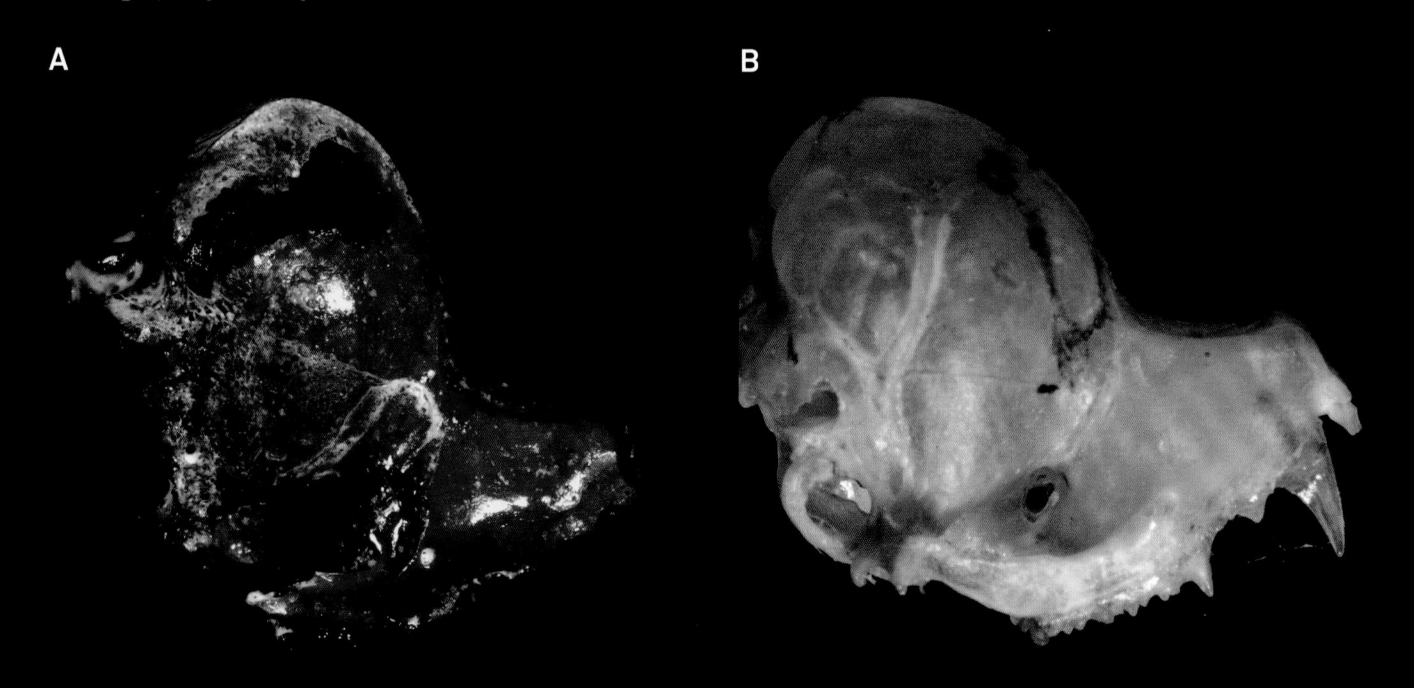

Figure 11.8.
Fossil Ghost-faced Bat from Cuba (**A**) compared with a recent specimen from the Yucatan Peninsula, Mexico (**B**).

Figure 11.9.
The range of the challenges of handling bats is illustrated by this comparison. In **A**, a Jamaican Fruit Bat illustrates the one glove technique. In **B**, the Large Flying Fox (*Pteropus vampyrus*) being held by Nancy's son Matthew. The bat, named Kamilah, was born at a zoo nineteen years ago and damaged her wings on a hard-sided caging before she could fly. She came to live at the Bat Zone at Cranbrook Institute of Science in 2002. Rob Mies, Executive Director of the Organization for Bat Conservation, travels widely with Kamilah and uses her to teach the public about bats. Photograph **B** by Nancy Simmons.

results of modeling and studies of genetics allow biologists to predict how bats may be affected by global warming and other changes. Meanwhile in the New World tropics, some species of New World Leaf-nosed Bats appear to be very sensitive to disturbances of their habitats. Species ranging from the large (200 g.) Spectral Bat to the tiny (7 g.) Little Big-eared (*Micronycteris megalotis*) Bat no longer occur at sites where the forest has been disturbed.

Scientists do not know exactly what the parameters to which these bats are sensitive, but they can document the changes in populations and distributions over time.

Perceptions of Bats

Bats have an image problem. Quite simply, many people don't like them or understand them, which can have negative effects on efforts to conserve bat populations. In Western popular culture and the press, bats are often associated with diseases, blood-feeding and other undesirable traits. But many positive perspectives about bats emerge from folklore in different cultures. (Chapter 10)

A comparison of the attitudes to bats expressed by five year olds and ten year olds living in urban areas reveals no innate aversion to bats, suggesting that negative attitudes about bats have to be learned. Five year olds tend to be curious about bats and fascinated by them. By age ten, girls tend to be "afraid" of bats, apparently the result of social conditioning. Boys remain interested and curious about bats.

The popularity of bats in recent years is attested to by the crowds gathered the Congress Avenue Bridge in Texas, as well as exhibits about bats at museums, zoos and conservation areas. High viewership indicates that people will take opportunities to see and to learn about bats if they are offered a chance. When we visit a classroom to talk about bats, elementary school students tend to be eager, well informed about the animals and full of questions. The response is even better if we can bring a bat with us. This is a widespread response, whether the class is in Australia, Canada, the United States, Belize or Jamaica. Evening presentations in the field about bats also attract many people. Although attendees at presentations on bats tend to be self-selected for interest in nature and animals, it nevertheless surprises us that many are quite well-informed about bats. While people in audiences ask about bats and rabies, they also express concerns about the impact that WNS and wind turbines have on bats. There is no doubt that public attitudes about bats have been changed by organizations such as Bat Conservation International, and through the efforts of bat researchers, dedicated teachers and professionals involved in public education.

There is an interesting, ongoing tension between bat researchers and public educators. Concerned about the possibility of bats being involved in the spread of diseases, some colleagues argue that people should not be allowed to touch bats. (Figure 11.9) The same concern extends to those who would strongly discourage anyone from showing a bat (or a picture of a bat) held in an ungloved hand.

Others, including us, believe that it is important to give people a chance to see and touch a real bat. We agree that it is important to pick the "right" bat, and hold it so that it cannot bite anyone but the holder. Touching a bat's wing or stroking its fur is an effective way to introduce a bat to others. If the holder makes the point about being careful to avoid being bitten and having rabies pre-exposure shots, then the net effect can be positive. Appearances of live bats from disease-free captive colonies, maintained and accustomed to public showing, can play a very important role in public education about bats. Speaking about diseases that may be associated with bats is important, as is emphasizing the fact that bats bite in self-defense. (See Chapter 9.) Even without a "real bat" in hand, appropriate images are vital. We choose with care the images that we present to give a balanced picture. Some bats can look very fierce, but not always. (See Figure 9.7.)

What People Can Do for Bats

For at least thirty years, people have been building and putting up "bat houses", artificial roosts for bats. Many garden centers and nature stores sell bat houses, and a simple internet search reveals a wealth of advice about how to build them and where to put them. Sometimes bat houses work to attract bats but often they do not. One large study in the United Kingdom suggested that only about 5 percent of bat boxes deployed in woodlands attracted resident bats. (See also Chapter 6.) Much depends on the location height above the ground, exposure to sunlight, nearby vegetation because bats are very picky about where they live. Bat researchers have used bat houses/boxes in their studies, demonstrating their effectiveness if situated correctly. In Texas, at least one large underground bat house has been built for and occupied by Brazilian Free-tailed Bats, a species well known for their opportunism when it comes to finding roosts. Work in Sweden illustrated how making roosts available brought several species of bats into pine plantations otherwise lacking bats. (See page 135.) In Israel, changing the ceiling surface in military bunkers, made the structures suitable for more bats.

Scientists need more basic knowledge about bats if they are to effectively conserve them. The survival of bats depends upon their having access to roosts and to food. Adequate food and roost resources usually translate into secure populations of bats—ones likely to survive as long as conditions remain the same or improve. The diversity of bats means that not all species use the same food and roosts. Thus a conservation plan may need to be adjusted by species and the particular habitat. Many details of the biology of a species are relevant to effectively protecting endangered species. How large is the geographic range of the species? How many individuals are there in the population? How long do individuals live? How often do they reproduce? What proportion of the young survive long enough to reproduce? What factors limit the size of the population? Unfortunately scientists do not have answers to most of these questions for the vast majority of the >1300 species of living bats.

Biologists use a range of basic biological information, including some of the features mentioned above, to assess the conservation status of a species. By broad consensus, most scientists use established criteria to label species as "Extinct" "Extirpated", "Endangered", "Threatened","Special Concern" or "Not at Risk." The category "Data Deficient" is used when scientists lack the necessary information to place a species in one of the other categories.

Details about the size of the population are essential when assigning conservation status, especially when populations have changed in size over time. Christmas Island Pipistrelles appear to be extinct, while Seychelles Sheath-tailed Bats remain Endangered. In 2012, the impact of WNS on populations of bat species that hibernate underground in northeastern North America was directly responsible for catastrophic declines in populations of species such as Little Brown Myotis and Northern Long-eared Bats. In the Northeastern United States, both of these species may be extirpated by 2020. WNS has not yet arrived in other parts of these bats' ranges and there populations of both species may appear to be secure.

Often scientists may be concerned about a species but lack the evidence to support the case that it should be considered Endangered and require protection. One challenge is to distinguish among species that are rarely encountered (because they are naturally rare or simply hard to detect or capture) and species whose survival is actually in question. This distinction is made more easily when there is quantitative evidence to consider. Sometimes the lack of data can move a species from Special Concern to Data Deficient. For example, Keen's Myotis is a species that occurs from coastal Alaska south through British Columbia and Washington. Although this species is wide-

Figure 11.10.
A New Zealand Lesser Short-tailed Bat. Note that the bat's face is covered in pollen. Photograph by David Mudge courtesy of Nata Manu Images and Stuart Parsons.

spread in coastal rainforest, most of what is known about it is from Hot Springs Island in Haida Gwaii National Park. At this location, crevices above the hot springs were used as roosts by Little Brown Myotis, Keen's Myotis and California Myotis (*Myotis californicus*). (See Figure 7.7.) The hot springs provided central heating for the bats. After, however, an earthquake in 2012, the hot springs dried up and it remains to be seen if the bats will continue to use the now unheated site. The change in the roost availability may affect the size of the population of Keen's Myotis at this site. How important this particular location is to the species as a whole is another question. Often catastrophic events that reduce a population at one site may be compensated for by population expansion elsewhere in the range of a species. This complicated dance means that researchers often need data from multiple locations to properly assess the status and heath of the overall population of a species.

Assessment of the conservation status of bat species can provide managers with useful information, but little is gained by listing a species of bat as Endangered if this is not supported by action. In the Seychelles, ending programs to control invasive plants around the roosts used by Seychelles Sheath-tailed Bats could result in the extinction of these animals even though they are listed and protected. In the case of WNS, it remains to be seen if listing Little Brown Myotis and Northern Long-eared Myotis as Endangered species in Canada will make any difference to the survival of these species there.

What are the prospects for reintroducing bats into areas where they used to occur? New Zea-

land Lesser Short-tailed Bats were much more widespread when the Maoris arrived in the 13th century than they are now. (Figure 11.10) In April 2005, twenty-five pregnant female New Zealand Lesser Short-tailed Bats were translocated to 165 Ha. Kapiti Island located off the southwest coast of the North Island of New Zealand. After capture at a site 40 km. from Kapiti Island, these pregnant females were housed in an aviary and provided with food (meal worms, honey and moths) and water. When twenty bats born to these females, the mothers and offspring were introduced to an aviary on Kapiti Island. There they had access to food, and then were free to leave the aviary. Initially, results of this project offered hope to those who would reintroduce some bats to areas from which their species had been extirpated. But an as yet unknown disease that caused hair loss and scab formation on the pinnae struck the introduced bats. They then had to be captured and returned to captivity.

Reintroductions of Little Brown Myotis, which have been decimated by WNS, back into locations in the northeastern United States was once thought to be a possible management strategy for dealing with the population declines wrought by the disease. The sources of bats might be the apparently isolated populations on Newfoundland and Haida Gwaii. But experiments by researchers in New York have showed that the spores of the fungus that causes WNS persist in roosts after local bat populations have been extirpated. This suggests that reintroductions are not a viable solution as long as natural roosts retain disease-causing spores even after the bats are gone. Other challenges with reintroductions is lack of knowledge about how bats learn their home ranges, the locations of roosts and foraging areas and how they establish their social networks.

Figure 11.11.
The nacelle, center point of the turbine blades, is 60 m. above the ground. Note the blurring of the tips of the blades indicating their movement.

Bats and WEFs (Wind Energy Facilities)

Virtually everywhere that someone has looked under the giant windmill turbines at wind energy facilities, they have found dead bats. Initially it appeared that bats are killed by colliding with the moving blades of turbines. (Figure 11.11) Then it emerged that embolisms in the lungs killed many of them. The embolisms are believed to have resulted from the negative pressure areas behind the blades.

One problem in attempting assess the threat that wind turbines pose to bats has been determining just how many bats are killed by them. Bat carcasses are typically small, decay quickly and are easily carried away by scavengers. It is easy to overlook a carcass, and it becomes easier the longer it has been there. It is even more difficult to identify remains by species (especially when they are not fresh), although distinct features such as their frosted fur make Hoary Bats easy to distinguish from others at sites in North America. Although some people have focused on deaths of migratory tree-roosting bats (Hoary Bats, Eastern Red Bats and Silver-haired Bats) at turbines, Little Brown Myotis represented 20 percent of the dead bats found beneath some turbines in the United States and Canada. By developing standardized protocols for searching for carcasses, researchers have increased the accuracy of estimates of bat mortality at wind turbines. Dogs trained to sniff out carcasses have been used in some cases. There remain several unanswered questions when trying to assess the effect of WEFs on bats. What is the size of the populations of bats affected? How significant are these losses to different species of bats? Lack of information makes it impossible to assess the impact of any mortality agent on populations of bats and complicates assigning conservation urgency to the problem.

By monitoring the echolocation calls of bats it is possible to assess and monitor the occurrence of bats in specific areas, and to evaluate the prevalence of particular species of interest (assuming that they have distinctive calls). Monitoring bat activity may help facilitate location of wind turbines away from sites with heavy bat activity. But turbines also kill birds, and sometimes the very features of a site that make it good for bats might make it bad for birds. Ongoing acoustic monitoring also could be used to determine when to shut down turbines at a particular site and time to minimize bats' exposure to the hazards associated with them.

Other unanswered questions lurk behind the wind turbine problem. Are bats attracted to turbines? If so, why? Could scientists find ways to warn bats away from them? These are topics of ongoing research. There will not be a single answer to these questions for any species, and there may be different answers for males versus females, or old versus young animals, for every species.

Change Over Time

Over repeated visits Brock, Nancy and their colleagues have documented the presence of about fifty species of bats in the area of Lamanai in Belize. About 1000 years ago (end of Late Classic Maya Period), the area shown in Figure 11.12 probably had been cleared to grow beans, squash and corn (the three sisters). Information is lacking about the bat faunas in the Maya centers, but land use patterns suggest the cultivation of fruit trees such as Breadnut (*Brosimum alicastrum*) and attendant large insect populations that could have provided food for some bats. Maya buildings also could have provided opportunities for roosting. By about 800 years ago, the large Maya complexes in the area were in disrepair, although people lived in some parts of the ruins through at least 1500 A.D. By this time, many of the Maya buildings were overgrown by the native dry tropical forest. Modern land clearing in Belize has yet again changed the picture in many areas. (See Figure 11.2.) Modern agriculture in the region does not appear to produce many crops sustainable over long periods. Within a few years, cleared land becomes pasture for cattle. The cattle, in turn, support thriving populations of Common Vampire bats. It is unlikely that vampires were as common during Maya times, when turkeys were the only domesticated animals. The habitats, roost sites and food sources available to bats around the Lamanai Maya complex have thus changed repeatedly over the last millennium. (See Figure 9.13.)

Change over time is one of the constants of life on Earth. On all continents, humans have simply speeded up the process. Although we tend to think of habitat losses and destruction as unidirectional, this is not always the case. In Africa, the savannah woodland had swallowed the ruins of 14th century Great Zimbabwe Empire by the time they were discovered by Europeans in the 16th century. Today the Great Zimbabwe ruins are home to quite a large community of bats. Protected areas can, and do, protect bat populations. The real question is whether or not this will be enough for many species.

Continued expansion of the human populations and the associated encroachment on natural systems surely means that some species of bats will dwindle and then disappear. Others, however, may prosper in the wake of human disruptions. We should remember that this is not a new development and that nature has repeatedly demonstrated its resilience. Nevertheless, human population growth and the expansion of modified landscapes throughout the world mean that bats are under increasing pressure every year that passes.

Figure 11.12.
Maya ruins at Lamanai, northern Belize, including the Jaguar Temple in the background.

12

What's Next in Bats?

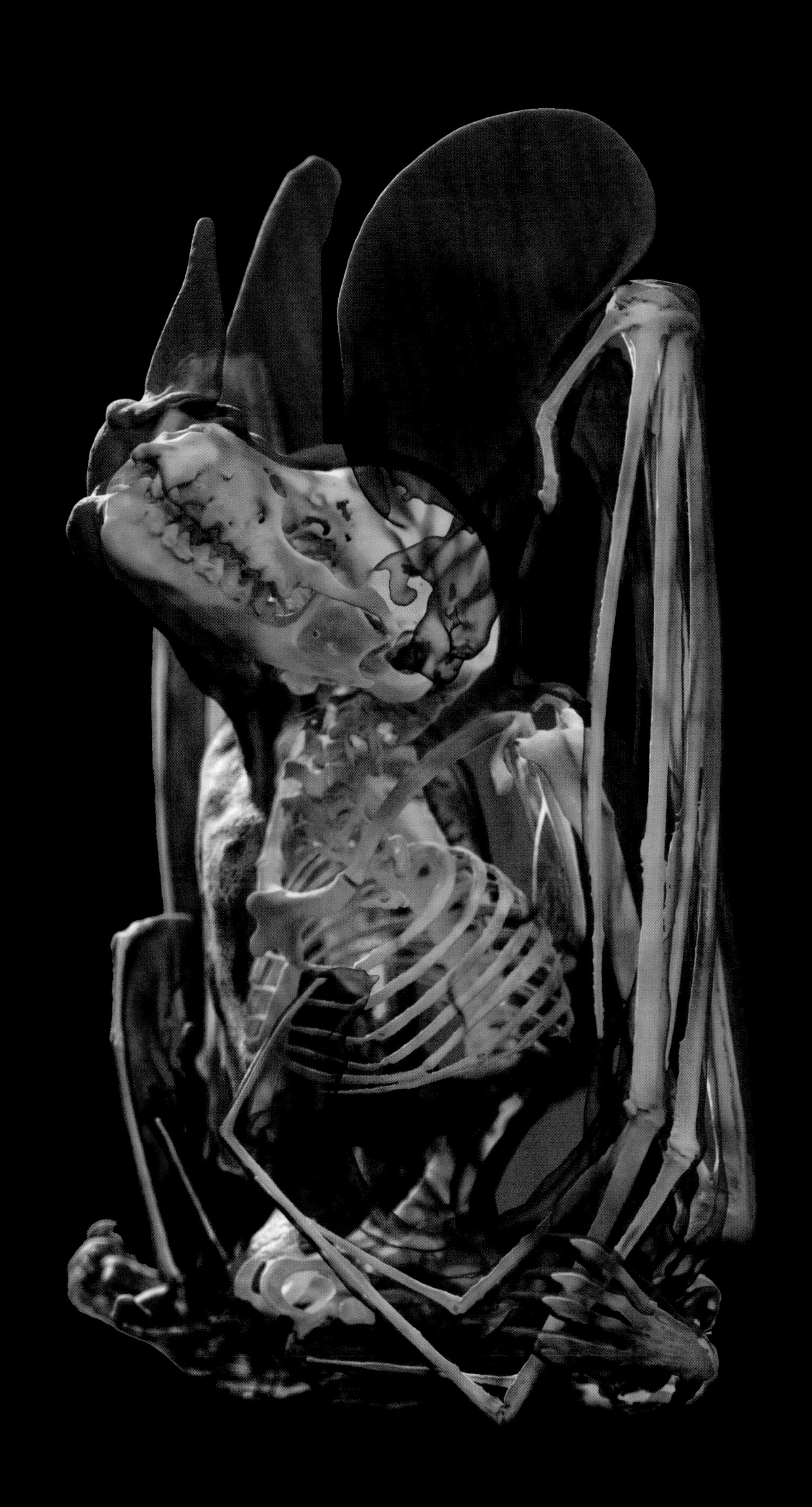

Figure 12.1.
Flying directly towards the camera, this Fringe-lipped Bat triggered an infrared beam, taking its own picture.

Nancy and Brock Photograph a Bat

The bat flying towards the camera epitomizes some of the challenges bats present to scientists. (Figure 12.1) At the time we took this picture, we agreed that it was a male, possibly a Fringe-lipped Bat. To confirm this, however, we had to catch one and examine it closely. Perhaps not surprisingly, the unusual distinctive faces of these bats have made it relatively easy to determine the geographic range of Fringe-lipped Bats, which extends from Mexico to Bolivia and Southeastern Brazil. But the Fringe-lipped Bats of Central America look a bit different from those that Nancy studied in South America, and genetic data has revealed that populations in different parts of their range, *e.g.*, in Belize versus Brazil, are quite genetically distinct. Are these different species or a single, variable species? This question has yet to be resolved, in part because while there are museum specimens from many areas, tissue samples are comparatively scarce. Genetic variation and additional genes in more intervening populations need to be examined before we will be sure how many species there may be in the complex of animals we call Fringe-lipped Bats.

The Unknown Known

What we know (and do not know) about bats is hard to describe even when one has a whole book in which to do it. For scientists, part of the allure of working with bats in the 21st century are the amazing questions that come to mind and the new methods that we have to address them. We know so much, and yet so little; but every year we learn more.

Looking closely at a single bat species exemplifies how scientists can simultaneously know quite a bit about an animal and still have remaining questions. Fringe-lipped Bats live in forests throughout their range, where they roost in hollows in trees or underground, usually near streams and pools. In Panama, controlled experiments by researchers at the Smithsonian Tropical Research Institute revealed that these bats listen for the mating calls of frogs, and use them to detect, locate and identify their targets. Fringe-lipped Bats unerringly identify poisonous frogs and toads by their calls and avoid them, while homing in on non-poisonous species, which they readily catch for food. Radio-tracking revealed that in Panama, Fringe-lipped Bats were most active along stream beds. Quickly known as "the frog-eating bat", it soon became obvious that Fringe-lipped Bats ate a wide range of insects in addition to frogs . (Figure 12.3) Furthermore, these bats watch others and quickly learn to identify prey from acoustic cues (songs, calls) they had not previously heard.

Looking at a portrait of this bat brings other questions to mind. (Figure 12.3) What functional roles do the noseleaf, tragus and wrinkles in the ears play? What is the function of the wart-like projections on the chin, lips and edges of the noseleaf? Biologists have speculated for decades that these warts might be sensory in nature, sensitive to either touch or chemical signals and perhaps used as last-minute warning devices to prevent the bats from capturing distasteful prey. This seems supported by observations of extensive innervation of these structures. However, nobody knows for sure what their function is, so the reason that

Fringe-Lipped Bats have fringe-like warts remains a mystery.

How do Fringe-lipped Bats integrate information collected by echolocation, listening to other sounds and by vision? It is easy to demonstrate that bats collect and attend to both echolocation and vision. For example, a Little Brown Myotis will fly around a room deftly avoiding all obstacles. Output from a bat detector confirms the flying bat produces echolocation calls. At some point, however, the bat likely will fly directly at, and crash into a window. While echolocation tells the bat the truth about window glass (that it represents a hard surface), vision does not. Bats like the Fringe-Lipped Bat see well, but biologists do not really understand the tradeoffs—for instance, under what conditions the benefits of vision outweigh those of echolocation, or vice versa. Biologists really do not understand when these bats switch between different sensory modalities in their normal nightly activities in the wild, let alone how much they use vision and echolocation in combination.

It will be challenging to extend observations and experiments about bats switching between vision and echolocation into the field. The work will require quite precise information about a bat's behavior associated with echolocation and vision, and achieving that will require some technological advances and some wonderful study sites. Bats, like people, probably continuously integrate input from different sensory channels. For bats on the wing, hearing, vision and input from other sensory receptors all seem to be very important as is hearing involved in prey-generated sounds as well as those used in echolocation. While some bats use echolocation to detect water surfaces, it is less clear what role vision plays in this behavior. The role of olfaction in the lives of bats largely remains a mystery as well. Scientists know that scents are very important for social communication—indeed, some species have specialized scent-producing glands on the chest or shoulders—but most species remain unstudied

The eyes of bats differ among species, suggesting that vision is more important to some bats than to others. In early studies of the homing behavior of bats, experimental animals were blinded or temporarily deprived of vision. As noted earlier, day-flying bats are quite vulnerable to attacks by predatory birds. (See page 177.) A blind-folded (or blinded) bay may not be able to tell if it is flying by day when it might have been an easy target for a predatory bird. The failure of a blinded or blind-folded bat to return home may reflect more about vulnerability to predators than about the cues that bats use in orientation.

Groups of bats clearly use and adjust their echolocation calls when flying in the same airspace with other members of the same species. The need to do this is apparent when large group of Fringe-lipped Bats emerge from their roost inside a Maya temple. How do they avoid crashing into each other and the walls of the narrow tunnel? Scientists do not yet know the details of how they manage this—just that they do it every night.

More impressive and perhaps more interesting is the behavior of nectar-feeding bats around feeders. (Figure 12.4) Several species of nectar-feeding bats may converge on a hummingbird feeder or a flower. The question of how they avoid mid-air collisions remains unanswered but likely involves some aspects of acoustic signals, namely echolocation calls. What is the significance of bats flying around with their tongues hanging out? These bats hydraulically extend their tongues by pumping blood into them. Does the photograph suggest that the bats take time to deflate their tongues?

Figure 12.2.
A portrait of a Fringe-lipped Bat showing prominent ears, noseleaf, tragus and wart-like projections around the mouth

Figure 12.3.
A Fringe-lipped Bat eating a katydid (Tettigoniidae).
Photograph by Beth Clare

Figure 12.2.

Figure 12.3.

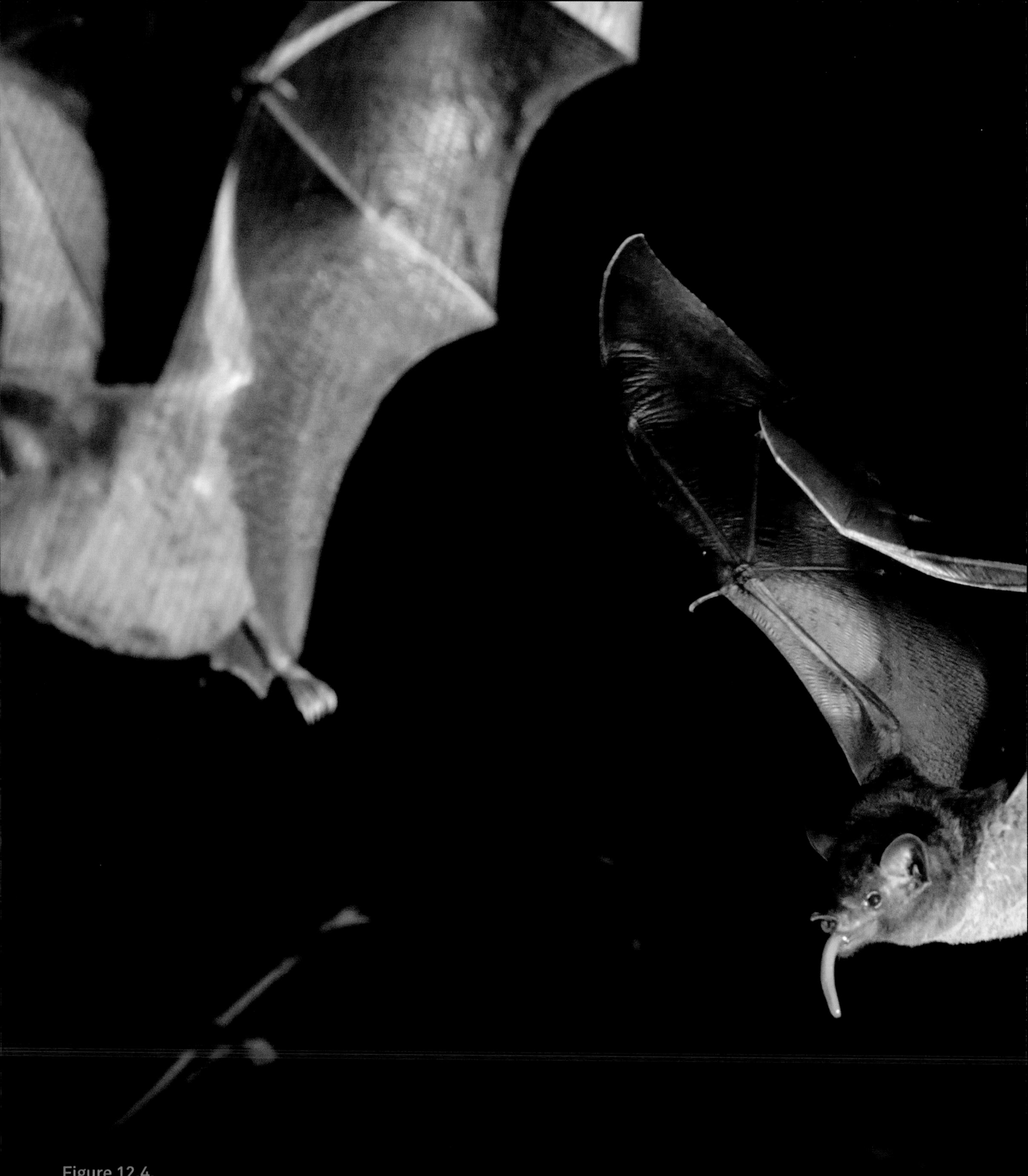

Figure 12.4.
Captive Pallas' Long-tongued Bats mill around a feeder in the Biodôme in Montreal, Quebec.

Figure 12.5.

Figure 12.5.
In May 2013 this group of mainly bat biologists met in Belize for an annual research retreat. Photograph by Sean Werle.

Figure 12.6.
Digital model of a complete Little Big-eared Bat (*Micronycteris megalotis*) based on a CT scan. Different structures are shown in context: (left) the brain inside the skeleton; (middle) the skeleton inside the skin; and (right) the skin surfaces of the specimen. Courtesy of Dale Webster.

Figure 12.6.

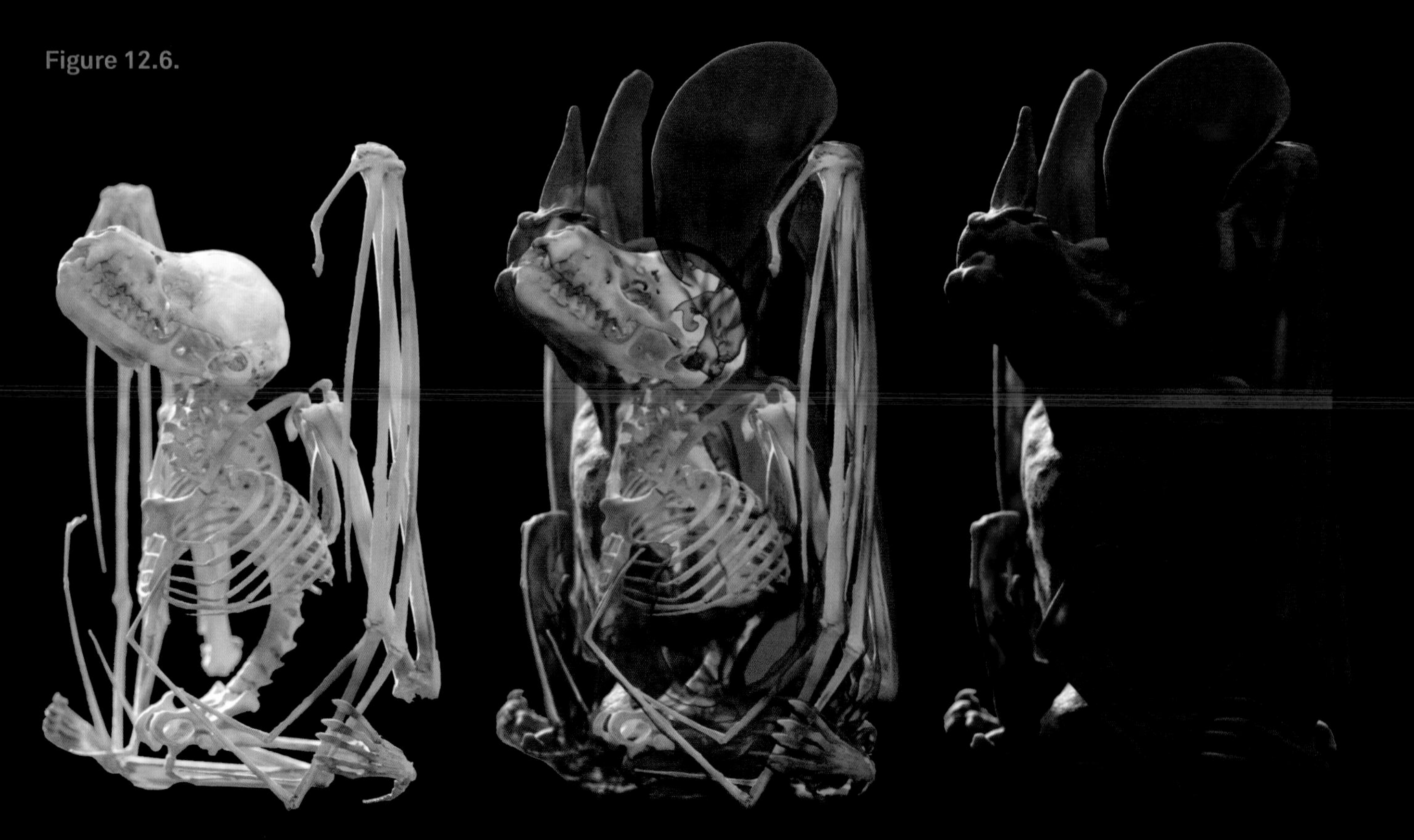

How We Progress

Scientists continue to make new discoveries about bats almost every day. These are often chiropterologists (biologists who focus their research on bats), but not always. Sometimes the most exciting discoveries emerge from teams of researchers including scientists who usually work in different fields or on different organisms. The importance of bringing to bear different points of view on a topic is illustrated by the expertise and interests of the people in Figure 12.5. In this group are people who study echolocation (Signe Brinklov, Paul Faure, Alan Grinnell, Toni Guillen-Servent, Katrina Hulgard, Cindy Moss, Jens Rydell, Jim Simmons, Mike Smotherman, Annemarie Surlykke), anatomy and flight (John Hermanson, Stuart Parsons, Scott Pedersen, Cosima Schunk, Sharon Swartz), behavior (Robert Barclay, Mark Brigham, Kim Briones, Hugh Broders, Zenone Czenze, Miranda Dunbar, Yvonne Dzal, Joel Jameson, Jennifer Krauel, David Johnston, Genni Spanjer, Cory Toth), evolution (Beth Clare, Kate Jones, Nancy Simmons), feeding behavior (Betsy Dumont, Sharlene Santana), as well as reproduction and development (Teri Orr, Karen Sears). Research teams working on bats may include engineers, botanists, geneticists, reproductive biologists and conservation biologists just to name a few. Meetings like this to exchange ideas and information often produces new discoveries about bats.

Technological advances in the last two decades have led to many new discoveries about bats. Digitization of preserved specimens allows preparation of new views of bats that do not require dissection. (Figure 12.6) Even the views provided by images taken with a digital camera can provide a closer look at details, *e.g.*, Fringe-lipped Bat. (Figure 12.7) These new techniques will reveal details about the roles of Ghost-faced Bat noseleaves and ear structures in echolocation. (See Chapter 4.)

Meanwhile, what is the function of the pebbled surfaces that appear on the ear and tragus of Greater Dog-like Bats and on the lower lips of Davy's Naked-backed Bats and Parnell's Moustached Bats? (See Figure 12.7.) It is tempting to associate these features with echolocation since the lips are involved in emission of sounds (which occurs through the mouth in these bats) and the ears are involved in receiving the returning echoes.

Computerized X-ray microtomography (CT scans) has provided yet another way to look at bats. This method allows researchers to digitally isolate and enlarge body parts for study, including everything from the skeleton to the tragus or margin of the ear. (See page .) In the latter case, application of engineering models to the data collected using CT imaging methods allows exploration of the importance of these structures in sound reception and transmission. Evolutionary modification of the ear margin and tragus probably strongly affects the ability of a bat to perceive echoes. By changing the facial features of a "virtual" bat, researchers can measure the role of features such as the tragus. (See page 13.) This approach offers a way for scientists to better understand the forces that have driven the remarkable modifications in ears and facial structures seen in bats. When it comes to bat faces, the Wrinkle-faced Bat clearly stands out—a story in structure and function waiting to be told. (See Figure 1.5D.)

By recording bat echolocation calls simultaneously on arrays of microphones, biologists can reconstruct flight paths of multiple bats and determine which bat said what, when and where. This approach to studying echolocation also allows measurement of the strength (intensity) of bat echolocation calls. This is an important advance that has revolutionized understanding of the echolocation behavior of bats. Donald Griffin had reported that while some bats produced very strong echolocation calls, others "whispered." The short duration of echolocation calls made it difficult to measure call intensity because the response time of the needle on sound level meters was longer than the calls. (See page 93.) Measures of call intensity allow better estimation of the range over which bats collect information by echolocation, now 20 m. to 30 m. versus earlier estimates of less than 10 m. Digitization of the recording and analysis of bat echolocation calls also has advanced the ability of scientists to recognize species by their calls. Research teams are assembling of libraries of bat calls, in turn allowing them to identify species of bats by their echolocation calls. (See Chapter 4.) Other advances in cell and molecular techniques have enhanced the ability scientists to explore other aspects of their biology, from physiology to evolution and social structure.

Evolution of Bats

The discovery of new fossils could strongly influence our view of bats and their evolution. Ad-di-tional fossil material could provide more definitive evidence about the mammals that gave rise to bats. (See Chapter 2.) What did the non-flying early members of the bat lineage look like? When and where did the transition from non-flying ancestors to bats occur? We expect that the transition occurred in a forest or woodland setting. At what point in the transition would we be able to say with certainty that the animal was a bat?

In 2013, analyses of genomic features confirmed convergences in genetic sequences among echolocating bats and toothed whales (dolphins and their relatives). Evolution of the ability to echolocate arose independently in these two groups, and convergence in genetic sequences is thought to reflect similar adaptations in very dif-ferent environments to meet the challenges of sound production and hearing. Interestingly, the genetic convergences extend to vision. Echolocating bats and dolphins see well but operate in environments where lighting is unpredictable, and hence may also have undergone convergent evolu-

Thinking about the origin and evolution of bats leads to other intriguing questions. What underlies the diversification of bats? Why are there so many species? The concept of "species" in biology is a basic premise, and biologists tend to believe that no two species living in the same place can fill the same niche. In other words, they cannot rely on exactly the same resources. The niche is the species' place in nature. Bats niches can roughly be defined based on some combination of where they live, what they eat, where they forage and where they roost. As discussed in Chapters 5 and 6, whether the focus is food, roosts or foraging areas, many species of bats are opportunistic. Is there competition among species of bats for access to food, foraging areas or roosts? The apparent abundance of resources raises the possibility that bat species—or at least many of them in any given area—do not directly compete with one another. An enticing challenge would be to design an experiment that would test the presence or absence of competition. It could also be argued that the very origin of bats was associated with an abundance of previously untapped resource available to those of the night sky.

Species of bat-eating bats occur in South and Central America, Africa, India, and Southeast Asia and Australia. Researchers do not know yet how much these bats rely on other bats as food. Nor do scientists know if any or all bat-eating bats use their echolocation to detect, track and identify bat prey, or if they sit silently and use the echolocation calls of their victims to locate targets. The answers to these questions may extend our knowledge about the advantages and disadvantages of echolocation.

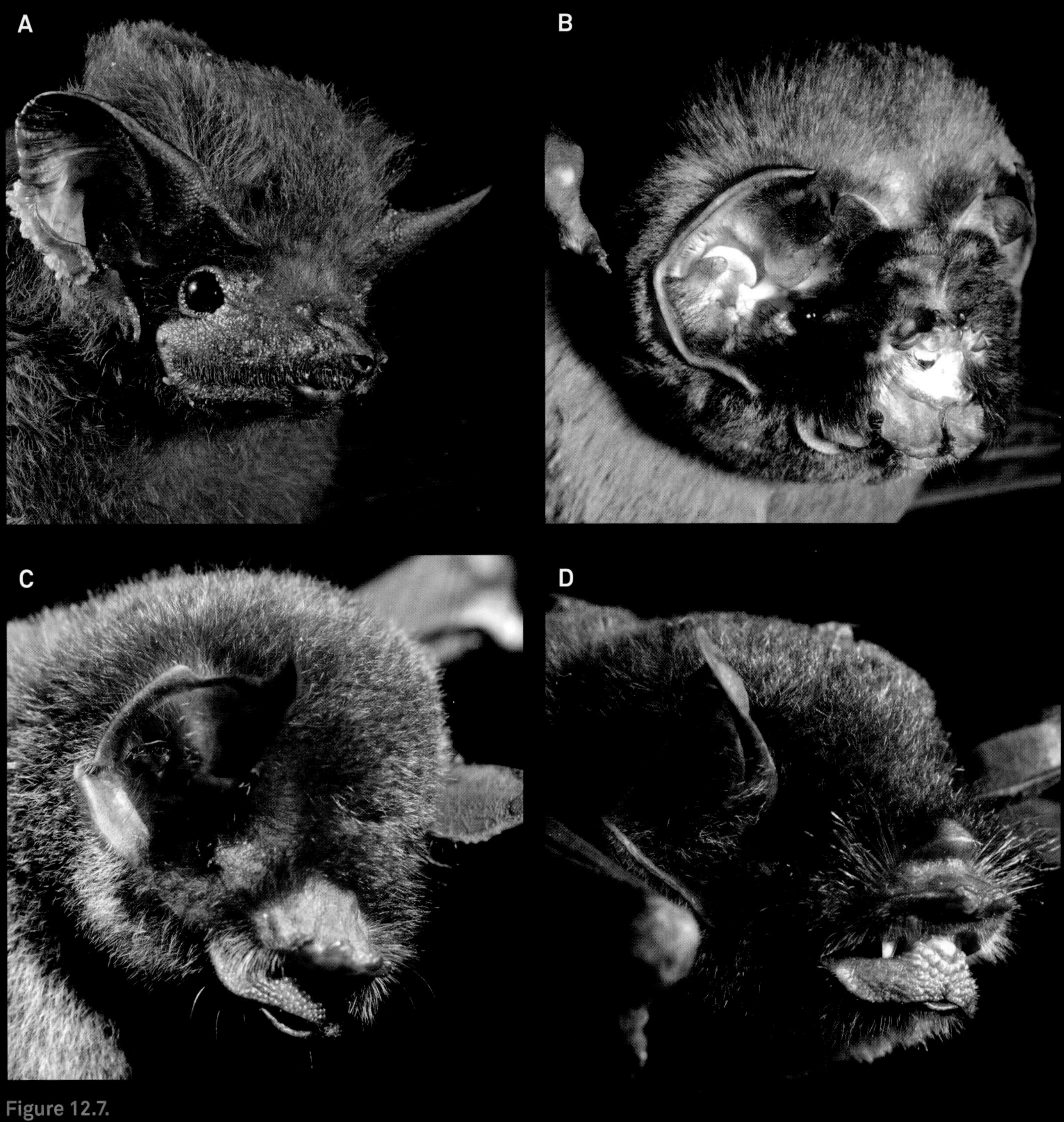

Figure 12.7.
Portraits of the four species. A Greater Dog-like Bat (**A**), a Ghost-faced Bat (**B**), a Davy's Moustached Bat (**C**) and a Parnell's Moustached Bat (**D**). Note the pebbled surface of the tragus in **A** and **B**, or the lower lip in **C** and **D**. Note the other details of the flaps of skin around the mouth in **B**, the features in the ear (pinna) and the ectoparasitic mites on the posterior surface of the ear in **A**. There is an orange streblid fly on the head in **D**.

Brock Encounters Bat-eating Bats

I remember the first time that I heard about a bat that eats other bats. When I was a graduate student at the Royal Ontario Museum, my supervisor, R.L. (Pete) Peterson, received some specimens from Panama, including one Spectral Bat. This large 200 g. bat had been caught in a mist net while it was trying to eat another bat entangled in the net. Then, in 1979, I heard about bats living in a disused water tower in Mana Pools National Park in Zimbabwe. (See Figure 5.3C.) Dolf Sasseen, who worked for the Zimbabwe Department of National Parks and Wildlife Management, had sent the skull of one of these bats to headquarters in Harare (then Salisbury) along with a report that the bats were eating frogs. When I visited the site in November 1979, I observed that the bats brought their prey back to the water tower and ate them there...with much lip smacking and crunching of bones. Among other prey, these 30–35 gram bats captured and ate 10 g. Egyptian Slit-faced Bats! In 1982 working with captive animals, Connie Gaudet, Dalhousie University biologist Marty Leonard and I established that Large Slit-faced Bats killed their bat prey with a powerful bite to the face, effectively smothering their victims. This is not "cannibalism"—which is defined as eating members of your own species, but instead represents the opportunistic habits of some carnivorous bats.

Nancy Captures More Bat-eating Bats

The very first specimens of Brosset's Big-eared Bats ever collected came out of a hollow tree in French Guiana. The tree was about a meter in diameter, and the hollow had a small opening (20 cm.) about a meter above the ground. I was working in the area with my husband, fellow mammalogist Rob Voss, who was helping me survey the bat fauna. We could hear bats inside the tree during the day, which was very exciting. What lived inside? We were dying to know. To catch the bats that were roosting in the tree hole, we built an enclosure out of mist nets (roofed by palm fronds) around the base of the tree, and then waited for dark. Much to our surprise, our bat trap did not just catch bats coming from inside the tree–it also

Figure 12.8.
This Striped Hairy-nosed Bat (*Mimon crenulatum*) has emerged from its roost ready to search for food. The large ears and noseleaf are striking features, but the bat's eyes are clearly visible. Note that its mouth is closed indicating that the bat transmits its echolocation calls through its nostrils.

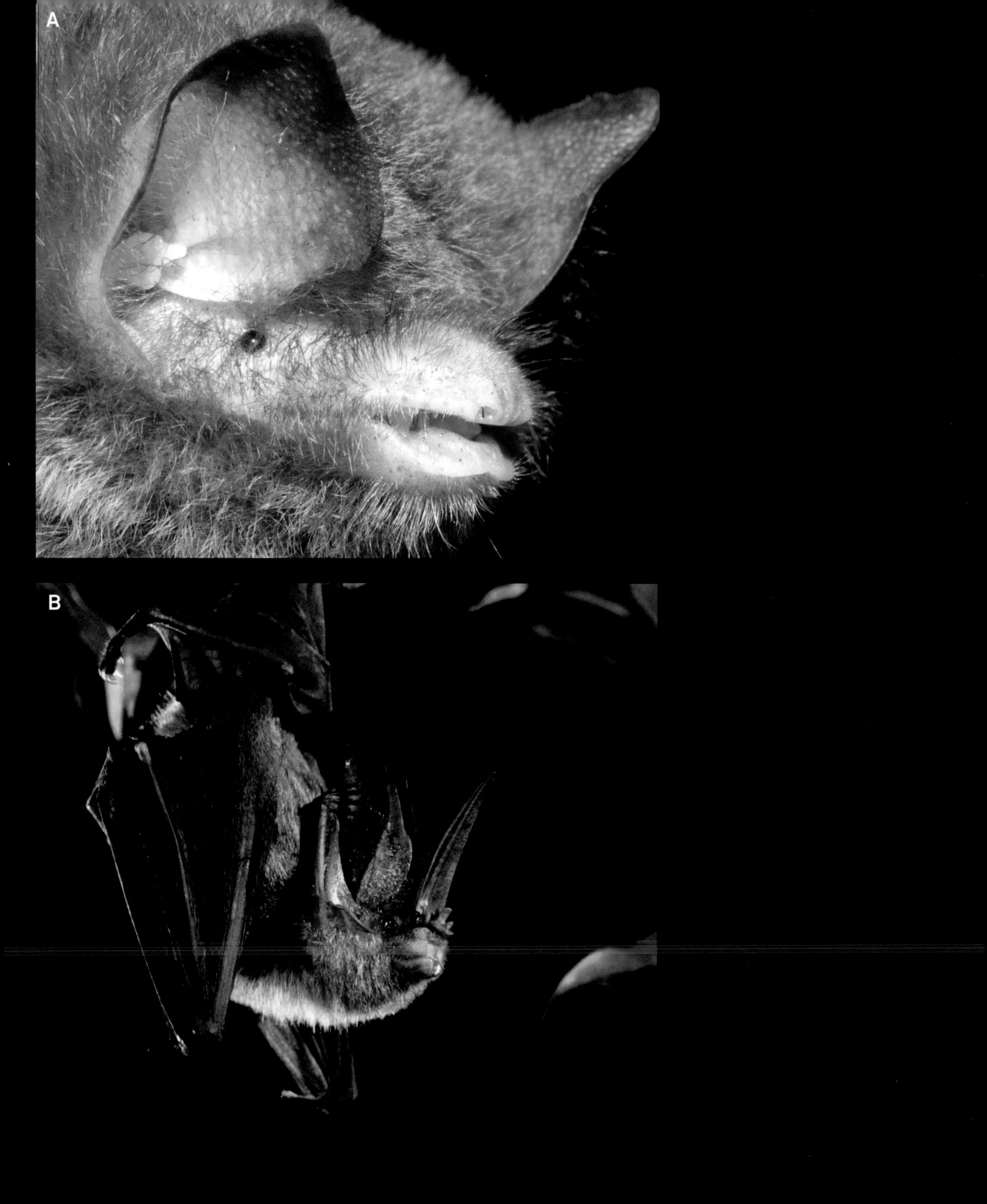
A
B

More Tantalizing Questions

It may be easy to wonder how anyone can spend their lives studying bats. Perhaps you now appreciate that the diversity of bats means an endless abundance of tantalizing questions for researchers Here are three more examples that especially interest Nancy and Brock:

Researchers do not know much about the biology of Striped Hairy-nosed Bat (*Mimon crenulatum*). (Figure 12.8) They appear to eat mainly animals, probably arthropods, and locate prey by sounds they produce. Nancy and her colleagues have noticed that they are among the first bats to emerge and become active as dusk falls in the Neotropics. Their echolocation calls are low in intensity, perhaps providing details about what is immediately before them. What role does their noseleaf play? Why is it serrated along the edges and hairy? Are their ears more attuned to the sounds of prey than to the bat's echolocation calls? Nobody yet knows the answers to these questions.

Another interesting group, Funnel-eared Bats (*Natalus spp.*), occur in parts of Central and South America, usually in lowland forest areas. These bats are small and delicately built. They roost underground and eat insects. Their flattened muzzles and funnel-shaped ears are distinctive, and their echolocation calls are low in intensity (Figure 12.9A). Beyond that our knowledge is sketchy. Bi-

Yet another amazing bat, Tome's Sword-nosed Bat has a noseleaf that lives up to its name. Again, however, beyond general descriptions about where it roosts (in hollows) and what it eats (insects), researchers know virtually nothing about this species. (Figure 12.9) How does it use its sword-like noseleaf to advantage? The evolutionary origins of such unusual appendages may have lessons to teach us that go far beyond bats to more general questions of how evolutionary novelties arise.

Figure 10.3C pictures a brightly colored bat on a snuff bottle from China. While most bats are not colorful, a pattern of red fingers on black wing membranes occurs in different Vesper Bats. (Figure 12.10) This prominent pattern appears in the Tacarcuna Bat (*Lasiurus castaneus*) of South and Central America, in Welwitsch's Bat (*Myotis welwitschii*) in Africa, and in various species of Woolly Bats (*Kerivoula*). The significance of the color pattern is high on our list of enticing questions about bats.

Figure 12.10.

Painted Bats like this one are among the most colorful, but striking blacks and reds also occur in Vesper Bats from the New World tropics, Africa and tropical Asia. The Painted Bat is depicted on a snuff bottle from China. (See Figure 10.3.) Photograph courtesy of Merlin Tuttle.

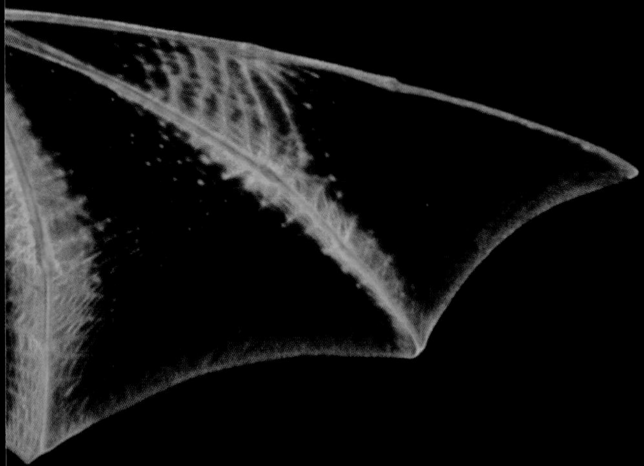

A Group Effort

Our collective knowledge about bats has mushroomed in the last forty years. Since 1970, the number of bat people attending the annual meetings of the North American Society for Bat Research (NASBR) has grown from about 30 to nearly 400 people. At the same time the number of oral presentations increased from 26 to almost 300. Nancy and Brock, along with literally hundreds of colleagues and their students, are the lucky ones. These bat people get to enjoy hearing about the great projects and discoveries of others, often including astonishing new details about bats. As science has become more collaborative in the age of the internet, scientific research has thrived in many fields, including chiropterology. We would love to be at the NASBR meetings in 2068 to see what has developed in the field of bats 100 years after the inaugural meeting!

Donald Griffin referred to echolocation as the "magic well" because every time he visited the topic, he discovered something new. Nancy and Brock agree, but think that the real "magic well" is the bats themselves. Endlessly fascinating and instructive about how nature and evolution work, bats provide scientific inspirations enough to last many lifetimes.

Annotated Bibliography

Adams, Rick A., and Scott C. Pedersen (eds.). *Ontogeny, Functional Ecology, and Evolution of Bats*. New York: Cambridge University Press, 2000.

______________. *Bat Evolution, Ecology, and Conservation*. New York: Springer-Verlag, 2013.

Altringham, John D. *Bats: From Ecology to Conservation*, 2nd ed. Oxford: Oxford University Press, 2011.

Barbour, Paul. *Vampires, Burial and Death: Folklore and Reality*. New Haven: Yale University Press, 1988.

Couffer, Jack. *Bat Bomb: World War II's Other Secret Weapon*. Austin: University of Texas Press, 1992.

Crichton, Elizabeth G., and Philip H. Krutzsch (eds.). *Reproductive Biology of Bats*. New York: Academic Press, 2000.

Csorba, Gábor, Peter Ujhelyi, and Nikki Thomas. *Horseshoe Bats of the World (Chiroptera: Rhinolophidae)*. Exeter, U.K.: Pelagic Publishing, 2003.

Denny, Mark. Blip, *Ping and Buzz: Making Sense of Radar and Sonar*. Baltimore: Johns Hopkins University Press, 2007.

Fenton, M. Brock. *Communication in the Chiroptera*. Bloomington: Indiana University Press, 1985.

Findley, James S. *Bats: A Community Perspective*. (Cambridge Studies in Ecology.) Cambridge: Cambridge University Press, 1993.

Fleming, Theodore H. *The Short-tailed Fruit Bat: A Study in Plant-Animal Interactions*. (Wildlife Behavior and Ecology Series.) Chicago: University of Chicago Press, 1988.

Fleming, Theodore H., and Harry Greene. *A Bat Man in the Tropics: Chasing El Duende*. (Organisms and Environments.) Chicago: University of Chicago Press, 2003.

Greenhall, Arthur M., and Uwe Schmidt. (eds.). *Natural History of Vampire Bats*. Boca Raton: CRC Press, 1988.

Gunnell, Gregg F., and Nancy B. Simmons (eds.). *Evolutionary History of Bats: Fossils, Molecules and Morphology*. (Cambridge Studies in Morphology and Molecules: New Paradigms in Evolutionary Bio.) Cambridge: Cambridge University Press, 2012.

Hill, John E., and James D. Smith. *Bats: A Natural History*. Austin: University of Texas Press, 1984.

Koopman, Karl F. "*Systematics: Chiroptera*" in *Handbook of Zoology*, vol. 8, part 60: Mammalia, Erwin Kulzer (ed.). Berlin: Walter de Gruyter & Co., 1994.

Kunz, Thomas H. (ed.). *Ecology of Bats*. New York: Plenum Press. 1982.

Kunz, Thomas H., and M. Brock Fenton (eds.). *Bat Ecology*. Chicago: University of Chicago Press, 2003.

Kunz, Thomas H., and Stuart Parsons (eds.). *Ecological and Behavioral Methods for the Study of Bats*, 2nd ed. Baltimore: Johns Hopkins University Press, 2009.

Kunz, Thomas H., and Paul A. Racey (eds.). *Bat Biology and Conservation*. Washington, D.C.: Smithsonian Institution Scholarly Press, 1998.

Lacki, Michael J., John P. Hayes and Allen Kurta (eds.). *Bats in Forests: Conservation and Management*. Baltimore: Johns Hopkins University Press, 2007.

Neuweiler, Gerhard. *The Biology of Bats*. Oxford: Oxford University Press, 2000.

Norberg, Ulla M. *Vertebrate Flight: Mechanics, Physiology, Morphology Ecology and Evolution*. (Zoophysiology.) New York: Springer-Verlag, 1990.

Novick, Ronald M. *Walker's Bats of the World*. Baltimore: Johns Hopkins University Press, 1994.

Pollak, George D., and John H. Casseday. *The Neural Basis of Echolocation in Bats*. (Zoophysiology.) New York: Springer-Verlag, 1989.

Popper, Arthur N., and Richard R. Fay (eds.). *Hearing by Bats. (Springer Handbook of Auditory Research.)* New York: Springer-Verlag. 1995.

Racey, Paul A., and Susan M. Swift (eds.). *Ecology, Evolution and Behaviour of Bats*. (Symposia of the Zoological Society of London 67.) Oxford: Oxford University Press, 1995.

Ransome, Roger. *The Natural History of Hibernating Bats*. (Christopher Helm Mammal Series.) Independence, Kentucky: Thomson Learning, 1990.

Richarz, Klaus, and Albred Limbrunner. *The World of Bats: The Flying Goblins of the Night*. Neptune, New Jersey: TFH Publications, 1993.

Richardson, Phil. *Bats*. Richmond Hill, Ontario: Firefly Books, 2011.

Robertson, James. *The Complete Bat*. London: Chato and Windus, 1990.

Ruff, Sue, and Wilson, Don E. *Bats. (Animal Ways.)* New York: Cavendish Square Publishing, 2001.

Ryberg, Olaf. *Studies on Bats and Bat Parasites, Especially with Regard to Sweden and Other Neighboring Countries of the North*. Stockholm: Svensk Natur, 1947.

Sales, Gillian, and David Pye. *Ultrasonic Communication by Animals*. New York: Springer Verlag, 1974.

Schober, Wilfried. *The Lives of Bats*. New York: Arco Publishing, 1984.

Schutt, Bill. *Dark Banquet: Blood and the Curious Lives of Blood-Feeding Creatures*. New York: Harmony Books, 2008.

Simmons, Nancy B. *"Taking Wing: Uncovering the Evolutionary Origins of Bats,"* Scientific American, December 2008: 96-103.

Thomas, Jeanette A., Cynthia F. Moss, and Marianne Vater (eds.). *Echolocation in Bats and Dolphins*. Chicago: University of Chicago Press, 2004.

Tupinier, Denise. *La chauve-souris et l'homme*. Paris: Editons L'Harmattan, 1989.

Tuttle, Merlin D. America's Neighborhood *Bats: Understanding and Learning to Live in Harmony with Them*. Austin: University of Texas Press, 1988.

Williams, Kim, Rob Mies, Donald Stokes, and Lillian Stokes. *Beginner's Guide to Bats*. New York: Little, Brown and Company, 2002.

Wilson, Don E. *Bats in Question: The Smithsonian Answer Book*. Washington, D.C., Smithsonian Books, 1997.

Zubaid, Akbar, Gary F. McCracken, and Thomas H. Kunz. (eds.). *Functional and Evolutionary Ecology of Bats*. Oxford: Oxford University Press, 2006.

Some Classic Books About Bats

Allen, Glover M. *Bats: Biology, Behavior and Folklore*. Cambridge: Harvard University Press, 1939.

Barbour, Roger W., and Wayne H. Davis. *Bats of America*. Lexington: University of Kentucky Press, 1969.

Griffin, Donald R. *Listening in the Dark: The Acoustic Orientation of Bats and Men*. New Haven: Yale University Press, 1958.

Marshall, Adrian G. *The Ecology of Ectoparasitic Insects*. London: Academic Press, 1981.

Miller, Gerrit S. Jr. *The Families and Genera of Bats*. U.S. National Museum Bulletin 57, 1907.

Novick, Alvin, and Nina Leen. *The World of Bats*. New York: Holt Rinehart and Winston, 1969.

Roeder, Kenneth D. *Nerve Cells and Insect Behavior*, 2nd ed. (Books in Biology Series.) Cambridge: Harvard University Press, 1967.

Wilson, Don E., and Alfred L. Gardner (eds.). *Proceedings of the Fifth International Bat Research Conference*. Lubbock, Texas: Texas Tech University Press, 1980.

Wimsatt, William A. (ed.). *Biology of Bats*, vol. 1. New York: Academic Press, 1970.

———. *Biology of Bats*, vol. 2. New York: Academic Press, 1970.

———. *Biology of Bats*, vol. 3. New York: Academic Press, 1977.

Yalden, Derek W., and P. A. Morris. *The Lives of Bats*. New York: A Demeter Press Book, 1975.

Some Examples of Regional Works

Baagøe, Hans J. "*Danish bats (Mammalia: Chiroptera): Atlas and analysis of distribution, occurrence and abundance.*" Steenstrupia, 26:1-117, 2001.

Barquez, Ruben M., Norberto P. Giannini, and Michael A. Mares. *Guide to the Bats of Argentina*. Norman, Oklahoma: Oklahoma Museum of Natural History, 1993.

Bates, Paul J. J., and David L. Harrison. *Bats of the Indian Subcontinent*. Sevenoaks, U.K.:Harrison Zoological Museum, 1997.

Emmons, Louise. *Neotropical Rainforest Mammals: A Field Guide*, 2nd ed. Chicago: University of Chicago Press, 1997.

Francis, Charles M. *A Field Guide to the Mammals of South-East Asia*. London: New Holland Publishers, 2008.

Kays, Ronald W., and Don E. Wilson. *Mammals of North America*. (Princeton Field Guides.) Princeton: Princeton University Press, 2009.

Kingston, Tigga, Boo L. Lim, and Zubaid Akbar. *Bats of Krau Wildlife Reserve*. Bangi, Malayia: Penerbit Universiti Kebangsaan Malaysia Press, 2006.

Menkhorst, Peter, and Frank Knight. *Field Guide to the Mammals of Australia*, 3rd ed. Oxford: Oxford University Press, 2010.

Monadjem, Ara, Peter J. Taylor, F. P. D. (Woody) Cotterill, and M. Corrie Schoeman. *Bats of Southern and Central Africa: A Biogeographic and Taxonomic Synthesis*. Johannesburg: Wits University Press, 2010.

Reid, Fiona. *A Field Guide to the Mammals of Central America and Southeast Mexico*, 2nd. Oxford: Oxford University Press, 2009.

Richards, Greg, and Leslie Hall. *A Natural History of Australian Bats: Working the Night Shift*. Canberra: CSIRO Press, 2012.

Acknowledgements

We start by recognizing our families for their long and continued support of us in our pursuit of bats. Brock thanks Eleanor, and Nancy thanks Rob, Nick, and Matt, all of whom have put up gracefully with our numerous long absences in the field.

Brock's research on bats has been supported by grants from the Natural Sciences and Engineering Research Council of Canada, the K.F. Molson Foundation and World Wildlife Fund Canada. Nancy's research on bats has been supported by the U.S. National Science Foundation, The National Geographic Society, and the American Museum of Natural History Taxonomic Mammalogy Fund. Without the support of these funding agencies, we would never have been able to complete the many studies of bats that have so inspired our careers.

We are grateful to the following colleagues who kindly permitted us to use pictures in this book they had taken or prepared, or put us in touch with people with the pictures we needed: Anne Brigham, Jorn Cheney, Beth Clare, Ray Crundwell, Jakob Fahr, Skanda de Saram, Yohami Fernandez Delgado, Ted Fleming, Dennis Francos, Simon Ghanen, Joerg Habersetzer, Tigga Kingston, Susan Koenig, Renate Matzke-Karasz, Liam McGuire, Greg McIntosh, David Mudge, Stuart Parsons, Sebastien Puechmaille, Sandra Peters, Mehefatiana Ralisata, John J. Rasweiler IV, Dan Riskin, Jens Rydell, Bill Scully, Mark Skowronski, Sharon Swartz, Merlin Tuttle, Paula Tuttle, Dane Webster, and Sean Werle. We particularly thank Beth Clare, Ted Fleming, I.L. (Naas) Rautenbach, Jens Rydell, and Merlin Tuttle for continuing discussions about the challenges of photographing bats.

We salute now-departed bat biologists who influenced us in many ways: James H. Fullard, Donald R. Griffin, Elisabeth K.V. Kalko, Karl Koopman, Gerhard Neuweiler, Pete (R.L.) Peterson, Kenneth D. Roeder, Björn Siemers, and Donald W. Thomas. These pioneers, colleagues, and friends are sorely missed, but their legacy has and will continue to influence all aspects of our understanding of bat biology as reflected in this book.

Numerous individuals contributed to the success and pleasures of our many field expeditions to study bats, including the trips from which the stories in this book were drawn. We are particularly grateful to Amanda Adams, Robert Barclay, Mark Brigham, Hugh Broders, Deanna Byrnes, Andrea Cirranello, Beth Clare, Neil Duncan, Yvonne Dzal, Erin Fraser, Paul Faure, Eli Kalko, Darrin Lunde, Liam McGuire, Emanuel Mora, John Ratcliffe, Nina Veselka and Robert Voss, with whom we spent many memorable nights and early mornings in the forest. We are also grateful to the colleagues who keep the North American Symposium on Bat Research (now the North American Society for Bat Research) alive and well. We also thank the wonderful field crews we have worked with over many years at Lamanai, Belize,

We thank Paul Faure, Gregg Gunnell, Sam Mubareka, and Sharon Swartz for reading and commenting on various parts of the manuscript, and two anonymous reviewers for their thorough reviews and helpful comments. We also thank Christie Hendry, Amy Krynak and Melinda Kennedy at the University of Chicago Press for their editorial support. We are grateful to Nicholas LiVolsi for his elegant book design. Finally, we thank our publisher and the inspiration for this book—Peter Névraumont, who read every word of text several times over, brought studies to our attention that we had missed, and generally guided the creation of this book from beginning to end. Thank you, Peter.

Angolan Free-tailed Bat	*Mops condylurus*	Free-tailed Bat	186
Antillean Fruit-eating Bat	*Brachyphylla cavernarum*	New World Leaf-nosed Bat	**112**, 124, 129, 256
Antillean Ghost-faced Bat	*Mormoops blainvillii*	Moustached Bat	110
Australian False Vampire Bat	*Macrotus gigas*	False Vampire Bat	107
Banana Pipistrelle	*Neoromica nanus*	Vesper Bat	150
Bent-winged Bat	*Miniopterus*	Bent-winged Bat	11, 55, 141, **143**, 164
Beuttikofer's Epauletted Bat	*Epomops beuttikoferi*	Old World Fruit Bat	116
Bicolored Round-leafed Bat	*Hipposideros bicolor*	Old World Leaf-nosed Bat	**97**
Big Brown Bat	*Eptesicus fuscus*	Vesper Bat	63, **66**, 67, 75, 98, **99** 110, 129, 140, 141, 168, 178, 180, 185, 186, **187**, 190, 197, 198, 199, 215, 216
Birdlike Noctule	*Nyctalus aviator*	Vesper Bat	107
Bismarck Masked Flying Fox	*Pteropus capistratus*	Old World Fruit Bat	168
Black Bonneted Bat	*Eumops auripendulus*	Free-tailed Bat	68
Black Flying Fox	*Pteropus alecto*	Old World Fruit Bat	163
Black Mastiff Bat	*Molossus rufus*	Free-tailed Bat	**81**, **144**, **154**
Black Myotis	*Myotis nigricans*	Vesper Bat	159
Blanford's Fruit Bat	*Sphaerias blanfordi*	Old World Fruit Bat	**42**
Brandt's Myotis	*Myotis brandti*	Vesper Bat	159
Brazilian Free-tailed Bat	*Tadarida brasiliensis*	Free-tailed Bat	**16**, 17, **88**, 110, 126, 134, 141, 153, **166**, 167, 178, 185, 192, 195, 213, 216, 238, 250, **252**, 255, 260
Brosset's Big-eared Bat	*Micronycteris brosseti*	New World Leaf-nosed Bat	10, 278
Brown Long-eared Bat	*Plecotus auritus*	Vesper Bat	159
Bulmer's Fruit Bat	*Aproteles bulmeri*	Old World Fruit Bat	256
Bumblebee Bat	*Crasonycteris thonglongyai*	Bumblebee Bat	11, **28**, 55
Butterfly Bat	*Glauconycteris variegata*	Vesper Bat	146
California Myotis	*Myotis californicus*	Vesper Bat	261
Carriker's Round-eared Bat	*Lophostoma carrikeri*	New World Leaf-nosed Bat	153
Chapin's Free-tailed Bat	*Chaerephon chapini*	Free-tailed Bat	177
Chestnut Short-tailed Bat	*Carollia castanea*	New World Leaf-nosed Bat	120
Christmas Island Flying Fox	*Pteropus melanotus*	Old World Fruit Bat	149
Christmas Island Pipistrelle	*Pipistrellus murrayi*	Vesper Bat	249, 260
Commissarisi's Long-tongued Bat	*Glossophaga commissarisi*	New World Leaf-nosed Bat	104
Common Big-eared Bat	*Micronycteris microtis*	New World Leaf-nosed Bat	96, **98**, 142
Common Pipistrelle	*Pipistrellus pipistrellus*	Vesper Bat	189, 255, 256
Common Vampire Bat	*Desmodus rotundus*	New World Leaf-nosed Bat	**26**, **27**, **29**, 78, **104**, **112**, **144**, 119, 120, 129, 137, **138**, 153, 159, 171, 185, 197, 213, **215**, **217**, 218, **221**, 231, 255
Commerson's Round-leafed Bat	*Hipposideros commersoni*	Old World Leaf-nosed Bat	99
Cuban Flower Bat	*Phyllonycteris poeyi*	New World Leaf-nosed Bat	**29**

Index

Brock getting ready to photograph bats leaving a cave in Cuba. Note liberal covering of bat guano on shirt and trousers.

Nancy relaxing in a portable hammock near mist nets set to catch bats at Lamanai Archaeological Reserve in Belize. Having a dry (and ant-free) place to sit can make it worth the effort of carrying and stringing up a hammock in the field.